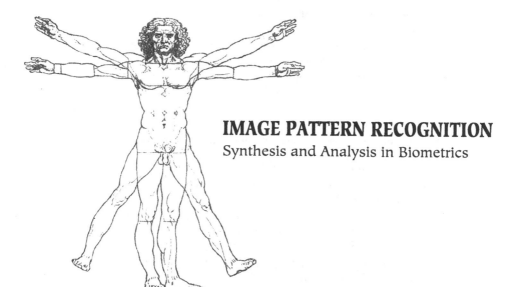

IMAGE PATTERN RECOGNITION
Synthesis and Analysis in Biometrics

SERIES IN MACHINE PERCEPTION AND ARTIFICIAL INTELLIGENCE*

Editors: **H. Bunke** (Univ. Bern, Switzerland)
P. S. P. Wang (Northeastern Univ., USA)

*For the complete list of titles in this series, please write to the Publisher.

Series in Machine Perception and Artificial Intelligence – Vol. 67

IMAGE PATTERN RECOGNITION
Synthesis and Analysis in Biometrics

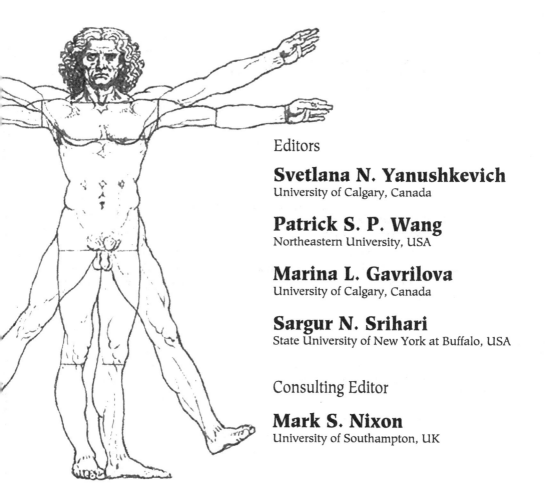

Editors

Svetlana N. Yanushkevich
University of Calgary, Canada

Patrick S. P. Wang
Northeastern University, USA

Marina L. Gavrilova
University of Calgary, Canada

Sargur N. Srihari
State University of New York at Buffalo, USA

Consulting Editor

Mark S. Nixon
University of Southampton, UK

 **World Scientific**

NEW JERSEY • LONDON • SINGAPORE • BEIJING • SHANGHAI • HONG KONG • TAIPEI • CHENNAI

Published by

World Scientific Publishing Co. Pte. Ltd.

5 Toh Tuck Link, Singapore 596224

USA office: 27 Warren Street, Suite 401-402, Hackensack, NJ 07601

UK office: 57 Shelton Street, Covent Garden, London WC2H 9HE

British Library Cataloguing-in-Publication Data
A catalogue record for this book is available from the British Library.

Series in Machine Perception and Artificial Intelligence — Vol. 67
IMAGE PATTERN RECOGNITION
Synthesis and Analysis in Biometrics

ISBN-13 978-981-256-908-0
ISBN-10 981-256-908-1

Printed in Singapore.

Preface

Computer Vision and *Computer Graphics* can be thought of as opposite sides of the same coin. In computer graphics we start, for example, with a three-dimensional model of a face, and we attempt to render or project this model onto a two-dimensional surface to create an image of the face. In computer vision we attempt to do the opposite — we start with a two-dimensional image of the face and we try to generate a computer model from a sequence of one or more such images. However, the two sides of the coin are by no means equal as far as the amount of research and development lavished upon them; computer graphics is a very advanced and developed field, whereas computer vision is still relatively in its infancy. This is largely because developments in computer graphics have been driven forwards by the multi-billion dollar markets for computer aided design, computer games, and the movie and advertising industry. It therefore makes a great deal of sense to try and exploit this powerful relationship between the two fields so that computer vision can benefit from the wealth of powerful techniques already developed for computer graphics.

In this book we apply this thinking to a field which whilst not exactly a sub-discipline of computer vision has a very great deal in common with it. This field is *Biometrics* where we attempt to generate computer models of the physical and behavioural characteristics of human beings with a view to reliable personal identification. It is not completely a sub-discipline of computer vision because the human characteristics of interest are not restricted to visual images, but also include other human phenomena such as odour, DNA, speech, and indeed anything at all which might help to uniquely identify the individual.

Although biometrics is at least as old as computer vision itself, research and development in this field has proceeded largely independently of computer graphics. We strongly believe that this has been a mistake in the past and we will attempt to redress this balance by developing the other side of the biometrics coin, namely *Biometric Synthesis* — rendering biometric phenomena from their corresponding computer models. For example, we could generate a synthetic face from its corresponding computer model. Such a model could include muscular dynamics to model the full gamut of human emotions conveyed by facial expressions.

We firmly believe that this will be a very fertile area of future research and development with many spin-offs. For example, much work has already been done on the information theory associated with computer graphics; just think of image and video compression — we can now fit a complete high quality $1\frac{1}{2}$ hour movie on a single 700MB CD using MPEG4 or DivX video compression. We should be able to exploit this valuable research to gain a much better understanding of the information theoretic aspects of biometrics which are not very well understood at present. This is just one example of how this powerful dual relationship between computer graphics and computer vision might be exploited.

This book is a collection of carefully selected chapters presenting the fundamental theory and practice of various aspects of biometric data processing in the context of pattern recognition. The traditional task of biometric technologies — human identification by analysis of biometric data is extended to include the new discipline, *Biometric Synthesis*— the generation of artificial biometric data from computer models of target biometrics. Some of new ideas were first presented at the

International Workshop on
"Biometric Technologies: Modeling and Simulation"

held in June 2004 in Calgary, Canada, and which was hosted by the research laboratory of the same name, from which the workshop took its title, "Biometric Technologies: Modeling and Simulation" at the University of Calgary.

The book is primarily intended for computer science, electrical engineering, and computer engineering students, and researchers and practitioners in these fields. However, individuals in other areas who are interested in these and related subjects will find it a most comprehensive source of relevant information.

Biometric technology may be defined as the automated use of physiological or behavioral characteristics to determine or verify an individual's identity. The word biometric also refers to any human physiological or behavioral characteristic[1] which possesses the requisite biometric properties. They are:

---

[1] A. Jain, R. Bolle, and S. Pankanti, Eds., *Biometrics: Personal Identification in a Networked Society*, Kluwer, 1999.

Universal (every person should have that
     characteristic),
Unique (no two people should be exactly the same in
     terms of that characteristic),
Permanent (invariant with time),
Collectable (can be measured quantitatively),
Reliable (must be safe and operate at a satisfactory
     performance level),
Acceptable (non-invasive and socially tolerable), and
Non-circumventable (how easily the system is fooled
     into granting access to impostors).

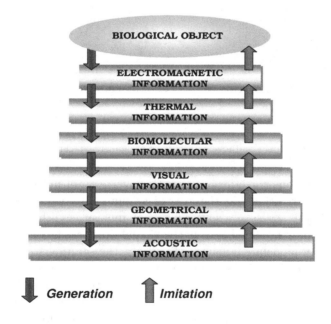

Fig. 1. A biological object as a generator of different types of information. Imitation of biometric data is the solution of the inverse problem.

Research and development in advanced biometrics techniques is currently proceeding in both directions simultaneously: analysis for identification or recognition of humans (direct problems), and synthesis of biometric information (inverse problems), see Fig. 1. The problem of analysis of biometric information has long been investigated. Many researchers have provided efficient solutions for human authentication based on signature, fingerprints, facial characteristics, hand geometry, keystroke analysis, ear, gait, iris and retina scanning. Active research is being conducted using both traditional and emerging technologies, to find better solutions to the problems of verification where claimants are checked against their specific biometric records in a database, and identification where a biometric database is searched to see if a particular candidate can be matched to any record. However, development of biometric simulators for generating synthetic biometric data has not yet been well investigated, except for the particular area of modeling of signature forgery, and voice synthesis, see Fig. 2.

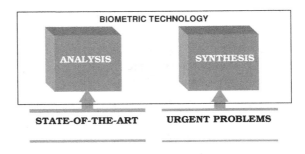

Fig. 2.   Direct and inverse problems of biometric technology.

Imitation of biometric information is the inverse problem to the analysis of biometric information. In the area of graphical image processing, for instance, synthesis serves as a source for many innovative technologies such as virtual simulation. The objects in the virtual world are modeled through a virtual reality modeling language. Similarly, the solution to the inverse problem in biometrics will foster pioneering applications, such as biometric imitators that reflect both psychological (mood, tiredness) and physical (normal vs. ultra-red light) characteristics. It will also alleviate the well known backlog problems in traditional biometric research, for example, by providing a novel approach to decision making based on the paradigm of relevance of information. Synthetic biometric data has been the focus of numerous previous studies, but these attempts were limited in that the synthesis was either not automated or semi-automated. Some examples are given in Table 1

Our aim has been to make this book both a comprehensive review and a suitable starting point for developing modeling and simulation techniques

Table 1. Synthetic biometrics.

BIOMETRIC	COMMENTS
Synthetic fingerprints	*Today's interest in automatic fingerprint synthesis addresses the urgent problem of testing fingerprint identification systems, training security personnel, biometric database security, and protecting intellectual property.*
Synthetic signatures	*Current interest in signature analysis and synthesis is motivated by the development of improved devices for human-computer interaction, which enable input of handwriting and signatures. The focus of this study is the formal modeling of this interaction.*
Synthetic irises	*The ocularist's approach to iris synthesis is based on the composition of painted primitives, and utilizes layering of semi-transparent textures built from topological and optic models. Vanity contact lenses are available with fake iris patterns printed onto them (designed for people who want to change eye colors). Colored lenses, i.e., synthetic irises, cause trouble for current identification systems based on iris recognition.*
Synthetic speech	*Synthetic speech has evolved considerably since the first experiments in the 1960s. New targets in speech synthesis include improving the audio quality and the naturalness of speech, developing techniques for emotional "coloring", and combining it with other technologies, for example, facial expressions and lip movement. Synthetic voice should carry information about age, gender, emotion, personality, physical fitness, and social upbringing. The synthesis of an individual's voice will be possible too, the imitation based upon the actual physiology of the person.*
Synthetic emotions and expressions	*Synthetic emotions and expressions are more sophisticated real world examples of synthesis. People often use their smile to mask sorrow, or mask gladness with a neutral facial expression. Such facial expressions can be thought of as artificial or synthetic in a social sense. In contrast to synthetic fingerprints and irises, the carrier of this synthetic facial information is a person's physical face rather than an image on the computer. The carrier of information can be thought of as facial topologies, indicative of emotions. To investigate the above problems, techniques for modeling facial expressions, i.e., the generation of synthetic emotions, must be developed. These results can be used, in particular, in a new generation of lie detectors. A related problem is how music or an instrument expresses emotions. To examine whether music produces emotions, a measuring methodology might be developed.*
Humanoid robots	*Humanoid robots are artificial intelligence machines that include challenging direct and inverse biometrics: language technologies, such as voice recognition and synthesis, speech-to-text and text-to-speech; face and gesture recognition of the "moods" of the instructor, following of cues; dialog and logical reasoning; vision, hearing, olfaction, tactile, and other sensing (artificial retinas, e-nose, e-tongue).*

in biometrics. To achieve this goal, the presentation is organized in three parts:

Part 1: Synthesis in Biometrics,
Part 2: Analysis in Biometrics, and
Part 3: Biometric Systems and Applications.

PART 1: SYNTHESIS IN BIOMETRICS

The first part is devoted to the inverse problems of biometrics — synthesis of biometric data. It includes four chapters and overviews existing and currently being developed approaches to synthesis of various biometric data.

Chapter 1 by Drs. *Yanushkevich, Shmerko, Stoica, Wang*, and *Srihari* "Introduction to Synthesis in Biometrics" introduces the inverse problems of biometrics — synthesis of biometric data. Synthetic biometric data provides for detailed and controlled modeling of a wide range of training skills, strategies, and tactics for forgeries of biometrics, thus enabling a better approach to enhancing system performance.

Chapter 2 by Dr *Popel*, "Signature Analysis, Verification and Synthesis in Pervasive Environments" introduces a system developed to identify and authenticate individuals based on their signatures or handwriting, or both. The issues of pervasive services are addressed by integrating unique data acquisition and processing techniques which are capable of communicating with a variety of off-the-shelf devices such as pressure sensitive pens, mice, and touch pads; by using sequence processing techniques (like matching, alignment or filtering) for signature analysis techniques and comparison; by using self-learning database solutions for achieving accurate results, and by utilizing signature synthesis techniques for benchmarking and testing.

Chapter 3 by Drs. *Samavati* and *Bartels*, and their student *Luke Olsen* "Local B-Spline Multiresolution with Examples in Iris Synthesis and Volumetric Rendering" studies *B*-splines and filtering techniques for image synthesis and reconstruction allowing high resolution of data.

Chapter 4 by Dr. *Gavrilova* "Computational Geometry and Image Processing Techniques in Biometrics: on the Path to Convergence" examines a unique aspect of the problem — the development of new approaches and methodologies for biometric identification, verification and synthesis, utilizing the notion of proximity and topological properties of biometric identifiers. The use of recently developed advanced techniques in computational geometry and image processing is examined with the purpose of finding the common characteristics between different biometric problems, and identifying the most promising methodologies. In particular, this chapter discusses applications of computational geometry methods such as constructing medial axis transforms, distance distribution computation, Voronoi diagrams and Delaunay triangulation, and topology-based approaches for feature extraction and pattern matching.

We also recommend the books by *S. Yanushkevich, A. Stoica, V. Shmerko* and *D. Popel* "Biometric Inverse Problems", Taylor & Francis/CRC Press 2005, and the book by *Z. Wen*, by *T. S. Huang* "3D Face Processing: Modeling, Analysis and Synthesis", Kluwer, 2004, and by *W. Zhao* and *R. Chellappa*, Eds., "Face Processing: Advanced Modeling and Methods", Elsevier, 2006, as an introduction to the problem of biometric data design.

PART 2: ANALYSIS IN BIOMETRICS

The second part includes six chapters on general methodology used in the traditional applications of biometric technologies — identification and verification of individuals. These methods are classified as image analysis and pattern recognition.

Chapter 5 by Dr. *Srihari* and his PhD student *Harish Srinivasan* "A Statistical Model for Biometric Verification", introduces a statistical learning methodology for determining whether a pair of biometric samples belong to the same individual. The proposed approach is applied to friction ridge prints (physical biometric) and and handwriting (behavioral biometric) and advantages over conventional methods are demonstrated (improved accuracy and a natural provision for combining with other biometric modalities).

Chapter 6 by Dr. *Parker* "Composite Systems for Handwritten Signature Recognition", introduces a simple way to reliably compare signatures in a quite direct fashion. Comparisons are made to other methods, and a four algorithm voting scheme is used to achieve over 99% success.

Chapter 7 by Dr. *Hurley* "Force Field Feature Extraction for Ear Biometrics", presents a new transformation and feature extraction technique using a force field model in the context of ear biometrics. This method automatically locates the potential wells and channels of an energy surface produced by the transform, which then forms the basis of characteristic ear features, which are subsequently used for ear recognition.

Chapter 8 by Drs. *You, Wang*, and their colleagues *Q. Chen* and *D. Zhang* "Nontensor-Product-Wavelet - Based Facial Feature Representation", introduces a method for facial feature representation by using a non-tensor product bivariate wavelet transform. Nontensor product bivariate wavelet filter banks with linear phase are constructed from the centrally symmetric matrices.

Chapter 9 by Drs. *Lu, Zhang, Kong*, and *Liao*, "Palmprint Identification by Fused Wavelet Characteristics" presents a novel method of feature extraction for palmprint identification based on the wavelet transform. This method is used to handle the textural characteristics of palmprint images at low resolution. The extraction of four sets of statistical features (the mean, energy, variance, and kurtosis), allows the achievement of high accuracy in identification.

Chapter 10 by Dr. *Traoré* and his PhD student *Ahmed Awad E. Ahmed* "Behavioral Biometrics for Online Computer User Monitoring" introduces an artificial neural network based techniques for analyzing and processing keystroke and mouse dynamics to achieve passive user monitoring. Keystroke dynamics recognition systems measure the dwell time and flight time for keyboard actions. Mouse dynamics are described as the characteristics of the actions received from the mouse input device for a specific user, while interacting with a specific graphical user interface.

We also recommend the book by R. C. Gonzalez, R. E. Woods, and S. L. Eddins *Digital Image Processing Using MATLAB*, Pearson Prentice Hall, 2004, as the basis of analysis of biometric data.

PART 3: BIOMETRIC SYSTEMS AND APPLICATIONS

The third part of the book includes five chapters in the broader context of systems and applications.

Chapter 11 by Drs. *Ratha, Bolle,* and *Pankanti,* "Large-Scale Biometric Identification: Challenges and Solutions" reviews the tools, terminology and methods used in large-scale biometrics identification applications. In large-scale biometrics, the performance of the identification algorithms need to be significantly improved, to successfully handle millions of persons in the biometrics database, matching thousands of transactions per day.

Chapter 12 by Dr. *Coello Coello,* "Evolutionary Algorithms: Basic Concepts and Applications in Biometrics" provides a short introduction to the main concepts related to evolutionary algorithms and several case studies on the use of evolutionary algorithms in both physiological and behavioral biometrics. The case studies include fingerprint compression, facial modeling, hand-based feature selection, handwritten character recognition, keystroke dynamics and speaker verification.

Chapter 13 by Dr. *Wang* "Some Concerns on the Measurement for Biometric Analysis and Applications", reexamines the "measurement" techniques essential for comparing the "similarity" of patterns. The chapter focuses on the concepts of "segmentation" and "disambiguation" from global semantic point of view, which are important in the context of pattern recognition and biometric-based applications.

Chapter 14 by Dr *Elliott* and his students *Eric Kukula,* and *Shimon Modi,* "Issues Involving The Human Biometric Sensor Interface", examines topics such as ergonomics, the environment, biometric sample quality, and device selection, and how these factors influence the successful implementation of a biometric system.

Chapter 15 by Drs. *Yanushkevich, Stoica,* and *Shmerko,* "Fundamentals of Biometrics-Based Training System Design", introduces the concept of biometric-based training for a wide spectrum of applications in the social sphere (airports and seaports, immigration service, border control, important public events, hospitals, banking, etc.). The goal of such training is to assist the officers in developing their skills for decision making, based on two types of information about an individual: biometric information collected during pre-screening or surveillance, and information collected during the authorization check itself.

We also recommend the book by A. Jain, S. Pankanti, and Bolle, R. Eds. *Biometrics: Personal Identification in Networked Society*, Kluwer, 1999.

There are a lot of new approaches that are not included in this book, for example, gait biometrics. The book by Dr *Nixon* et al. "Human ID based on Gait", Springer, 2006, can help the reader to get familiar with this biometric. Note that gait biometrics are very useful in security systems which use screening and the early warning paradigm (see Chapter 15).

This collection of selected papers introduces the activity of several internationally recognized research centres:

Analytical Engines Ltd., UK, represented by Dr. *D. J. Hurley*

Biometric Research Center directed by Dr. *D Zhang*, Hong Kong Polytechnic University, Hong Kong

Biometrics Standards, Performance, and Assurance Laboratory directed by Dr. *S. J. Elliott*, Purdue University, U.S.A.

Biometric Technology Group directed by Dr. *D. V. Popel*, Baker University U.S.A.

Biometric Technology Laboratory: Modeling and Simulation directed by Dr. *S. N. Yanushkevich* and Dr. *M. L. Gavrilova*, University of Calgary, Canada

Center of Excellence for Document Analysis and Recognition (CEDAR) directed by Dr.*S. N. Srihari*, State University of New York at Buffalo, U.S.A.

Department of Information Security, Office of National Statistics, UK, directed by Eur. Ing. *Phil Phillips*

Digital Media Laboratory directed by Dr. *J. Parker*, University of Calgary, Canada

European Center for Secure Information and Systems, Iasi, Romania, represented by Drs. *A. Stoica, S. Yanushkevich*, and *V. Shmerko.*

Humanoid Robotics Laboratory directed by Dr. *A. Stoica*, California Institute of Technology, Jet Propulsion Laboratory, National Aeronautics and Space Agency (NASA), U.S.A.

Evolutionary Computation Group, directed by Dr. *Carlos A. Coello Coello*, CINVESTAV-IPN, Mexico

IBM Thomas J. Watson Research Center, Exploratory Computer Vision Group, IBM, N.Y. U.S.A., represented by Drs. *N. K. Ratha, R. M. Bolle,* and *S. Pankanti*

Image Processing Group directed by Dr. *P. S. Wang*, Northeastern University, U.S.A.

Information: Signals, Images, System Research Group, School of Electronics and Computer Science at the University of Southampton, UK, represented by Dr. *M. S. Nixon*

Information Security and Object Technology Laboratory, directed by Dr. *I. Traoré*, University of Victoria, Canada

Visual Information Processing Laboratory directed by Dr. *Q. M. Liao*, Graduate School at Shenzhen, Tsinghua University, China

Acknowledgments

All phases of the organization of the first forum on synthetic biometrics, Workshop "Biometric Technologies: Modeling and Simulation" (BT'2004), were supported by

Dr. *Vlad Shmerko*, Biometric Technologies: Modeling and Simulation Laboratory, University of Calgary, Canada

Eur. Ing. *Phil Phillips*, Head of the Information Security Department for the UK Government, and

Mr. *Bill Rogers*, Publisher of Biometric Digest and Biometric Media Weekly, U.S.A.

They presented their vision of state-of-the-art biometrics, which one can find in the Workshop Proceedings. We thank *Vlad, Phil,* and *Bill* for their unselfish contribution.

The work by *Dr. P. Wang* was partially supported by the Alberta Informatics Circle of Excellence (iCore) when he was a visiting Professor at the Biometric Technologies Laboratory at the University of Calgary in 2006-2007. The activity of Drs. *S. Yanushkevich* and *M. Gavrilova* was partially supported by the Canadian Foundation for Innovations (CFI), the Government of Alberta, and BCT TELUS, Canada.

The Editors gratefully acknowledge the authors and the important support given by our reviewers. They gave their valuable time and made incredible efforts to review the papers.

This volume should be especially useful to those engineers, scientists, and students who are interested in both aspects - analysis and synthesis - of biometric information and their applications. It is our hope that this book will inspire a new generation of innovators to continue the development of biometric-based systems.

Svetlana N. Yanushkevich
BIOMETRIC TECHNOLOGIES: MODELING AND SIMULATION
LABORATORY, University of Calgary, Canada

Patric S. Wang
IMAGE PROCESSING GROUP, Northeastern University, MA, U.S.A.

Sargur N. Srihari
CENTER OF EXCELLENCE FOR DOCUMENT ANALYSIS AND
RECOGNITION, State University of New York at Buffalo, NY, U.S.A.

Marina L. Gavrilova
BIOMETRIC TECHNOLOGIES: MODELING AND SIMULATION
LABORATORY, University of Calgary, Canada

Mark S. Nixon
INFORMATION: SIGNALS, IMAGES, SYSTEM RESEARCH GROUP,
School of Electronics and Computer Science at the University of
Southampton, UK

Calgary
Boston
New York
Southampton

Contents

PART 1
SYNTHESIS IN BIOMETRICS

PART 1

SYNTHESIS IN BIOMETRICS

- *Direct and inverse biometric problems*
- *Basic tools*
- *Modeling and synthesis*
- *Synthetic biometrics*
- *Applications*
- *Ethical and social aspects*

SYSTEM FOR SIGNATURE ANALYSIS, VERIFICATION, AND SYNTHESIS

- *Signature representation*
- *Analysis and verification*
- *Signature synthesis*

MULTIRESOLUTION IN IMAGE SYNTHESIS

- *Wavelets and multiresolution*
- *B-splinemultiresolution filters*
- *Biometric data synthesis*

COMPUTATIONAL GEOMETRY FOR SYNTHESIS BIOMETRICS

- *Computational geometry*
- *Voronoi diagram techniques*
- *Delaunay triangulation*
- *Applications in biometrics*

Chapter 1

Introduction to Synthesis in Biometrics

S. Yanushkevich[*], V. Shmerko[†], A. Stoica[‡], P. Wang[§], S. Srihari[¶]

Biometric Technology Laboratory: Modeling and Simulation,
University of Calgary, Canada
[*]*syanshk@ucalgary.ca*
[†]*vshmerko@ucalgary.ca*

Humanoid Robotics Laboratory, California Institute of Technology,
NASA Jet Propulsion Laboratory, California Institute of Technology,
Pasadena, CA, USA
[‡]*adrian.stoica@jpl.nasa.gov*

Image Processing Group, Northeastern University, Boston, MA, USA
[§]*pwang@ccs.neu.edu, patwang@mit.edu*

Center of Excellence for Document Analysis and Recognition,
State University of New York at Buffalo, Amherst, NY, USA
[¶]*srihari@cedar.buffalo.edu*

The primary application focus of biometric technology is the verification and identification of humans using their possessed biological (anatomical, physiological and behavioral) properties. Recent advances in biometric processing of individual biometric modalities (sources of identification data such as facial features, iris patterns, voice, gait, ear topology, etc.) encompass all aspects of system integration, privacy and security, reliability and countermeasures to attack, as well as accompanying problems such as testing and evaluation, operating standards, and ethical issues.

Contents

1.1. Introduction

The typical application scenario of biometric technologies involves the interaction of different levels of physical access control with different levels of data sensors. The human user of a the biometric system is at the center of this interaction. The centre in this interaction is the human user of a biometric system. The system must assist the user by providing high quality biometric data to ensure optimal system operation, e.g. to minimize false rejection errors or to provide early warning information.

This chapter addresses important questions of protecting against an attack on biometric systems. It focuses on studying the extent to which artificial, or synthetic biometric data (e.g. synthesized iris patterns, fingerprint, or facial images) can be useful in this task. Artificial biometric data are understood as biologically meaningful data for existing biometric systems. Synthetic biometric data can be useful for:

(a) Testing biometric devices with "variations" or "forgeries" of biometric data,
(b) Simulation of biometric data on computer-aided tools (decision-making support, training systems, etc.)

Testing biometric devices is of urgent importance [32,56] and can be accomplished by providing variations on biometric data mimicking unavailable or hard to access data (for example, modeling of badly lit faces, noisy iris images, "wet" fingerprints etc.) Synthetic data can also be used for "spoofing" biometric devices with "forged" data. We argue that synthetic biometric data can:

Improve the performance of existing identification systems. This can be accomplished by using automatically generated biometric data to create statistically meaningful sets of data variations (appearance, environmental, and others, including "forgeries").

Improve the robustness of biometric devices by modeling the strategies and tactics of forgery.

Improve the efficiency of training systems by providing the user-in-training with the tools to model various conditions of biometric data acquisition (non-contact such as passive surveillance, contact, cooperative, non-cooperative), environmental factors (light, smog, temperature), appearance (aging, camouflage).

Therefore, synthetic biometric data plays an important role in enhancing the security of biometric systems. Traditionally, security strategies (security levels, tools, etc.) are designed based on assumptions about a *hypothetical* robber or forger. Properly created artificial biometric data provides for detailed and controlled modeling of a wide range of training skills, strategies and tactics, thus enabling a better approach to enhancing the system's performance. This study aims to develop new approaches for the detection of attacks on security systems. Figure 1.1 introduces the basic configuration for inverse biometric problems.

1.1.1. *Basic Paradigm of Synthetic Biometric Data*

Contemporary techniques and achievements in biometrics are being developed in two directions:

Analysis for identification and recognition of humans (direct problems) and
Synthesis of biometric information (inverse problems) (Fig. 1.1).

Note that synthesis also means modeling. In general, modeling is considered to be an inverse problem in image analysis [9]. However, there are several differences between synthesis of realistic biometric data and modeling used in analysis.

Synthesis of artificial biometric data is the inverse task of analysis, performed as a part of the process of verification and identification. The crucial point of modeling in biometrics is the *analysis-by-synthesis* paradigm. This paradigm states that synthesis of biometric data can verify the perceptual equivalence between original and synthetic biometric data, i.e. synthesis

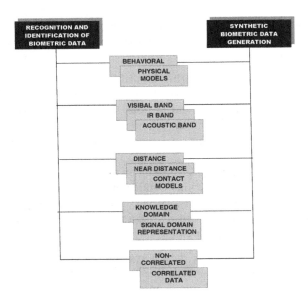

Fig. 1.1.   Basic tools for inverse biometric problems include facilities for generation of synthetic data and its analysis.

based feedback control. For example, facial analysis can be formulated as deriving a symbolic description of a real facial image (Fig. 1.2). The aim of face synthesis is to produce a realistic facial image from a symbolic facial expression model [25].

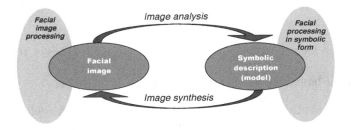

Fig. 1.2.   Analysis-by-synthesis approach in facial image.

In Table 1.1, the basic terminology of synthetic biometrics is given. In particular, the term "face reconstruction" indicates a class of problems aimed at synthesizing a model of a static face. The term "mimics animation" is defined as the process of modeling facial appearance and facial topology, a behavioral characteristic called facial expression, the visible result of synthetic emotion.

We consider synthetic single and synthetic multi-biometric data: eyes, hands, and face are multi-biometric objects because of the different topologies of the iris and retina of the eye; fingerprints, palmprints, and hand-geometry; face geometry which is a highly dynamic topological structure (smile, lip, brow, and eye movements). Synthetic signatures [20], [37], voice, and ears [21], [22] are examples of synthetic biometrics.

Table 1.1. Direct and inverse problems of biometric technology.

DIRECT PROBLEM	INVERSE PROBLEM
Signature identification	Signature *forgery*
Handwriting character recognition	Handwritten text *forgery*
face recognition	Face *reconstruction* and mimics animation
Voice identification and speech recognition	Voice and speech *synthesis*
Iris and retina identification	Iris and retina *image synthesis*
Fingerprint identification	Fingerprint *imitation*
hand geometry identification	Hand geometry *imitation*
Infrared identification	infrared image *reconstruction*
Gait identification	Gait *modeling*
Ear identification	Ear-print *imitation*

1.2. Synthetic Approaches

There are two approaches to synthetic biometric data design [39]:

(a) Image synthesis-based, and
(b) Statistical physics-based.

Both approaches use statistical models in the form of equations based on underlying physics or empirically derived algorithms, which use pseudo-random numbers to create data that are statistically equivalent to real data. For example, in face modeling, a number of ethnic or race models can be used to represent ethnic diversity, the specific ages and genders of individuals, and other parameters for simulating a variety of tests.

1.2.1. *Image Synthesis*

The image synthesis-based approach falls into the area of computer graphics, a very-well explored area with application from forensics (face reconstruction) to computer animation.

In generating physiological biometric objects (faces, fingerprints), the physics-based approach overlaps with the image-based approach, as it tries to model visual appearance and the physical properties and topology of the objects (including physics-based models to control physical form, motion and illumination properties of materials).

1.2.2. *Physics-Based Modeling*

Physics-based models attempt to mimic biometric data through the creation of a pattern similar to that acquired by biometric sensors, using knowledge of physical processes and sensor measurement.

1.2.3. *Modeling Taxonomy*

A taxonomy for the creation of physics-based and empirically derived models for the creation of statistical distributions of synthetic biometrics was first attempted in [5]. There are several factors affected the modeling biometric data: behavior, sensor, and environmental factors.

Behavior, or appearance, factors are best understood as an individual's presentation of biometric information. For example, a facial image can be camouflaged with glasses, beards, wigs, make-up, etc.

Sensor factors include resolution, noise, and sensor age, and can be expressed using physics-based or geometry-based equations. This factor is also relevant to the skills of the user of the system.

Environmental factors affect the quality of collected data. For example, light, smoke, fog, rain or snow can affect the acquisition of visual-band images, degrading the biometric facial recognition algorithm. High humidity or temperature can affect infrared images. This environmental influence affects the acquisition of fingerprint images differently for different types of fingerprint sensors.

1.3. Synthetic Biometrics

1.3.1. *Synthetic Fingerprints*

Albert Wehde was the first to "forge" fingerprints in the 1920s. Wehde "designed" and manipulated the topology of synthetic fingerprints at the physical level. The forgeries were of such high quality that professionals could not recognize them [11], [33]. Today's interest in automatic fingerprint synthesis addresses the urgent problems of testing fingerprint identification systems, training security personnel, biometric database security, and protecting intellectual property [7,60].

Traditionally, two possibilities of fingerprint imitation are discussed with respect to obtaining unauthorized access to a system: (i) the authorized user provides his fingerprint for making a copy, and (ii) a fingerprint is taken without the authorized user's consent, for example, from a glass surface (a classic example of spy-work) by forensic procedures.

Cappelli et al. [6,7] developed a commercially available synthetic fingerprint generator called SFinGe. In SFinGe, various models of fingerprints are used: shape, directional map, density map, and skin deformation models (Fig. 1.3). To add realism to the image, erosion, dilation, rendering, translation, and rotation operators are used.

Fig. 1.3. Synthetic fingerprint assembly (growth) generated by SFinGe.

Methods for continuous growth from an initial orientation map, a new synthesized orientation map (as a recombination of segments of the orientation map)) using a Gabor filter with polar transform (Fig. 1.4) have been reported in [60]. These methods alone can be used to design fingerprint benchmarks with rather complex structural features.

Fig. 1.4. Synthetic fingerprint assembly (growth) using a Gabor filter with polar transform.

Kuecken [26] developed a method for synthetic fingerprint generation based on natural fingerprint formation and modeling based on state-of-the-art dermatoglyphics, a discipline that studies epidermal ridges on finger-prints, palms, and soles (Fig. 1.5).

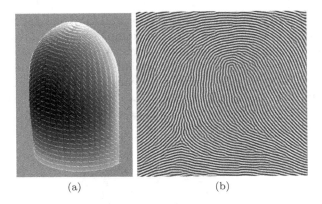

(a)   (b)

Fig. 1.5. Synthetic 3D (a) and 2D (b) fingerprint design based on physical modeling (reprinted with permission from Elsevier).

The methods of fingerprint synthesis can also be applied to synthetic palmprint generation. Note that the generation of synthetic hand topologies is a trivial problem.

1.3.2. *Synthetic Signatures*

Current interest in signature analysis and synthesis is motivated by the development of improved devices for human-computer interaction which enable input of handwriting and signatures. The focus of this study is the formal modeling of this interaction [4], [10], [20], [23], [37], [42], [44], [57].

Similarly to signature imitation, the imitation of human handwriting is a typical inverse problem of graphology. Automated tools for the imitation of handwriting have been developed. It should be noted that more statistical data, such as context information, are available in handwriting than in signatures.

The simplest method of generating synthetic signatures is based on geometrical models. Spline methods and Bezier curves are used for curve approximation, given some control points. Manipulations of control points give variations on a single curve in these methods [47,60].

The following evaluation properties are distinguished for synthetic signatures [60]: *statistical*, *kinematical* (pressure, speed of writing, etc.), *geometric*, also called *topological*, and *uncertainty* (generated images can be intensively "infected" by noise) properties.

An algorithm for signature generation based on deformation has been introduced in [37]. Hollerbach [23] has introduced the theoretical basis of handwriting generation based on an oscillatory motion model. In Hollerbach's model, handwriting is controlled by two independent oscillatory motions superimposed on a constant linear drift along the line of writing. There are many papers on the extension and improvement of the Hollerbach model.

To generate signatures with any automated technique, it is necessary to consider: (a) the formal description of curve segments and their kinematical characteristics, (b) the set of requirements which should be met by any signature generation system, and (c) possible scenarios for signature generation.

Various characteristics are used in so-called *off-line* (static) analysis and *on-line* (kinematic) analysis of the acquired signature.

A model based on combining shapes and physical models in synthetic handwriting generation has been developed in [57]. The so-called *delta-log normal model* was developed in [42]. This model can produce smooth connections between characters, but can also ensure that the deformed characters are consistent with the models. In [10], it was proposed to generate character shapes by Bayesian networks. By collecting handwriting examples from a writer, a system learns the writers' writing style.

An example of a combined model based on geometric and kinematic characteristics (in-class scenario) is illustrated by Fig. 1.6.

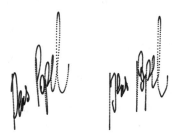

Fig. 1.6. In-class scenario: the original signature (left) and the synthetic one (right), courtesy of Prof. D. Popel, Baker University.

1.3.3. *Synthetic Retina and Iris Images*

Iris recognition systems scan the surface of the iris to compare patterns [13]. Retina recognition systems scan the surface of the retina and compare nerve patterns, blood vessels and such features. To the best of our knowledge, automated methods of *iris* and *retina image reconstruction*, or *synthesis* have not been developed yet, except for an approach based on generation of iris layer patterns [47,60].

Iris pattern painting has been used by ocularists in manufacturing glass eyes or contact lenses for sometime. The ocularist's approach to iris synthesis is based on the composition of painted primitives, and utilized layered semi-transparent textures built from topological and optic models [27]. These methods are widely used by today's ocularists: vanity contact lenses are available with fake iris patterns printed onto them (designed for people who want to change eye colors). Other approaches include image processing and synthesis techniques such as PCA combined with super-resolution [13], and random Markov field [46].

A synthetic image can be created by combining segments of real images from a database. Various operators can be applied to deform or warp the original iris image: translation, rotation, rendering etc. Various model of the iris, retina, and eye can be used to improve recognition, and can be found in [8,34,48]. In [2], a cancelable iris image design is proposed for the problem as follows. The iris image is intentionally distorted to yield a new version. For example, a simple permutation procedure is used for generating a synthetic iris.

An alternative approach is based on synthesis of patterns of the iris layers [60] followed by superposition of the layers and the pupil (black centre).

Below is an example of generating posterior pigment epithelia of the iris using the Fourier transform on a random signal. A fragment of the FFT signal is interpreted as a grey-scaled vector: the peaks in the FFT signal represent lighter shades and valleys represent darker shades. This procedure is repeated for other fragments as well. The data plotted in 3D, a 2D slice of the data, and a round image generated from the slice using the polar transform, are shown in Fig. 1.7.

Other layer patterns can be generated based on wavelet, Fourier, polar, and distance transforms, and Voronoi diagrams [60]. For example, Fig. 1.8 illustrates how a synthetic collarette topology has been designed using a Bezier curve in a cartesian plane. It is transformed into a concentric pattern, and superimposed with a random signal to form an irregular boundary curve.

Superposition of the patterns of various iris layers form a synthetic iris pattern. Figure 1.9 illustrates three different patterns obtained by this method.

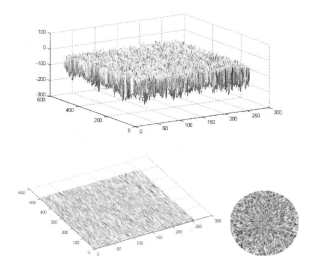

Fig. 1.7. A 3D grey-scale interpretation of the Fourier transform of a random signal, a slice of this image and the result of polar transform for a synthetic posterior pigment epithelia of the iris

Fig. 1.8. Synthetic collarette topology modelled by Bezier curves and a randomly generated curve.

Fig. 1.9. Synthetic iris patterns.

1.3.4. *Synthetic Speech and Voice*

Synthetic speech and voice have evolved considerably since the first experiments in the 1960s. New targets in speech synthesis include improving the audio quality and the naturalness of speech, developing techniques for emotional "coloring" [12,16,52,58], and combining it with other technologies, for example, facial expressions and lip movement [16,38,58]. Synthetic voice should carry information about age, gender, emotion, personality, physical fitness, and social upbringing. A closely related but more complicated problem is generating a synthetic singing voice for training singers, studying the famous singers' styles, and designing synthetic user-defined styles combining voice with synthetic music.

1.3.5. *Gait Modeling*

Gait recognition is defined as the identification of a person through the pattern produced by walking [14,35]. The potential of gait as a biometric was encouraged by the considerable amount of evidence available, especially in biomechanics literature [51,41]. A unique advantage of gait as biometrics is that it offers potential for recognition at a distance or at low resolution, when other biometrics might not be perceivable. As gait is behavioural biometrics there is much potential for within-subject variation [3]. This includes footwear, clothing and apparel. Recognition can be based on the (static) human shape as well as on movement, suggesting a richer recognition cue. Model-based techniques use the shape and dynamics of gait to guide the extraction of a feature vector.

Gait signature derives from bulk motion and shape characteristics of the subject, articulated motion estimation using an adaptive model and motion estimation using deformable contours; examples of all of these processes can be seen in Fig. 1.10.

The authors of [61] propose to use the gait biometrics in the pre-screened area of a security system.

1.3.6. *Synthetic Faces*

Face recognition systems detect patterns, shapes, and shadows in the face. The reverse process — face reconstruction — is a classical problem of criminology.

Many biometric systems are confused when identifying the same person smiling, aged, with various accessories (moustache, glasses), and/or in badly lit conditions (Fig. 1.11). Facial recognition tools can be improved by training on a set of synthetic facial expressions and appearance/environment

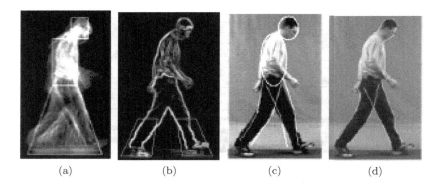

(a)   (b)   (c)   (d)

Fig. 1.10. Parameter extraction in gait model: shape estimation (a), period estimation (b), adaptive model (c), and deformable countours (d), courtesy of Prof. M. Nixon, University of Southampton, UK.

variations generated from real facial images.

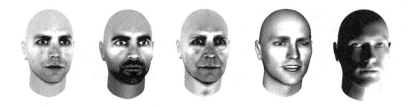

Fig. 1.11. Modeling of facial accessories, aging, drunk, and a badly lit faces.

A face model is a composition of various sub-models (eyes, nose, etc.) The level of abstraction in face design depends on the particular application. Traditionally, at the first phase of computer aided design, a generic (master) face is constructed. At the next phase, the necessary attributes are added. The composition of facial sub-models is defined by a global topology and generic facial parameters. The face model consists of the following facial sub-models: *eye* (shape, open, closed, blinking, iris size and movement, etc.), *eyebrow* (texture, shape, dynamics), *mouth* (shape, lip dynamics, teeth and tongue position, etc.), *nose* (shape, nostril dynamics), and *ear* (shape). Figure 1.12 illustrates one possible scheme for automated facial generation [60].

Usage of databases of synthetic faces in a facial recognition context has been reported in [55].

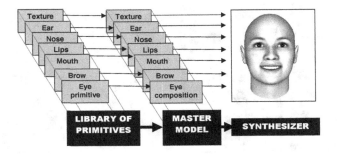

Fig. 1.12. Partitioning of the face into regions in the model for facial analysis and synthesis.

1.3.6.1. *Animation as Behavioral Face Synthesis*

An example of a direct biometric problem is identifying speech given a video fragment without recorded voice. The inverse problem is mimicry synthesis (animation) given a text to be spoken (synthetic narrator) [37,52,58]. Behavioral biometric information can also be used in evaluation of the truth in answers to questions, or the truth of a person speaking [17].

Facial expressions are formed by about 50 facial muscles that are controlled by hundreds of parameters. Psychologists distinguish two kinds of short-time facial expressions: *controlled* and *non-controlled* facial expressions [40]. Controlled expressions can be fixed in a facial model by generating control parameters, for example, a type of smile. Non-controlled facial expressions are very dynamic and are characterized by short time durations[a]. The difference between controlled and non-controlled facial expressions can be interpreted in various ways. The example below illustrates how to use short-term facial expressions in practice.

In Fig. 1.13, a sample of two images taken two seconds apart shows the response of a person to a question [60]. The first phase of the response is a non-controlled facial expression that is quickly transformed into another facial expression corresponding to the control face. The *facial difference* of topological information Δ, for example, in mouth and eyebrow configurations, can be interpreted by psychologists based on the evaluation of the first image as follows

---

[a]Visual pattern analysis and classification can be carried out in 100 *msec* and involves a minimum of 10 synaptic stages from the retina to the temporal lobe (see, for example, Rolls ET. Brain mechanisms for invariant visual recognition and learning, *Behavioural Processes, 33:113–138, 1994.*

$$\text{Mouth} = \begin{cases} \text{Irritation;} \\ \text{Aggression;} \\ \text{Discontent.} \end{cases} \qquad \text{Brows} = \begin{cases} \text{Unexpectedness;} \\ \text{Astonishment;} \\ \text{Embarrassment.} \end{cases}$$

Decision making is based on analysis of facial expression change while the person listens and responds to the question. More concretely, the *local facial difference* is calculated for each region of the face that carries short-term behavioural information. The local difference is defined as a change in some reliable topological parameter. The sum of weighted local differences is the *global facial difference*.

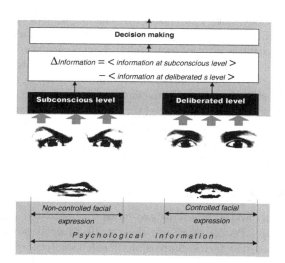

Fig. 1.13. The controlled and non-controlled phases of facial expressions.

1.3.6.2. *Caricature as Synthetic Face*

Caricature is the art of making a drawing of a face which makes part of its appearance more noticeable than it really is, and which can make a person look ridiculous. A caricature is a synthetic facial expression, where the distances of some feature points from the corresponding positions in the normal face have been exaggerated (Fig. 1.14).

Exaggerating the Difference from the Mean (EDFM) is widely accepted among caricaturists to be the driving factor behind caricature generation. The technique of assigning these distances is called a *caricature style*, i.e. the

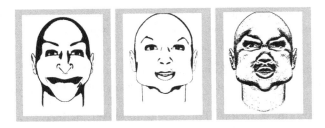

Fig. 1.14.   Three caricatures automatically synthesized given some parameters.

art-style of the caricaturist. The reason why the art-style of the caricaturist is of interest for image analysis, synthesis, and especially facial expression recognition and synthesis is as follows [19,28]. Facial caricatures incorporate the most important facial features and a significant set of distorted features. Some original features (coordinates, corresponding shapes, and the total number of features) in a caricature are very sensitive to variation, however the rest of the features can be distorted significantly. Restoration of a facial image based on caricatures is an inverse problem itself [29]. Various benefits are expected in identification, recognition and matching techniques, if the art-style of the caricaturist can be understood.

1.3.6.3. *Synthetic Emotions and Expressions*

Synthetic emotions and expressions are more sophisticated real world examples of synthesis. People often use their smile to mask sorrow, or mask gladness with a neutral facial expression. Such facial expressions can be thought of as artificial or synthetic in a social sense. Facial topologies are carriers of information, that is, emotions. Visual-band images along with thermal (infrared) images can be used in this task [38,54]. These results have been used, in particular, in a new generation of lie detectors [17,43].

1.4.   Examples of Usage of Synthetic Biometrics

In this section, we consider several problems where synthetic data is useful.

1.4.1. *Testing*

The commercially available synthetic fingerprints generator [6,7] has been used, in particular, in the Fingerprint Verification Test competition since 2003. An example of a tool used to create databases for fingerprints is SFinGe, developed at the University of Bologna (http://bias.csr.unibo.it/research/biolab/sfinge.html). The generated databases were entered in the Fingerprint Verification Competition FVC2004 and performed just as well as real fingerprints [18].

1.4.2. *Databases of Synthetic Biometric Information*

Collection of large databases of biometric data, such as fingerprints, is troublesome for many researchers due to the protection of personal information. Imitation of biometric data allows the creation of databases with tailored biometric data without expensive studies involving human subjects [59].

Usage of databases of synthetic faces in a facial recognition context has been reported in [55].

A simulator of biometric data is understood as a system for modeling specific conditions of intake and processing of biometric data. An example of such a system is a simulator for training bank officers (supported by signature imitation and handwriting character imitation), or a simulator for training customs officers (supported by a signature imitator, face imitator, and fingerprint imitator) [59]. The multi-biometric system constitutes the core of the simulator. It provides the user identification based on the traditional methods of biometrics. This basic configuration inherits all the pitfalls of current biometric systems. In the development of training simulators, such infrastructure can be configured to meet specific customer requirements.

It is difficult to satisfy the requirements of different standard testing methodologies because of the limited set of standard tests. Hence, developing a methodology for generating biometric tests (standard as well as special customer requirements) is an urgent problem.

1.4.3. *Humanoid Robots*

Humanoid robots are anthropomorphic robots (have human-like shape) that include also human-like behavioral traits. The field of humanoid robotics includes various challenging direct and inverse biometrics. Examples include:

Language technologies such as voice identification and synthesis, speech-to-text (voice analysis) and text-to-speech (voice synthesis)

Face and gesture recognition, to recognize and obey the "master", also to recognize the "moods" of the instructor, following of cue and to act intelligently depending on the mood context.

Vision, hearing, olfaction, tactile, (implemented through artificial retinas, e-nose, and e-tongue, etc.) provide senses analogous to those of humans, and allow the robot an analysis of the world and humand with whom it interacts.

On the other hand, in relation to inverse biometrics, robots attempt to generate postures, poses, face expressions to better communicate their human masters (or to each other) the internal states [53]). Robots such as Kismet express calm, interest, disgust, happiness, surprise, etc. (see (MIT, http://www.ai.mit.edu/projects/humanoid-robotics-group/kismet/). More advanced aspects include dialogue and logical reasoning similar to those of humans. As more robots would enter our society it will become useful to distinguish them among each other by robotic biometrics.

1.4.4. *Cancelable Biometrics*

The issue of protecting privacy in biometric systems has inspired the area of so-called *cancelable biometrics.* It was first initiated by the Exploratory Computer Vision Group at IBM T.J. Watson Research Center and published in [2]. Cancelable biometrics aim to enhance the security and privacy of biometric authentication through generation of "deformed" biometric data, i.e. synthetic biometrics. Instead of using a true object (finger, face), the fingerprint or face image is intentionally distorted in a repeatable manner, and this new print or image is used. If, for some reason, the old print or image is "stolen", an essentially "new" print can be issued by simply changing the parameters of the distortion process. This also results in enhanced privacy for the user since his true print is never used anywhere, and different distortions can be used for different types of accounts.

1.4.5. *Synthetic Biometric Data in the Development of a New Generation of Lie Detectors*

The features of the new generation of lie detectors include [17,43,60]: (a) Architectural characteristics (highly parallel configuration), (b) artificial intelligence support of decision making, and (c) New paradigms (non-contact testing scenario, controlled dialogue scenarios, flexible source use, and the possibility of interaction through an artificial intelligence supported machine-human interface). The architecture of the new generation of lie detectors includes (Fig. 1.15): an interactive machine-human interface, video and infrared cameras, and parallel hardware and software tools.

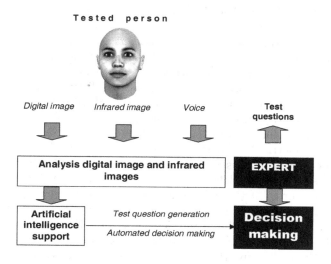

Fig. 1.15.　The next generation of non-contact lie detector system.

1.4.6.　*Synthetic Biometric Data in Early Warning and Detection System Design*

The idea of modeling biometric data for decision making support enhancement at checkpoints is explored, in particular, at the Biometric Technologies Laboratory at the University of Calgary (http://enel.btlab.ucalgary.ca) [61]. A facial model of a tested person is captured, and can be manipulated to mimic changes in visual and infrared bands caused by physiological changes during the questioning period. These can be compared against generic models of change based on statistical data. This approach can also be used to train personnel involved in questioning procedures. The decision making support system utilizes synthetic models for modeling of biometric data to support decision making.

Simulators of biometric data are emerging technologies for educational and training purposes (immigration control, banking service, police, justice, etc.). They emphasize decision-making skills in non-standard and extreme situations. For instance, the simulator for immigration control officer training must include various scenarios of generation of biometric data. This information is the system's input data. The system then analyzes the current biometric data (collates the passport photo with the present image, verifies signatures, analyzes handwriting, etc.).

1.5.   Biometric Data Model Validation

Data generated by various models are classified as *acceptable* or *unaccept-able* for further processing and use in various applications. The application-specific criteria must provide a *reasonable level of acceptability*. Accept-ability is defined as a set of characteristics which distinguish original and synthetic data. A model that approximates original data at reasonable lev-els of accuracy for the purpose of analysis is not considered a generator of synthetic biometric information.

Artificial biometric data must be verified for their meaningfulness. Sta-tistical model verification is accomplished by solving the equations that describe physics-based models, and obtaining the correct values. Model validation must prove if the equations that describe the model are right. Comparing the statistical distributions of real biometrics to the statistical distributions from empirical and physics-based models for a wide range of operational conditions validates these models for the range of conditions provided by the real biometric samples [30]. A simple method for validat-ing these distributions is via visual comparison of overlapped distributions. For example, Daugman used plots for comparing the hamming distances for 9.1 million iris comparisons to the Beta-binomial distribution, showing that the data fit the distribution remarkably well [15].

The MITRE research project [39] used synthetically generated faces to better understand the performance of face recognition systems. If a person's photo in the system's database was taken 10 years ago, is it possible to identify the person today? A pose experiment was also conducted with synthetic data to isolate and measure the effect of camera angle in one-degree increments.

The modeling technique will provide an effective, more structured basis for risk management in a large biometric system. This will help users choose the most effective systems to meet their needs in the future.

1.6.   Ethical and Social Aspects of Inverse Biometrics

Ethical and social aspects of inverse biometrics include several problems, in particular, the prevention of undesirable side-effects, and targeting of areas of social concern in biometrics. Prevention of undesirable side effects aims at studying the potential negative impacts of biometrics, as far as important segments of society are concerned, and how can these be prevented. The undesirable ethical and social effects of the solutions of inverse biometrics have not been studied yet. However, it is possible to predict some of them.

The particular examples of negative impact of synthetic biometrics are as follows:

(a) Synthetic biometric information can be used not only for improving the characteristics of biometric devices and systems, but also can be used by forgers to discover new strategies of attack.
(b) Synthetic biometric information can be used for generating multiple copies of original biometric information.

1.7. Conclusion

The concept of inverse biometrics arose from the *analysis-by-synthesis* paradigm, and has become an integral part of the modeling and simulation of human biometrics in many applications. Data generated by various models are used as databases (for example, databases of synthetic fingerprint) of synthetic biometrics for testing biometric hardware and software. The other application is biometric-based decision making support systems for security, banking, and forensic applications. A generator of synthetic biometric information (for example, an aging or surgically changed face), is a vital component of such systems. Yet another recently emerging application is the creation of simulators for training highly qualified personnel in biometric-based physical access control systems such as airport gates, hospital registration, and others. The ability to increase the reliability and accuracy of these systems while probing their vulnerabilities under numerous environmental conditions is critical, as biometrics becomes an essential part of law enforcement and security communities.

Acknowledgment

The authors acknowledge the help and suggestions of Dr. M. S. Nixon and Dr. D. J. Hurley.

Bibliography

1. Boles, W. and Boashash, B. (1998). A human identification technique using images of the iris and wavelet transform, *IEEE Trans. Signal Processing*, **46**, 4, pp. 1185–1188.
2. Bolle, R., Connell, J., Pankanti, S., Ratha, N. and Senior, A. (2004). *Guide to Biometrics*, Springer.

3. Boulgouris, N. V., Hatzinakos, D. and Plataniotis, K. N. (2005). Gait recognition: A challening signal processing technology for biometric identification, *IEEE Signal Processing Magazine*, November, pp. 78–90.
4. Brault, J. J. and Plamondon, R. (1993). A complexity measure of handwritten curves: modelling of dynamic signature forgery, *IEEE Trans. Systems, Man and Cybernetics*, **23**, pp. 400–413.
5. Buettner, D. J. and Orlans, N. M. (2005). A taxonomy for physics based synthetic biometric models, *Proc. 4th IEEE Workshop on Automatic Identification Advanced Technologies*, pp. 10–14.
6. Cappelli, R. (2003). Synthetic fingerprint generation, In D. Maltoni, D. Maio, A. K. Jain, and S. Prabhakar, Eds., *Handbook of Fingerprint Recognition*, pp. 203–232, Springer.
7. Cappelli, R. (2004). SFinGe: Synthetic fingerprint generator, *Proc. Int. Workshop Modeling and Simulation in Biometric Technology*, Calgary, Canada, pp. 147–154.
8. Can, A., Steward, C. V., Roysam, B. and Tanenbaum, H. L. (2002). A feature-based, robust, hierarchical algorithm for registering pairs of images of the curved human retina, *IEEE Trans. Analysis and Machine Intelligence*, **24**, 3, pp. 347–364.
9. Chalmond, B. (2003). Modeling and Inverse Problems in Image Analysis, *Applied Mathematical Sciences*, vol. 155, Springer.
10. Choi, H., Cho, S. J. and Jin Kim, J. H. (2003). Generation of handwritten characters with bayesian network based on-line handwriting recognizers, In *Proc. 17th Int. Conf. Document Analysis and Recognition*, Edinburgh, Scotland, pp. 995–999.
11. Cole, S. A. (2001). *Suspect Identities – A History of Fingerprinting and Criminal Identification,* Harvard University Press.
12. Cook, P. R. (2002). *Real Sound Synthesis for Interactive Applications*, A K Peters, Natick, MA.
13. Cui, J., Wang, Y., Huang, J., Tan, T., Sun, Z. and Ma, L. (2004). An iris image synthesis method based on PCA and super-resolution, *Proc. Int. Conf. Pattern Recognition*.
14. Cunado D., Nixon, M.S., and Carter, J.N. (2003). Automatic extraction and description of human gait models for recognition purposes, *Computer Vision and Image Understanding*, **90**, 1, pp. 1–14.
15. Daugman, J. (2003). The importance of being random: Statistical principles of iris recognition, *Pattern Recognition*, **36**, 2, pp. 279–291.
16. Du, Y. and Lin, X. (2002). Realistic mouth synthesis based on shape appearance dependence mapping, *Pattern Recognition Letters*, **23**, pp. 1875–1885.
17. Ekman, P. and Rosenberg, E. L., Eds. (1997). *What the Face Reveals: Basic and Applied Studues of Spontaneouse Expression Using the Facial Action Coding System (FACS)* Oxford University Press.
18. The Fingerprint Verification Competition FVC2004 http://bias.csr.unibo.it/fvc2004/databases.asp
19. Fujiwara, T., Koshimizu, H., Fujimura, K., Kihara, H., Noguchi, Y. and Ishikawa, N. (2001). 3D Modeling System of Human Face and Full 3D Fa-

cial Caricaturing. In *Proc. 3rd IEEE Int. Conf. 3D Digital Imaging and Modeling,* Canada, pp. 385–392.

20. Guyon, I. (1996). Handwriting synthesis from handwritten glyphs, *Proc. 5th Int. Workshop Frontiers of Handwriting Recognition,* Colchester, UK, pp. 309-312.

21. Hurley, D. J. (2006). Synthetic ear biometric based on force field modeling, *Private Communication,* University of Southampton, UK.

22. Hurley, D. J. Nixon, M. S. and Carter, J. N. (2005). Force field feature extraction for ear biometrics, *Computer Vision and Image Understanding,* **98**, pp. 491–512.

23. Hollerbach, J. M. (1981). An oscilation theory of handwriting, *Biological Cybernetics,* **39**, pp. 139–156.

24. Jain, A. K. Ross, A. and Prabhakar, S. (2004). An introduction to biometric recognition, *IEEE Trans. Circuit and Systems for Video Technology,* **14**, 1, pp. 4–20.

25. Koch, R. (1993). Dynamic 3-D scene analysis through synthesis feedback control, *IEEE Trans. Pattern Analysis and Machine Intelligence,* **15**, 6, pp. 556–568.

26. Kuecken, M. U. and Newell, A. C. (2004). A model for fingerprint formation, *Europhysics Letters* **68**, 1, pp. 141–146.

27. Lefohn, A., Budge, B., Shirley, P., Caruso, R. and Reinhard, E. (2003). An ocularists approach to human iris synthesis, *Computer Graphics and Applications,* IEEE Magazine, **23**, 6, pp. 70–75.

28. Luo W. C., Liu, P. C. and Ouhyoung, M. (2002). Exaggeration of Facial Features in Caricaturing, *Proc. Int. Computer Symposium,* China.

29. Luo, Y., Gavrilova, M. L., Sousa, M. C., Pivovarov, J. and Yanushkevich, S. (2005). Morphing Facial Expressions from Artistic Drawings, In T. Simos, G. Maroulis, Eds., *Advanced in Computational Methods in Sciences and Engineering. Lecture Series on Computer and Computational Sciences,* Brill Academic Publishers, The Netherlands, Vol. 4, pp. 1507–1511.

30. Ma, Y., Schuckers, M. and Cukic, B. (2005). Guidelines for appropriate use of simulated data for bio-authentication research, *Proc. 4th IEEE Workshop Automatic Identification Advanced Technologies,* Buffalo, New York, pp. 251–256.

31. Manolakis, D. and Shaw, G. (2002). Detection algorithms for hyperspectral imaging applications, *IEEE Signal Processing Magazin,* **19**, pp. 29–43.

32. Mansfield, A. and Wayman, J. (2002). Best Practice Standards for Testing and Reporting on Biometric Device Performance, *National Physical Laboratory of UK.*

33. Matsumoto, H., Yamada, K. and Hoshino, S. (2002). Impact of Artificial Gummy Fingers on Fingerprint Systems, In *Proc. SPIE, Optical Security and Counterfeit Deterrence Techniques* IV, **4677**, pp. 275–289.

34. Moriyama, T., Xiao, J., Kanade, T. and Cohn, J. F. (2004). Meticulously Detailed Eye Model and its Application to Analysis of Facial Image. *Proc. IEEE Int. Conf. Systems, Man, and Cybernetics,* pp. 629–634.

35. Nixon, M. S., Carter, J. N., Grant, M. G., Gordon, L. and Hayfron-Acquah,

J. B. (2003). Automatic recognition by gait: progress and prospects, *Sensor Review* **23**, 4, pp. 323-331,

36. Okabe, A. Boots, B. and Sugihara, K. (1992). *Spatial Tessellations. Concept and Applications of Voronoi Diagrams.* Wiley, New York.

37. Oliveira, C., Kaestner, C. Bortolozzi, F. and Sabourin, R. (1997). Generation of signatures by deformation, *Proc. the BSDIA97*, pp. 283-298, Curitiba, Brazil.

38. Oliver, N. Pentland, A. P. and Berard, F. (2000). LAFTER: a real-time face and lips tracker with facial expression recognition, *Pattern Recognition*, **33**, 8, pp. 1369-1382.

39. Orlans, N. M., Buettner, D.J. and Marques, J. (2004). A Survey of Synthetic Biometrics: Capabilities and Benefits, *Proc. Int. Conf. Artificial Intelligence*, CSREA Press, vol. I, pp. 499-505.

40. Pantic, M. and Rothkrantz, L. J. M. (2000). Automatic analysis of facial expressions: the state-of-the-art, *IEEE Trans. Pattern Analysis and Machine Intelligence*, **22**, 12, pp. 1424-1445.

41. Pappas, I.P.I., Popovic, M.R., Keller, T., Dietz, V., and Morari, M. (2001). A reliable gait phase detection system, *IEEE Transaction on Neural System Rehabilitation Engineering*, June, **9**, 2, pp. 113-125.

42. Plamondon, R. and Guerfali, W. (1998). The generation of handwriting with delta-lognormal synergies, *Bilogical Cybernetics*, **78**, pp. 119-132.

43. *The Polygraph and Lie Detection.* The National Academies Press, Washington, DC, 2003.

44. Popel, D. (2006). Signature analysis, verification and synthesis in pervasive environments, This issue.

45. Reed, I. S. and Yu, X. (1990). Adaptive multiple-band CFAR detection of an optical pattern with unknown spectral distribution, *IEEE Trans. Acoustic, Speech and Signal Processing.* **38**, pp. 1760-1770.

46. Makthal, S. and Ross, A. (2005). Synthesis of iris images using Markov random fields, *Proc. 13th European Signal Processing Conf.*, Antalya, Turkey.

47. Samavati, F. F., Bartels, R. H., and Olsen, L. (2006). Local B-spline multiresolution with example in iris synthesis and volumetric rendering, this issue.

48. Sanchez-Avila, C. and Sanchez-Reillo, R. (2002). Iris-based biometric recognition using dyadic wavelet transform, *IEEE Aerospace and Electronic Systems Magazine*, October, pp. 3-6.

49. Shmerko, V., Phil Phillips, Kukharev, G., Rogers, W. and Yanushkevich, S. (1997). Biometric technologies, *Proc. Int. Conf. The Biometrics: Fraud Prevention, Enchanced Service*, Las Vegas, Nevada, pp. 270-286.

50. Shmerko, V., Phil Phillips, Rogers, W., Perkowski, M. and Yanushkevich, S. (2000). Bio-technologies, *Bulletin of Institute of Math-Machines*, Warsaw, Poland, ISSN 0239-8044, **1**, pp. 7-30.

51. Sloman, L., Berridge, M., Homatidis, S., Dunter, D., and Duck, T. (1982). Gait patterns of depressed patients and normal subjects, *American Journal of Psyhiatry*, **139**, 1, pp. 94-97.

52. Sproat, R. W. (1997). *Multilingual Text-to-Speech Synthesis: The Bell Labs*

Approach, Kluwer.

53. Stoica, A. (1999). Learning Eye-Arm Coordination Using Neural and Fuzzy Neural Techniques, In H. N. Teodorescu, A. Kandel, and L. Jain, Eds., *Soft Computing in Human-Related Sciences*, pp. 31–61, CRC Press, Boca Raton, FL.

54. Sugimoto, Y., Yoshitomi, Y. and Tomita, S. (2000). A Method for detecting transitions of emotional states using a thermal facial image based on a synthesis of facial expressions, *Robotics and Autonomous Systems*, **31**, pp. 147–160.

55. Sumi, K., Liu, C. and Matsuyama, T. (2006). Study on Synthetic Face Database for Performance Evaluation, *Proc. Int. Conf. Biometric Authentication*, pp. 598–604, LNCS-3832, Springer.

56. Tilton, C. J. (2001). An Emergin Biometric Standards, *IEEE Computer Magazine*. Special Issue on Biometrics, **1**, pp. 130–135.

57. Wang, J., Wu, C., Xu, Y. Q., Shum, H. Y. and Li, L. (2002). Learning Based Cursive Handwriting Synthesis, *Proc. 8th Int. Workshop Frontiers in Handwriting Recognition*, Ontario, Canada, pp. 157–162.

58. Yamamoto, E, Nakamura, S. and Shikano, K. (1998). Lip Movement Synthesis From Speech Based on Hidden Markov Models, *Speech Communication* **26**, 1,2, pp. 105–115

59. Yanushkevich, S. N., Stoica, A. Srihari, S. N., Shmerko, V. P. and Gavrilova, M. L. (2004). Simulation of Biometric Information: The New Generation of Biometric Systems, In *Proc. Int. Workshop Modeling and Simulation in Biometric Technology*, Calgary, Canada, pp. 87–98.

60. Yanushkevich, S. N., Stoica, A., Shmerko, V. P. and Popel, D. V. (2005). *Biometric Inverse Problems*, Taylor & Francis/CRC Press, Boca Raton, FL.

61. Yanushkevich, S. N., Stoica, A. and Shmerko, V. P. (2006). Fundamentals of Biometric-Based Training System Design, this issue.

Chapter 2

Signature Analysis, Verification and Synthesis in Pervasive Environments

Denis V. Popel

Biometric Technology Group,
Computer Science Department, Baker University,
P.O. Box 65, Baldwin City, KS 66006, USA
denis.popel@bakeru.edu

Despite the wide appreciation of biometric principles in security applications, biometric solutions are far from being affordable and available "on demand" anytime and anywhere. Many security biometric solutions require dedicated devices for data acquisition delaying their deployment and limiting the scope. The chapter focuses primarily on analysis of data taken from a human signatures for his/her authentication or identification. Also the chapter introduces a system developed to identify and authenticate individuals based on their signatures and/or handwriting. The issues of pervasive services are addressed (i) by integrating unique data acquisition and processing techniques which are capable of communicating with a variety of off-the-shelf devices such as pressure sensitive pens, mice, and touch pads, (ii) by using sequence processing techniques (like matching, alignment or filtering) for signature analysis techniques and comparison, (iii) by using the self-learning database solutions for achieving accurate results, and (iv) by utilizing signature synthesis techniques for benchmarking and testing.

Contents

2.1. Introduction

The research in signature biometrics is traditionally interdisciplinary and includes such fields as data retrieval and scanning, sequence processing and understanding physical properties, image processing and pattern recognition, and even physiology and psychology. Signature processing deals with behavioural human characteristics, and can be classified as *behavioural biometrics* [5]. These characteristics can be captured by various input devices such as scanners and tablets. Currently there is no universal standard to represent signature biometrics. The variety of input devices with their own formats complicates the situation for data storage and processing. Human involvement with societal and legal issues, such as protection of privacy and legality of information gathering, makes the situation even more complicated.

There are many techniques used (see [3,5,10,9,15,16] for details) to acquire and process signatures and/or handwriting depending on the target application area (e.g. user authentication or document automation), signature and handwriting producer (e.g. written by an ink pen or gyro-pen) and image acquisition devices (e.g. flat scanners or pressure sensitive tablets). Further, there is no standard representation for this wide variety of signature images. However, our preliminary study [8,11] suggests that this question is answerable by an appropriate architecture of the system which integrates mathematical models, formats, and self-learning databases. Our research demonstrates that the majority of contour images should be modelled using graph-based approaches, or, at the very least, using approaches derived from continuous or multiple-valued (as opposed to binary) logic [14]. Some known modeling systems discard graph information by preserving only sequential representations. This explains many obvious limitations of current systems such as the lack of a unified approach for image transformation operations, strong dependence on the resolution of image acquisition systems, and the inability of software to make use of the originally graph-based nature of some images. By using proper formats, the tasks below can be accomplished faster, easier, and at a lower cost.

Some biometric characteristics, such as irises, retinas, and fingerprints, do not change significantly over time, and thus they have low in-class variation. The in-class variations for signatures can be large meaning that signatures from a single individual tend to be different. Even forensic experts with long-term experience cannot always make a decision whether a signature is authentic or forged [1]. Signatures are easier to forge than, for example, irises or fingerprints. However, the signature has a long standing tradition in many commonly encountered verification tasks, and signature verification is accepted by the general public as a non-intrusive procedure while hand prints and fingerprint verification, on the other hand, are associated with criminal investigations. It is assumed in automated signature processing that an individual's signature is unique. The state-of-the-art tools for automated signature analysis and synthesis rely partially or totally on methods of artificial intelligence. Some of these methods are mentioned in this chapter.

There are two types of automated signature processing: *off-line* and *on-line*. Off-line signature processing is used when a signature is available on a document. It is scanned to obtain its digital representation. On-line signature processing relies on special devices such as digitizing tablets with pens and/or gyropens to record pen movements during writing. In on-line processing, it is possible to capture the dynamics of writing, and, therefore, it is difficult to forge. Even with an image of the signature available, dynamic information (kinematics) pertinent to a signature is not as readily available to a potential forger as is the shape of signature (geometry). Skilled forgers can reproduce signatures to fool the unskilled eye. Human experts can distinguish genuine signatures from the forged ones. Modeling the invariance in the signatures and automating the signature recognition process pose significant challenges.

Aiming to solve some critical problems in pervasive signature processing, this chapter describes a system built to analyze and simulate human signatures. It is accomplished by (i) introducing some widely used signature descriptions, (ii) outlining *a method for aligning sequences of multi-valued and continuous data* for signature verification, and (iii) describing *a methodology of automated signature generation* for testing. It also introduces an efficient algorithm based on matching principles which is capable of measuring distances between signatures and aligning them to fulfil maximum correlation requirement. Overall, the system exploits the complete cycle of signature processing which includes representation, analysis, modeling, matching, and simulation.

This chapter is structured as four main blocks: an overview of signature representation methods is given in Section 2.2, signature comparison methods and algorithms for sequence alignment and matching are outlined in Section 2.3, some techniques for signature synthesis are introduced in Section 2.4, the overall architecture of the system as well as implementations issues are given in Section 2.5. Section 2.6 concludes the chapter and presents possible extensions.

2.2. Signature Representation

Many types of images like signatures can be classified as contour images or *line objects* stored on paper or electronic media. Line objects carry both *geometric* and *kinematic* information and cannot be handled in the same manner as *raster* or 2D images. In signatures, the line shape as well as its thickness, and other characteristics along the contour, which can be retrieved from the completed image, represent geometric information, while characteristics such as speed of writing or exerted pressure along the line drawn can only be retrieved in the process of creating the image and thus represent kinematic information.

Example 2.1. Both geometric and kinematic characteristics can be merged into a single 3D representation. The 3D representation of a signature is given in Fig. 2.1.

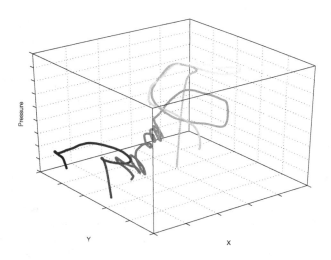

Fig. 2.1.   3D view of the on-line signature: the plain curve is given by the two-tuple (X, Y), the pressure is associated with the Z axis, the speed of writing is depicted by the shade of the curve, where darker is slower speed (Example 2.1).

The simplest form of the handwriting is represented by a plane curve. There are three commonly used descriptions of plane curves:

Implicit form: $f(x, y) = 0$;
Explicit form: $y = f(x)$, considered as a function graph;

Parametric form: $\mathbf{r}(t) = (x(t), y(t))$, using Cartesian parametric equations $x = x(t)$ and $y = y(t)$.

The latter form has been selected for further plane curve description of handwriting. Such a selection is justified by the nature of data which is acquired in a parameterized format. We use the following characteristics of the plane curve representing handwriting with an arbitrary parameter t (Fig. 2.2): position, velocity, speed, and acceleration.

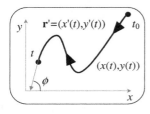

Position: $\mathbf{r}(t) = (x(t), y(t))$;

Velocity:
$$\mathbf{r}'(t) = (x'(t), y'(t));$$

Speed:
$$|\mathbf{r}'(t)| = |x'(t), y'(t)|;$$

Acceleration:
$$\mathbf{r}''(t) = (x''(t), y''(t)).$$

Fig. 2.2. Characteristics of the plane curve representing handwriting with an arbitrary parameter t.

If $t \in [t_0, t] \mapsto \mathbf{r}(t)$ with velocity $\mathbf{v}(t) = \mathbf{r}'(t)$ and speed $|\mathbf{v}(t)|$, then

$$s(t) = \int_{t_0}^{t} |\mathbf{v}(t)| \mathrm{d}t \tag{2.1}$$

is called the arc lengthof the curve. For a plane curve:

$$s(t) = \int_{t_0}^{t} \sqrt{x'(t)^2 + y'(t)^2} \tag{2.2}$$

The arc length is independent of the parameterization of the curve. The value $\frac{\mathrm{d}s(t)}{\mathrm{d}t} = s'(t) = \sqrt{x'(t)^2 + y'(t)^2}$ indicates the rate of change of arc length with respect to the curve parameter t. In other words, $s'(t)$ gives the speed of the curve parameterization.

For convenience, we also parameterize the curve with respect to its arc length s: $\mathbf{r}(s) = (x(s), y(s))$. The parameterization defines a direction along the curve on which s grows. Let $\mathbf{t}(s)$ be the *unit tangent vector*associated with the direction. Denote by $\varphi(s)$ the angle between the tangent at a point $(x(s), y(s))$ and the positive direction of the axis x. The *curvature* measures the rate of bending as the point moves along the curve with unit speed and can be defined as

$$k(s) = \frac{\mathrm{d}\varphi(s)}{\mathrm{d}s}. \tag{2.3}$$

The curvature is the length of the acceleration vector if $\mathbf{r}(t)$ traces the curve with constant speed 1. A large curvature at a point means that the curve is strongly bent. Unlike the acceleration or the velocity, the curvature does not depend on the parameterization of the curve (Fig. 2.3).

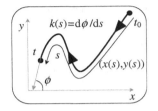

Unit tangent vector:
$$\mathbf{t}(t) = \frac{(r)'(t)}{|\mathbf{r}'(t)|}$$

Curvature: $k(t) = \frac{|\mathbf{t}'(t)|}{|\mathbf{r}'(t)|}$

$$k(t) = \frac{|\mathbf{r}'(t) \times \mathbf{r}''(t)|}{|\mathbf{r}'(t)|^3}$$

Fig. 2.3.   The curvature measures.

Curvatures are used in the statistical analysis of some contour shapes [4]. Curvature possesses the useful beneficial features for signature processing:

- Invariant to shifting (transition) and rotation;
- Gives a unique representation of the curve up to rigid motion; and
- Inversely proportional to scaling: the curve $c\alpha(t)$ has the curvature $\frac{k(t)}{c}$ for $c > 0$.

But the most important feature is that the computation of curvature is a reversible operation: by performing the reverse sequence of steps (numerical integration), it is possible to restore the original image. This fact has been used in the comparison method presented below where signature matching is performed by graph comparison and sequence matching and alignment.

2.3.   Signature Comparison

There are two approaches to signature comparison: static and dynamic. In the static signature comparison, only geometric (shape) features such as curvatures and angles are used for authentication of identity. Typically, the signature impressions are normalized to a known size and decomposed into simple components (strokes). The shapes and stroke relationships are used as features forming a graph structure. In the dynamic signature comparison, not only the shape features are used for authentication of the signature but the dynamic features like pressure and velocity are also employed. The signature impressions are processed as in the static signature comparison. Invariants of the dynamic features augment the static features, making forgery difficult since the forger must know not only the impression of the signature but also the way the impression was made. The following sequential characteristics recorded along the trajectory are used for sequence matching: curvature (geometric), pressure (kinematic), and velocity (kinematic).

2.3.1. *Sequence Alignment*

Let us introduce the following notation that will represent the sequence (vector) of n values $A = \{a_1, a_2, \ldots, a_n\}$. Another vector of size m that will be used for comparison is $B = \{b_1, b_2, \ldots, b_m\}$. The parametric arrangement of these sequences is based on an order index (e.g., a time stamp).

An *alignment* of two sequences, A and B, is a pairwise match between the characters of each sequence. Three arrangements are possible for comparing two sequences: *global alignment* which returns the similarity score for entire vectors; *local alignment* which finds the longest common subsequence; and *semiglobal alignment* which compares a short sequence and a longer sequence without considering both ends of the sequence. The focus of this section is on global alignment methods only. Three kinds of changes can occur at any given position within a sequence: (i) a mutation that replaces one value with another; (ii) an insertion that adds one or more positions; or (iii) a deletion that eliminates one or more positions. Since there are no common rules for value insertion or deletion in the sequences of curvatures or velocities, *gaps* in alignments are commonly added to reflect the occurrence of this type of change. The pairwise alignment for the sequences A and B results in a two-row matrix built in a way such that removing all gaps ('-') from the first row gives A, and removing gaps from the second row gives B. A column with two gaps is not allowed.

The problem of finding the optimal matching or alignment can be solved using dynamic programming, which is described by using a matrix M with n rows and m columns. One sequence is placed along the top of the matrix and the other sequence is placed along the left side. The process of finding the optimal matching between two sequences is the process of finding an optimal path from the top left corner to the bottom right corner of the matrix. Any step in any path can only be done to the right, down or diagonal. Every diagonal move corresponds to match or mismatch, whereas a right move corresponds to the insertion of a gap in the vertical sequence, and a down move corresponds to the insertion of a gap in the horizontal sequence.

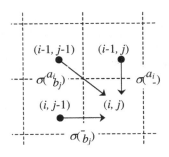

More formally, there are three possibilities: $\begin{pmatrix} a \\ b \end{pmatrix}$, $\begin{pmatrix} - \\ b \end{pmatrix}$, and $\begin{pmatrix} a \\ - \end{pmatrix}$. The figure on the left depicts the cell (i, j) and possible scores. For each aligned pair $\begin{pmatrix} \mathcal{A} \\ \mathcal{B} \end{pmatrix}$, where $\mathcal{A}$ and $\mathcal{B}$ are either normal sequence entries or gaps, there is an assigned score $\sigma\begin{pmatrix} \mathcal{A} \\ \mathcal{B} \end{pmatrix}$. The total score of a pairwise alignment is defined to be the sum of the σ values of all aligned pairs.

Objective function. The central problem in sequence matching is to find a mathematical function, so called objective function, which will be able to measure the quality of an alignment. Such an objective function should incorporate everything that is known about the sequences including their structure, functionality and a priory data. These data are rarely known and usually being replaced with sequence similarity based on metrics and scoring.

Metrics and scoring. The main idea of metrics is to get a universal mechanism of comparing two sequences. Metrics can vary depending on the objective function, they can embrace distance measures and/or correspond to statistical computations. The scoring is determined by the amount of credit an alignment receives for each aligned pair of identical entities (the *match* score), the penalty for aligned pairs of non-identical entities (the *mismatch score*), and the *gap penalty* for the alignment with gaps. A simple alignment score can be computed as follows:

$$\sum_{k=1}^{K} \begin{cases} gap\ penalty;\ if\ a_i =' -'\ or\ b_j =' -', \\ match\ score;\ if\ no\ gaps\ and\ a_i = b_j, \\ mismatch\ score;\ if\ no\ gaps\ and\ a_i \neq b_j. \end{cases}$$

Data alphabet. The alphabet for sequence processing in case of curvature, velocity and pressure are represented originally by continuous values and cab be encoded by integers forming a multiple-valued representation. This encoding procedure can be based on learning rules and/or discretization principles for continuous sequences [13].

Example 2.2. Let us apply the alignment technique described above for two sequences $\{1, 2, 1, 4, 2, 1, 4\}$ and $\{1, 3, 2, 3, 4\}$ assuming that the gap penalty is -1, the match score is $+1$, and the mismatch score is 0. The global alignment of the two sequences is equivalent to a path from the upper left corner of the matrix to the lower right. A horizontal move in the matrix represents a gap in the sequence along the left axis. A vertical move represents a gap in the sequence along the top axis, and a diagonal move represents an alignment of nucleotides from each sequence. To fill in the matrix, we take the maximum value of the three choices (moves).

		1	3	2	3	4
	0	−1	−2	−3	−4	−5
1	−1					
3	−2					
1	−3					
4	−4					
2	−5					
1	−6					
4	−7					

		1	3	2	3	4
	0↘	−1	−2	−3	−4	−5
1	−1	1↘	0	−1	−2	−3
3	−2	0	2↓	1	0	−1
1	−3	−1	1↓	2	1	0
4	−4	−2	0↘	1	2	2
2	−5	−3	−1	1↘	1	2
1	−6	−4	−2	0	1↘	1
4	−7	−5	−3	−1	0	**2**

Using the global alignment algorithm, we obtain the following alignment:
$\left\{ \begin{matrix} 1 & 3 & - & - & 2 & 3 & 4 \\ 1 & 3 & 1 & 4 & 2 & 1 & 4 \end{matrix} \right\}$. The total score is 2.

Extending the technique presented above, let us outline a new practical method for aligning multiple-valued and continuous sequences to calculate the degree of similarity for signatures. The standard algorithms of Needleman-Wunsch [7] and of Smith-Waterman [17] align sequences by maximizing a score function that favors matching element pairs over mismatches and gaps. They tend to become sensitive to the choice of scoring parameters and therefore less reliable with increasing distance between sequences. Therefore, for aligning multiple-valued and continuous sequences we selected the scoring mechanism that depends on data only.

New objective function. Since each element of either multiple-valued or continuous sequence is described by a numerical value, we can use a limited set of arithmetic operations on them. Intuitively, the objective function for the alignment process should minimize the differences between two given sequences. Moreover, there is no distinction between matches and mismatches in aligning of multiple-valued and continuous sequences, and gaps can be replaced by values resulted from interpolation or averaging.

New metrics and scoring. The following scoring based on differences in values is suggested:

$$\min \begin{cases} M(i, j-1) + \Delta \\ M(i-1, j-1) + \Delta, \ where \ \Delta = |a_i - b_j|. \\ M(i-1, j) + \Delta \end{cases}$$

We use the scoring system which helps to minimize the differences between two arbitrary sequences. Thus, an optimal matching is an alignment with the lowest score. The score for the optimal matching is then used to find the alignment which possesses the following properties:

Property 1. The sequences can be shifted left or right, and the alignment will not be affected.

Property 2. Gaps are allowed to be inserted into the middle, beginning or end of sequences. The gap is represented by a value from the affected sequence (multiple-valued case), or the average/interpolated value from the neighborhood (continuous case).

Alignment of continuous sequences

There are few existing techniques for measuring correlation between two continuous sequences, like Cramér-von Mises [2], Kuiper [6] and Watson [19], but none of them gives the sequence of aligned segments as the result of performed measurements. In the proposed method, both a correlation score and two aligned sequences are returned.

Example 2.3. Let us consider two continuous sequences $\{0.3, 0.1, 0.5, 0.7, 0.0\}$ and $\{0.1, 0.5, 0.5, 0.0, 0.1\}$. These sequences are placed along the top and left margins (left panel). Following the dynamic programming technique and new metrics previously explained, we obtain both the total score and the alignment (right panel).

	0.3	0.1	0.5	0.7	0.0
0.1					
0.5					
0.5					
0.0					
0.1					

	0.3	0.1	0.5	0.7	0.0
0.1	**0.2→**	**0.2↘**	0.6	1.2	1.3
0.5	0.4	0.6	**0.2↘**	0.4	0.9
0.5	0.6	0.8	0.2	**0.4↘**	0.9
0.0	0.9	0.7	0.7	0.9	**0.4↓**
0.1	1.1	0.7	1.1	1.3	**0.5**

The resulted aligned sequences with gaps and interpolated values are:

$$\left\{\begin{array}{cccccc} 0.3 & 0.1 & 0.5 & 0.7 & 0.0 & - \\ - & 0.1 & 0.5 & 0.5 & 0.0 & 0.1 \end{array}\right\}, \quad \left\{\begin{array}{cccccc} 0.3 & 0.1 & 0.5 & 0.7 & 0.0 & 0.0 \\ 0.1 & 0.1 & 0.5 & 0.5 & 0.0 & 0.1 \end{array}\right\}.$$

The total alignment score is 0.5.

The method has been evaluated on sample signature data from an image processing system [14]. The alignment is done for two continuous sequences which represent the speed of writing along the trajectory for the same individual. The upper panel of Fig. 2.4 gives the original unaligned sequences, the panel below shows the output of the alignment.

　　　Exploring various formats of data, the extensions based on cumulative and difference data processing are outlined below.

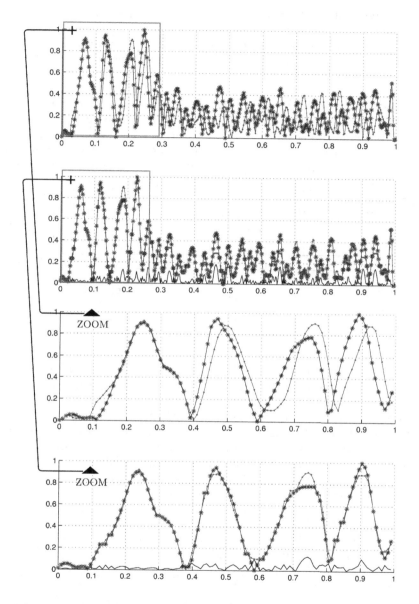

Fig. 2.4.   Unaligned and aligned continuous sequences (from top to bottom): unaligned sequences, aligned sequences, zoomed view of unaligned sequences, and zoomed view of aligned sequences. The panels show the difference by the solid black line.

Cumulative data. A "rough" alignment can be performed linearly by using cumulative distributions. Such a "rough" alignment can be used in two possible scenarios: as a preliminary step to identify regions that will be further aligned by the dynamic alignment algorithm; and as a real-time alignment if necessary.

Example 2.4. (Continuation of Example 2.3) Let us consider two sequences $\{0.3, 0.1, 0.5, 0.7, 0.0\}$ and $\{0.1, 0.5, 0.5, 0.0, 0.1\}$, and cumulative counterparts $\{0.3, 0.4, 0.9, 1.6, 1.6\}$ and $\{0.1, 0.6, 1.1, 1.1, 1.2\}$. Following the dynamic programming technique and new metrics introduced previously, we build the alignment matrix and restore the aligned sequences.

	0.3	0.4	0.9	1.6	1.6
0.1					
0.6					
1.1					
1.1					
1.2					

	0.3	0.4	0.9	1.6	1.6
0.1	**0.2**╲	0.5	1.2	2.4	3.1
0.6	0.4	**0.4**╲	0.7	1.7	2.7
1.1	1.2	1.1	**0.6**╲	1.1	1.6
1.1	2.0	1.8	0.8	**1.1**╲	1.6
1.2	2.9	2.6	1.1	1.2	**1.6**╲

The resulted aligned sequences with the total alignment score of 1.6 are:
$\left\{ \begin{matrix} 0.3 & 0.4 & 0.9 & 1.6 & 1.6 \\ 0.1 & 0.6 & 1.1 & 1.1 & 1.2 \end{matrix} \right\}$, but this does not reflect the alignment of original sequences.

It is illustrated in Fig. 2.5 that the restoration of initial sequences based on aligned sequences of cumulative data leads to misalignment.

Finite differences. The same method works for the alignment of finite differences. Thus, aligning the differences gives some clues about the dynamics of the process, not just statical characteristics. Sequences aligned by using their finite differences depict where the speed (of changing the values) is identical. If necessary, the normalization takes place to perform level adjustments.

Example 2.5. (Continuation of Example 2.3) Let us consider the sequences $\{0.3, 0.1, 0.5, 0.7, 0.0\}$ and $\{0.1, 0.5, 0.5, 0.0, 0.1\}$, and their difference counterparts $\{0.0, -0.2, 0.4, 0.2, -0.7\}$ and $\{0.0, 0.4, 0.0, -0.5, 0.1\}$.

	0.0	-0.2	0.4	0.2	-0.7
0.0					
0.4					
0.0					
-0.5					
0.1					

	0.0	-0.2	0.4	0.2	-0.7
0.0	**0.0**→	**0.2**╲	0.6	0.8	1.5
0.4	0.4	0.6	**0.2**╲	0.4	1.5
0.0	0.4	0.6	0.6	**0.4**╲	1.1
-0.5	0.9	0.7	1.5	1.1	**0.6**↓
0.1	1.0	1.0	1.0	1.1	**1.4**

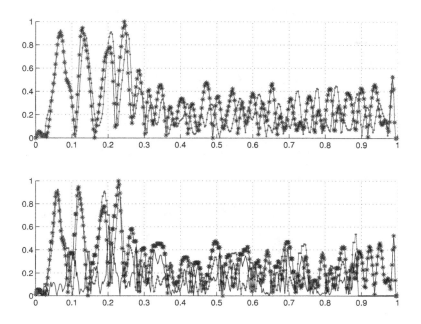

Fig. 2.5. Unaligned sequences (top) and misaligned sequences restored based on cumulative data processing (bottom).

The total alignment score is 1.4. The resulted aligned sequences with gaps and interpolated values are:

$$\left\{\begin{array}{cccccc} 0.0 & -0.2 & 0.4 & 0.2 & -0.7 & - \\ - & 0.0 & 0.4 & 0.0 & -0.5 & 0.1 \end{array}\right\} \text{ and } \left\{\begin{array}{cccccc} 0.0 & -0.2 & 0.4 & 0.2 & -0.7 & -0.7 \\ 0.0 & 0.0 & 0.4 & 0.0 & -0.5 & 0.1 \end{array}\right\}.$$

It reflects the alignment of original sequences.

Figure 2.6 depicts two unaligned (upper panel) and aligned (lower panel) sequences of finite differences. It is shown in Fig. 2.7 that the restoration of initial sequences based on aligned sequences of difference data gives well aligned sequences.

2.3.2. *Algorithm and Experimental Results*

Since the exhaustive search for all possible alignments is an NP-complete problem, it is impossible to compute in reasonable amount of time any real sequences. By applying dynamic programming principles, the computational complexity is in $O(n \times m)$. The algorithm works in four steps:

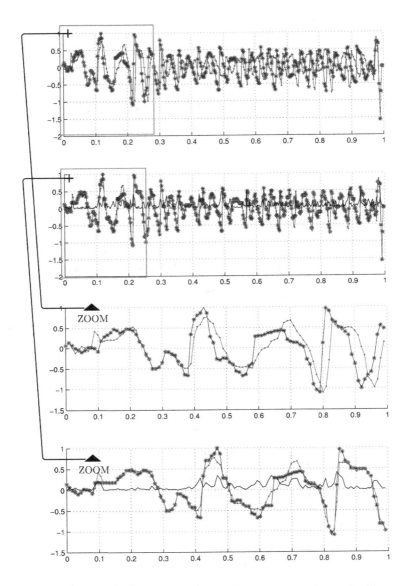

Fig. 2.6.   Unaligned and aligned sequences of finite differences (from top to bottom): unaligned sequences, aligned sequences, zoomed view of unaligned sequences, and zoomed view of aligned sequences. The panels show the difference by the solid black line.

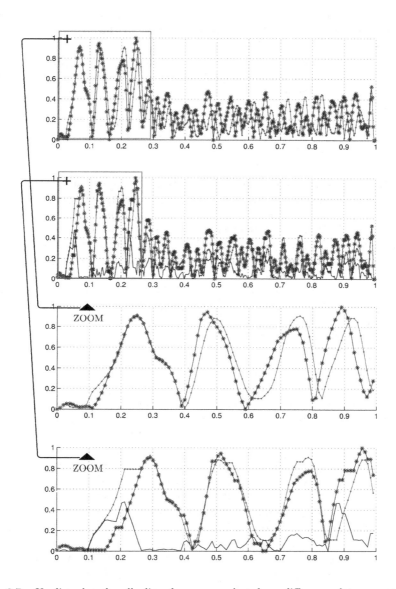

Fig. 2.7.   Unaligned and well aligned sequences based on difference data processing (from top to bottom): unaligned sequences, aligned sequences, zoomed view of unaligned sequences, and zoomed view of aligned sequences. The panels show the difference by the solid black line).

Step 1. The sequences A and B are placed along the left margin and on the top of the matrix M. There is no need for initialization because the scoring is based on data only.

Step 2. Other elements of the matrix are obtained by computing the difference $\Delta = |a_i - b_j|$ and finding the maximum value among the following three values:

$$M(i, j) = min \begin{cases} M(i, j - 1) + \Delta \\ M(i - 1, j - 1) + \Delta. \\ M(i - 1, j) + \Delta \end{cases}$$

Step 3. The dynamic programming algorithm propagates scores from the matching start point (upper-left corner) to the destination point (lower-right corner) of the matrix. The score that ends up in the lower-right corner is the optimal sequence alignment score. After finding the final score for the optimal alignment, the final similarity between the two sequences is computed by considering the final optimal score and the length of the two sequences.

Step 4. The optimal path is then achieved through back propagating from the destination point to the starting point. In all given examples, the optimal path found through back propagating is connected by arrows. This optimal path tells the best matching pattern.

Multiple-valued and continuous sequences are widely used in the area of pattern recognition and signature comparison, where the values describe certain features like curvatures, angles, velocity, etc. and the alignment of sequences often corresponds to the decision making procedure. Biometrical applications are good examples of sequence processing. For example, by aligning and comparing sequences from handwritten signatures, it is possible to build a reliable and robust verification system (see Fig. 2.8).

Sequence alignment and matching is an important task in multiple-valued and continuous systems like biometrics in order to group similar sequences and identify trends in functional behavior. Accurate processing and sequence comparison depend on good correlation measures between sequences. This section gave an outline of methods used in sequence processing and their applications to signature comparison. We introduced a new similarity measure based on sequence alignment and extended this approach to various data formats. Having a robust mechanism for signature analysis and comparison, there is a need in getting enough distinct signatures for testing and benchmarking. The next section introduces some methodological principles and techniques for signature modeling and synthesis.

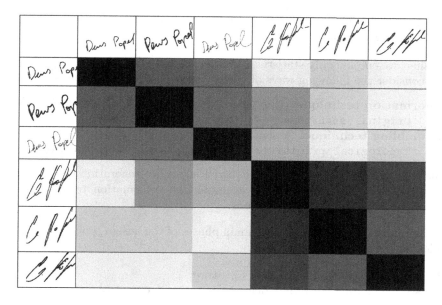

Fig. 2.8. Alignment scores for continuous sequences in signature verification. The range of colors corresponds to the white, showing the minimum similarity score, and gradually changing to the black, representing the maximum similarity score.

2.4. Signature Synthesis

The modeling and simulation in biometrics, or inverse problems of biometrics have not been investigated until recently [18]. However, the demand for synthetic biometric data is now led to many practical and important applications. Thus, in signature processing, synthetic signatures are used to test the system and evaluate the performance of signature recognition systems.

2.4.1. *Signature Synthesis Techniques*

This section focuses on the basics of signature synthesis techniques. Known approaches to analysis and synthesis of signatures can be divided into two distinct classes:

- *Statistical* approaches use the fact that a signature shape can be described by a limited set of features. These features are statistically characterized.
- *Rule-based* approaches assume that a signature is the composition of a limited number of basic topological primitive which can be formally

described. This composition can be implemented by a set of rules and/or grammars.

Further, we employ only statistical approaches to signature synthesis. One can consider the following ways of generating synthetic signatures:

Deformation technique. Applying reversible transformations
    `<Original signature>` ⇔ `<Synthetic signature>`
Assembling technique. Assembling topological primitives
    `<Topological primitives>` ⇒ `<Synthetic signature>`

The choice of an appropriate technique for signature generation is dictated by a particular application. For example, the deformation technique is suitable for testing in-class variations.

Deformation technique. The main phases of the design process are the following:

Phase 1. Preprocess the original signature.
Phase 2. Choose appropriate deformation rules and control parametersfor the deformation.
Phase 3. Transform the original signature into its deformable equivalent.
Phase 4. Use the synthesized signature for matching.

Assembling technique. The main phases of the design process are the following:

Phase 1. An arbitrary set of two topological primitives is taken from the database of primitives.
Phase 2. Based on appropriate assembling rules, a macroprimitive is created by merging the two initial primitives.
Phase 3. The next topological primitive is taken, and then added to the macroprimitive by applying appropriate assembling rules.
Phase 4. Control parameters direct the process of "growing" the macro-primitive. This process is stopped by a predefined criterion from the set of control parameters.

2.4.2. *Statistically Meaningful Synthesis*

Some methods in signature synthesis rely on statistical data collected from pre-existing databases of signatures. The first step in a statistically-driven synthesis is *learning*. Learning from the pre-existing collection of signatures (or signature database) allows the accumulation of a priori knowledge about image structures, detailed curve behaviors, and other important characteristics, including those of kinematic nature if it is an on-line signature. A pre-existing online database of signatures can be used to generate this statistical information.

Preprocessing: Learning. The learning stage results in multiple distributions with distribution function $\Phi(z)$ giving the probability that a standard normal variate assumes a value z. The following statistically meaningful characteristics can be acquired through the learning process from the pre-existing database of original signatures (Fig. 2.9):

- Distribution of the number of segments,
- Distribution of the segments' length,
- Distribution of the length and orientation of inter-segment connections, and
- Distribution of the angle $\Phi(\alpha)$ and curvature $\Phi(k)$ characteristics along the contingent segments.

The distributionof curvature can be replaced with the distribution of length's intervals $\Phi(\Delta s)$. To improve the visual quality of the curve, it is important to consider sequential distributions in the form of Markov chains or n-grams.

Other distributions of various kinematic characteristics include pressure $\Phi(P)$, time intervals $\Phi(\Delta t)$, etc. These distributions are used to complement different geometric characteristics.

2.4.3. *Geometrically Meaningful Synthesis*

The main question in a geometry-driven synthesis is how to preserve the geometric structure of the curve. One possible solution is based on learning principles and n-grams which model information about the curve structure. Curvature is defined by (see Eq. (2.3)).

$$k = \frac{\mathrm{d}\phi(s)}{\mathrm{d}s}.$$

For a plane curve,

$$k = \frac{(x(t)'^2 + y(t)'^2)^{3/2}}{x(t)'y(t)'' - x(t)''y(t)'}$$

The curvature in the equations above can be either positive or negative.

The representation of the curve is based on its intrinsic properties such as its arclength s, and its curvature k. For every curve there is a dependance of the form $F(s, k) = 0$. Such an equation is called the *natural equation* of the curve. This representation does not depend on any coordinate system and relates two geometric properties of the curve.

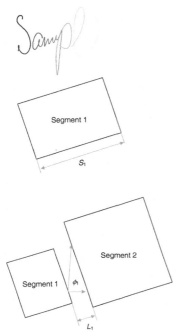

Original signature from the database of signatures. Distribution of the number of segments $\Phi(N_{\text{segments}})$. It is a simple and common observation that visual parts (segments) determine the overall visual appearance of the signature. For example, the original signature can be segmented to $N_{\text{segments}} = 1$ segments.

Distribution of the segment's length $\Phi(S_{\text{segment}})$. In addition to the distribution of the segment's length across all segments $\Phi(S_{\text{segment}})$, it is necessary to collect information about the dependence of the segments' length on the segment's number C_{segment}: $\Phi(S_{\text{segment}}, C_{\text{segment}})$. In this case, the segmented signature has $S_{\text{segment}1} = S_1$.

Distribution of the length $\Phi(L_{\text{inter-segment}})$ and orientation $\Phi(\phi_{\text{inter-segment}})$ of inter-segment connections. In addition to the distribution of characteristics of all inter-segment connections for the entire image, it is necessary to collect information about the dependence of these characteristics on the inter-segments' number $C_{\text{inter-segment}}$: $\Phi(L_{\text{inter-segment}}, C_{\text{inter-segment}})$ and $\Phi(\phi_{\text{inter-segment}}, C_{\text{inter-segment}})$. In this example, there are no other segments.

Fig. 2.9. Statistically meaningful characteristics are acquired through the learning process from the pre-existing database of original signatures.

Two curves with the same natural equation can differ only by their location in the plane so that the shape of the curve is determined uniquely by its natural equation (see [18] for the complete justification of this statement).

The preprocessing part of the synthesis method uses sequence processing techniques to gather essential information about the sequential nature of the curve. In the natural equation, $F(s, k) = 0$, it is important to recognize some patterns of how the curve behaves and changes. In digitized cases, a curve is described by a selected parameterization technique, for example, the unit tangent vector can be a function of t: $\phi(t)$. In such cases, we can compute the curvature using the chain rule:

$$\frac{\mathrm{d}\phi}{\mathrm{d}t} = \frac{\mathrm{d}\phi}{\mathrm{d}s}\frac{\mathrm{d}s}{\mathrm{d}t}.$$

Basic method of curve generation. The *basic method* of generating curves based on various distributions (given below) is controlled by the length of the curve. Figure 2.10 shows the distribution of interval lengths and angles for the database of signatures used in our study.

Step 1. Generate lengths of intervals along the trajectory so that the total length is the given length of the curve.

Step 2. For each interval with its length s, generate the curvature k or angle ϕ based on the statistical distribution of curvatures, their derivatives and the spatial distributions of $k(s)$. Repeat Steps 1–2 until the desired total arc length is obtained.

Example 2.6. Figure 2.10 gives an example of statistical distributions generated from the preexisting database of original signatures.

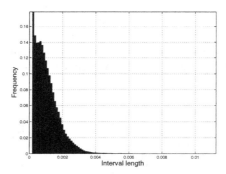

Fig. 2.10.   Distributions of interval lengths for the preexisting signature database.

Method based on the n-gram model. The basicmethod outlined above has been extended to improve the visual appearance of the generated signatures using n-gram structures and modeling. The sequential nature of curve characteristics, length intervals, angles and curvatures allows us to design a model which encompasses the probability of possible combinations of values and their transition from one to another. Thus, in our study we used digram (a pattern of two adjacent values) and trigram (a pattern of three adjacent values) models to simulate the sequential behavior of curves. Models based on digrams proved to be more efficient than those based on hidden Markov models in terms of storage and run-time. In addition to these general properties assessed by the final outcome, the digram models can be built with less computational resources than hidden Markov models.

The generation method produces an angle value ϕ_i based on digram model statistics and a previously generated value ϕ_{i-1}. The same methodology is applied to other sequential characteristics, such as curvature and length intervals. Formally, the method can be expressed as:

Step 1. Randomly select the initial values for a length interval and angle (or curvature).

Step 2. Build the distribution of values for length intervals and angles (or curvatures) from the corresponding digram models taking previous values as reference parameters.

Step 3. Randomly generate current values for length intervals and angles (or curvatures) based on distributions from Step 2, and repeat Steps 2–3 until the total arc length is reached.

Example 2.7. Some samples of generated signatures are given in Fig. 2.11. The digram model $\Theta(\phi, A)$ for angle sequences is defined on the alphabet A of angle values (in our experiments $A = 0 \ldots 359$) and contains the frequencies of all pairs $< \phi_{i-1}\ \phi_i >$ acquired at the learning stage. Figure 2.12 shows the angle digram model for the entire signature database. Evidently, only a discrete set of angles were permitted.

Fig. 2.11.   Three signatures generated by the digram model.

Method based on vector recombination.   In order to generate images with a high percentage of natural looking parts, we introduce a method based on repositioning image sub-parts randomly selected from the pool of training images. Since the curve structure is described in sequential forms, the repositioning implies vector recombination. Thus the resulting signature with the number of segments N_{segments} and the their lengths S_{segment} can be composed of N_{segments} pieces taken from different signatures with the total length S_{segment}. This method is outlined in the following series of steps:

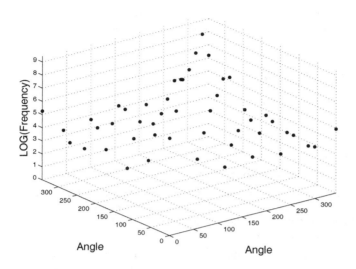

Fig. 2.12. An angle digram model for the preexisting signature database.

Step 1. The sequential representation of interval lengths and angles (or curvatures) is treated as a pool of sequences.

Step 2. Randomly select the start point within the pool of sequences and select as many points as the length of the current segment. This guarantees the mutual correspondence between interval lengths and angles (or between interval lengths and curvatures).

Step 3. Repeat Step 2 for the given number of segments.

Example 2.8. Some samples of generated signatures are given in Fig. 2.13. Experimental results demonstrate that the method based on vector recombination is fast, and visually accurate, but allows inconsistencies of the inter-segment parts.

Fig. 2.13. Three synthetic signatures generated based on the recombination model.

One can observe that the segment-to-segment connection is predominantly straight without significant changes in inter-segment angles.

2.4.4.   *Algorithm and Experimental Results*

The final algorithm consists of the following steps:

Step 1. Select a sample (training) set of signatures from the database.

Step 2. Generate statistical data from the sample set.

Step 3. Applying the random number generator and considering various distributions from the sample set, determine the number of segments, the length and orientation of inter-segment intervals, etc.

Step 4. Applying the random number generator and considering one of the outlined scenarios, generate the interval's lengths and curvatures (or the interval's lengths and angles) along the trajectory.

Step 5. Applying the random number generator and considering various distributions from the sample set, determine kinematic characteristics along the trajectory.

Step 6. Perform postprocessing validation of the obtained image.

Input data are signatures selected from either the entire database for the intra-class scenario or class-specific signatures for the in-class scenario. The basic algorithm implements the statistically meaningful synthesis by analyzing a sample set of signatures (Step 1) in order to build necessary statistical distributions (Step 2). The set of statistical distributions includes distributions of (i) the number of segments, (ii) the segment's length, (iii) the length and orientation of inter-segment connections, (iv) the angle and curvature characteristics along the contingent segments; and (v) the pressure and time intervals. The surrogate generator is used to randomize various signature characteristics (Steps 3–5). Adding a noise (an optional step in the surrogate generator) is aimed to increase the variability of signature characteristics. The result is a generated signature described geometrically and kinematically. Such a signature possesses similar statistical characteristics as the sample set. The postprocessing validation is necessary to evaluate deviations between the generated signature and the existing database of signatures.

The preexisting database of handwritten signatures has been used to generate necessary statistical distributions.[a] Experiments reveal the following characteristics:

Accuracy of visual appearance. All three generators embedded into the signature generation algorithm produced acceptable results in terms of visual appearance. Strokes have not only straight but also cursive shapes. Their relationships are also natural. Since the generation has been based on statistical data only, curves and shapes do not contain any information about language.

Accuracy of postsynthesis comparison. Both intra-class and in-class scenarios have been checked through the validation stage with the following outcome: 6% of signatures were rejected for the intra-class scenario, and 4% of were rejected for the in-class scenario. A similarity threshold of 95% was set up to reject signatures for the intra-class scenario, and accept signatures for the in-class scenario.

Speed. It is capable of generating more than 10 signatures a second (workstation with CPU 2Ghz and RAM 256Mb has been used in our experiments), which results in more than a million synthetic signatures a day.

Adding extra parameters such as speed, acceleration, curvature, etc. results in an improvement in accuracy characteristics of the system. Biometric fusion of several methods (fingerprint, voice, handwriting, etc.) can improve the results of verification and identification. Even combining methods within the same biometric method, like the fusion of on-line and off-line signature processing, can make the system more robust and efficient.

The system architecture that embraces signature comparison and signature synthesis methods and techniques is given in the next section.

2.5. System Architecture

In many instances, current data processing methods provide high accuracy of comparison and are often sufficient for the decision making. However, the increased variety of input devices and formats of storage and the large volume of data lead to the increased risk of forgeries, and require usage of new mathematical models which are capable of structuring, analyzing,

---

[a]The database contained 120 signatures collected from 25 different individuals. The signatures were acquired through pen based interfaces (i.e. pressure sensitive tablets) that provide 100 samples per second containing values for pen coordinates, pressure, and time stamps.

and mining large volumes of signature biometrics. Most current state-of-the-art solutions require proprietary hardware and are incompatible with legacy systems (i.e. unable to verify the bitmap images of the signatures). Moreover, none of the existing products solve the problem of identification for large sets of data.

To address those problems, we created a system for signature verification, identification and synthesis with flexible mechanisms of database organization. The system offers the following enhancements over existing packages in pervasive environments:

Advantage 1. An extended applicability to numerous security application areas, e.g. personal security, financial transactions and document automation.

Advantage 2. Rapid retrieval of information based on the attributes stored with the image and/or on relevant visual attributes.

Advantage 3. A unified approach to the design of image processing tools that are independent of image resolution, size or orientation.

Advantage 4. A software package which is able to process human biometrics of different types (e.g., handwriting, signatures, voice, etc.).

Advantage 5. An easy-to-use environment for signature verification and most importantly identification available "on demand" and accessible online.

The functionality of the system is enabled through the integration of three main components: (1) a unified data format with supporting mathematical models, (2) an image processing toolbox with image generation capabilities, and (3) a database of images with self-learning capabilities. The diagrammatic structure of the system is depicted in Fig. 2.14.

Data acquisition. Signatures can be obtained by various image acquisition devices. Dynamic signature acquisition records kinematic characteristics of human handwriting such as pressure, speed, acceleration, duration of writing, etc. Dynamic acquisition devices include numerous tablets, writing pads, etc. Pen-based interfaces are becoming more integrated in human-computer interfaces because of overall convenience of pen enabled devices vs. traditional keyboards and mice. There are various pen-enabled devices ranging from pressure sensitive tablets to gyropens. Recently developed devices include handheld computers (PDAs and smart phones), and Tablet PCs. In the study presented here, we use a Wacom® compatible pressure sensitive tablet. It is set to provide 100 samples per second containing values for pen coordinates, pressure,

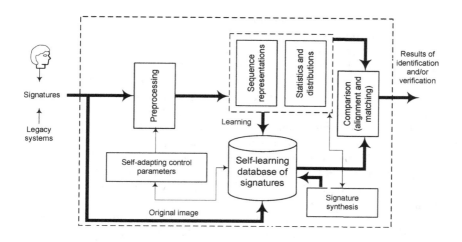

Fig. 2.14. The architecture of the system.

and time stamps. On the contrary, scanned signatures lose all kinematic characteristics retaining only geometric ones such as shape and color intensity. Scanners, cameras, etc. are examples of *static* acquisition devices.

Signatures collected by various image acquisition devices have different resolutions, unspecified orientation, different sizes and/or different file formats. These unconformities are addressed by the system's unified approach, which ultimately enhances the robustness of the image database.

Image Processing Toolbox. The processing of signature images should ideally be invariant to transformations of coordinates such as translation, rotation, and simple uniform scaling. The system implements a range of image processing tools based on the graph structures. These tools cover basic transformations which are (i) geometric, such as rotation, scaling, cropping, or translation; and (ii) image-processing, such as compression and filtering. The toolbox also includes a raster-to-vector format transformation, which consists of the following stages: (1) image thinning (skeletization); (2) image representation as a graph; (3) shape smoothing; (4) graph minimization; and (5) shape compression. These tools extend overall operability to scanned signatures and signatures from legacy systems.

Signature comparison engine. The signature comparison engine is a key component of the software package. It includes the tools for generating multiple distributions, the self-learning database, scoring mechanisms, and decision making tools. The engine is a robust and scalable technology designed to support behavioural authentication mechanisms based on handwritten electronic signatures for identification and verification.

The core of the engine is a database which adjusts the levels of accuracy and the speed of processing depending on the amount and quality of inserted data. The database integrates designed input/output formats, improves searching and matching capabilities, and establishes intermodular connections among signature processing tools. At the testing stage, the database is populated with artificially generated signatures (see Section 2.4). This permits a complete look at the robustness of the signature processing toolbox and accuracy of matching assuming that noise and various resolutions or writing conditions could adversely affect the overall performance of the system.

Most methods for detecting images in the database that are similar to a query image check every image in the database for similarity, a process that is very slow for very large data sets. While techniques do exist that speed up the search by indexing the database, they fail to restrict the search to a small portion of the database when applied to our problem. Such an approach to image searching has two major strengths. First, the measures of image similarity are robust to the common transformations performed on images, e.g. rotation, translation, scaling, cropping, filtering, compression, etc. Second, the measure of similarity is adjustable to gain maximum selectivity. New comparison metrics are created to demonstrate a compromise between robustness and selectivity. The database includes advanced indexing techniques. The self-learning database allows us to perform multi-factor optimization for speed and accuracy.

Online implementation. The online implementation of the system includes (i) signature acquisition tools, (ii) a searchable signature database (the engine), and (iii) an online interface. The system currently supports pressure sensitive tablets which allow recording both geometric (signature contours, shapes, etc.) and kinematic/dynamic characteristics (pressure, time stamps, etc.), touch pads and mice for collecting human handwriting and strokes. This implementation has been successfully tested on PDA devices and smart-phones where signatures are produced by a stylus pen. Overall, the system is a flexible and viable solution for many pervasive environments.

The web-based interface has six basic modules: `login`, `upload`, `list`, `draw` and `search`, `verify`, and `identify`. The database is protected against any unauthorized access by the login module. After the successful login, the user is given administrative rights to the upload and list functions. The upload module allows the user to upload a signature image providing a descriptive keyword (e.g. a person's name), and to choose a file type from the drop down list (see Fig. 2.15, first panel). After clicking `submit`, the web script updates the database and generates all the necessary distributions for the given image.

The `list` script creates a table, listing all the data from the database. For signature images, the data are listed in the form of thumbnails (see Fig. 2.15, second panel). A button labelled `regenerate` is also available for administrative users to automatically regenerate distributions for all signatures. This is especially useful when a new classification feature is added to the engine. By clicking `regenerate`, all previously stored data are recalculated for every signature in the database. Images can be inspected and deleted when necessary.

The only functions accessible to non-administrative users are `draw and search`, `verify` and `identify`, because they do not alter the database. `Identify` is a module that allows the user to upload a signature image, generate distribution data, and compare the generated data against the data of all images in the database (see Fig. 2.16, first panel). The verification module collects the keyword label from the user and compares the generated data against a limited set of images. Both modules create a table displaying the testing signature and listing the top ten signatures from the database along with similarity ratings (see Fig. 2.16, second panel). The draw and search module allows to use real-time signature acquisition from the browser window.

2.6. Concluding Remarks and Future Work

In this section, we provided an outline of a complete cycle of signature processing techniques ranging from various signature representations, signature analysis and comparison, as well as signature synthesis. This cycle is implemented in a system used for identification and authentication of signatures and human handwriting in pervasive environments. The system changes the mechanism of data collection and processing, and concentrates on a very specific area of personal authentication and identification based on signature input. The system can verify, compare, and organize both the visual appearance of a signature and the behavioural characteristics of the signature including the speed of writing and pressure. Our proprietary technology is based on a unique mixture of sequence processing and statistical techniques.

In general, the system uses drivers to integrate with many off-the-shelf image acquisition devices and standardized software platforms, and connectors to interface with legacy and commonly used authentication

Fig. 2.15.   Screenshots of the upload and list modules.

systems and applications. The system will have applicability in all areas where signature identification or verification is desirable or required. Using the system, businesses, practitioners and researchers will gain access to a robust, automated and scalable system that renders behavioural biometrics available for analysis and will be able to build a dynamically configurable knowledge base for the design of future pervasive systems or data exploration. Additionally, the system can complement existing software packages in security technologies and enhance their functionality across

Fig. 2.16. Screenshots of the identification and **draw and search** modules.

a wide range of products including personal access, online transactions, banking and financial transactions, secure document automation, etc.

Possible extensions can include voice processing and fusion of voice and image biometrics. An information-theoretic approach can be utilized to speed-up the prediction of possible alignments for continuous sequences (see, for example, [12]).

Acknowledgments

We express our appreciation to Drs. Vlad P. Shmerko and Svetlana N. Yanushkevich for valuable discussions, as well as to Joshua Strohm and Scott Alexander, former students, for critical comments and data preparation.

Bibliography

1. Brault, J. and Plamondon, R. (1993). Complexity measure of handwritten curves: modelling of dynamic signature forgery, *IEEE Trans. Systems, Man and Cybernetics*, 23, pp. 400–413.
2. Darling, D. (1957). The Kolmogorov-Smirnov, Cramér-von Mises tests. *Ann. Math. Stat.*, 28, pp. 823–838.
3. Dimauro, G., Impedovo, S., Modugno, R. and Pirlo, G. (2002). Analysis of stability in hand-written dynamic signatures, *Proc. 8th International Workshop on Frontiers in Handwriting Recognition*, pp. 259–263.
4. Dryden, I. and Mardia, K. (1998). *Statistical Shape Analysis*, John Wiley & Sons, Chichester.
5. Jain, A., Griess, R. and Connell, S. (2002). On-line signature verification, *Pattern Recognition*, 35, 12, pp. 2963–2972.
6. Kuiper, N. (1962). Tests concerning random points on a circle, *Proceedings of the Koninklijke Nederlandse Akademie van Wetenschappen*, A63, pp. 38–47.
7. Needleman, S. and Wunsch, C. (1970). A general method applicable to the search for similarities in the amino acid sequence of two sequences, *J. Molecular Biology*, 48, pp. 443–453.
8. Nikitin, A. and Popel, D. (2004). signmine algorithm for conditioning and analysis of human handwriting, *Proc. Int. Workshop on Biometric Technologies*, Calgary, Canada, pp. 171–182.
9. Parker, J. R. (2006). Composite Systems for Handwritten Signature Recognition, *this issue*.
10. Plamondon, R. and Srihary, S. (2000). Online and off-line handwriting recognition: a comprehensive survey, *IEEE Trans. Pattern Analysis and Machine Intelligence*, 22, pp. 63–84.
11. Popel, D. (2002). A compact graph model of handwritten images: Integration into authentification and recognition, *Proc. IAPR Int. Workshop on Structural and Syntactic Pattern Recognition, Lecture Notes in Computer Science 2396*, pp. 262–270.

12. Popel, D. (2003). Conquering uncertainty in multiple-valued logic design, *Artificial Intelligence Review, Kluwer*, 20, pp. 419–443.
13. Popel, D. and Popel, E. (2004). Controlling uncertainty in discretization of continuous data, *Proc. IEEE Int. Symp. on Multiple-Valued Logic*, pp. 288–293.
14. Popel, D. and Popel, E. (2005). bioGlyph: Biometric identification in pervasive environments, *Proc. IEEE Int. Symp. on Multimedia*, pp. 713–718.
15. Shmerko, V., Phil Phillips, Kukharev, G., Rogers, W., Yanushkevich, S. Biometric technologies, *Proc. Int. Conf. The Biometrics: Fraud prevention, Enchanced Service*, Las Vegas, Nevada, pp. 270–286, 1997.
16. Shmerko, V., Phil Phillips, Rogers, W., Perkowski, M., Yanushkevich, S. Bio-Technologies, *Bulletin of Institute of Math-Machines*, Warsaw, Poland, ISSN 0239-8044, no.1, pp. 7–30, 2000
17. Smith, T. and Waterman, M. (1981). Identification of common molecular subsequences, *Journal of Molecular Biology*, 147, pp. 195–197.
18. Yanushkevich, S., Shmerko, V., Stoica, A. and Popel, D. (2005). *Inverse Problems of Biometrics*, CRC Press/Taylor & Francis Group, Boca Raton, FL.
19. Watson, G. (1961). Goodness-of-fit tests on the circle, *Biometrika*, 48, pp. 109–114.
20. Zhang, D. (2000). *Automated Biometrics: Technologies and Systems*, Kluwer, Dordrecht.

Chapter 3

Local B-Spline Multiresolution with Example in Iris Synthesis and Volumetric Rendering

Faramarz F. Samavati*,‡, Richard H. Bartels†,§, Luke Olsen*

*Department of Computer Science,
University of Calgary, Calgary, Alberta,
T2N 1N4, Canada
†School of Computer Science,
University of Waterloo, Waterloo, Ontario,
N2L 3G1, Canada
‡samavati@cpsc.ucalgary.ca
§rhbartel@uwaterloo.ca

Multiresolution has been extensively used in many areas of computer science, including biometrics. We introduce local multiresolution filters for quadratic and cubic B-splines that satisfy the first and the second level of smoothness respectively. For constructing these filters, we use a reverse subdivision method. We also show how to use and extend these filters for tensor-product surfaces, and 2D/3D images. For some types of data, such as curves and surfaces, boundary interpolation is strongly desired. To maintain this condition, we introduce extraordinary filters for boundaries. For images and other cases in which interpolating the boundaries is not required or even desired, we need a particular arrangement to be able to apply regular filters. As a solution, we propose a technique based on symmetric extension. Practical issues for efficient implementation of multiresolution are discussed. Finally, we discuss some example applications in biometrics, including iris synthesis and volumetric data rendering.

Contents

3.1. Introduction

Multiresolution provides a tool for decomposing data into a hierarchy of components with different scales or resolutions. This hierarchy can be used for noise removal, compression, synthesizing and recognition of the objects. An efficient and economical multiresolution is associated with wavelets. Therefore, this kind of multiresolution has been employed in many areas such as image processing [13], biomedical [21] and computer graphics [18].

Noise removal, data synthesis and feature recognition are common problems in biometrics. Multiresolution techniques have been employed in iris identification [7,9,23], iris synthesis [22], feature extraction from fingerprints and irises [3,10,14], and for detecting faces in images [24].

To date, Haar wavelets have primarily been used in biometric applications due to their simplicity. Haar wavelets are based on zero-degree B-splines. Unfortunately, zero-degree B-spline functions and their wavelet functions are not even continuous and consequently they are not well suited when we need a "smooth multiresolution representation" for data.

In this chapter, we introduce multiresolution representations for quadratic and cubic B-spline that satisfy the first and the second level of smoothness respectively. To keep the simplicity and efficiency of the resulting techniques, we build and describe *local multiresolution filters* from a condition for *biorthogonal wavelets* using the general approach of *reverse subdivision.* Consequently, the purpose of this work is not only describing the construction method (Sections 3.3–3.5), but also discussing practical implementation issues (Sections 3.6–3.8).

3.2. Wavelets and Multiresolution Background

Multiresolution operations are specified by a set of filter matrices $\mathbf{A}^k$, $\mathbf{B}^k$, $\mathbf{P}^k$ and $\mathbf{Q}^k$. Consider a given discrete signal C^k, expressed as a column vector of samples. A lower-resolution sample vector C^{k-1} is created by a down-sampling filter on C^k. This process can be expressed as a matrix equation

$$C^{k-1} = \mathbf{A}^k C^k \ .$$

The *details* D^{k-1} lost through the down-sampling are captured using $\mathbf{B}^k$

$$D^{k-1} = \mathbf{B}^k C^k \ .$$

The pair of matrices $\mathbf{A}^k$ and $\mathbf{B}^k$ are called *analysis filters* and the process of splitting a signal C^k into C^{k-1} and D^{k-1} is called *decomposition.* Recovering the original signal C^k is called *reconstruction.* It involves refinement of the low-resolution sample C^{k-1} and details D^{k-1} using the *synthesis filters* $\mathbf{P}^k$ and $\mathbf{Q}^k$, which reverse the operations of $\mathbf{A}^k$ and $\mathbf{B}^k$

$$C^k = \mathbf{P}^k C^{k-1} + \mathbf{Q}^k D^{k-1} \ .$$

The matrices $\mathbf{A}^k$, $\mathbf{B}^k$, $\mathbf{P}^k$ and $\mathbf{Q}^k$ form the core of the multiresolution approach, and the efficiency of the resulting techniques depends on the structure of these matrices. For an efficient and useful representation, the following properties are desired:

- All matrices should be banded, with repetitive row/column entries.
- C^{k-1} is a good approximation for C^k.
- The storage requirement for storing C^{k-1} and D^{k-1} is not more than that of C^k.
- The time required to decompose and reconstruct the signal is linearly dependent on the size of C^k.

Decomposition and reconstruction operations take place in some underlying function spaces $\mathcal{V}^{k-1} \subset \mathcal{V}^k$ wherein C^k defines some function $f^k = \sum_i c_i^k \phi_i^k$ in the large space, C^{k-1} defines an approximation $f^{k-1} = \sum_i jc_j^{k-1} \phi_j^{k-1}$ to that function in the smaller space, and D^{k-1} defines the difference $g^{k-1} = \sum_i jd_j^{k-1} \psi_j^{k-1}$ in the complement space $\mathcal{V}^k \setminus \mathcal{V}^{k-1}$. The basis functions ψ_i^{k-1} are conventionally called *wavelets* and the ϕ_j are called *scale functions*.

Wavelet systems are usually classified according to the relationship between the wavelets and the scaling functions. Stollnitz et al. provide an excellent overview of wavelet classifications[18], which we summarize here.

Orthogonal wavelets. An orthogonal wavelet system is one in which "the scaling functions are orthogonal to one another, the wavelets are orthogonal to one another, and each of the wavelets is orthogonal to every coarser scaling function." In such a setting, the determination of the multiresolution filters is quite easy. Unfortunately, orthogonality is difficult to satisfy for all but the most trivial scaling functions.

Semiorthogonal wavelets. Semiorthogonal wavelets relax the orthogonality conditions, only requiring that each wavelet function is orthogonal to all coarser scaling functions. By relaxing the constraints on the wavelets, it is easier to derive a $\mathbf{Q}^k$ filter (note that there is no unique choice of $\mathbf{Q}^k$, but there are some choices that are better than others). The drawback of semiorthogonal wavelets is that while $\mathbf{P}^k$ and $\mathbf{Q}^k$ will be sparse matrices (meaning that reconstruction can be done in linear time), the decomposition filters $\mathbf{A}^k$ and $\mathbf{B}^k$ offer no such guarantee. It often turns out that the decomposition filters are full matrices while these matrices are very simple and banded in the case of Haar wavelets [18].

Biorthogonal wavelets. Finally there are biorthogonal wavelets, which have many of the properties of semiorthogonal wavelets but enforce no orthogonality conditions. The only condition in a biorthogonal setting is that $\left[\mathbf{P}^k | \mathbf{Q}^k\right]$ is invertible, which implies that the decomposition filters $\mathbf{A}^k$ and $\mathbf{B}^k$ exist such that

$$\begin{bmatrix} \mathbf{A}^k \\ \mathbf{B}^k \end{bmatrix} \begin{bmatrix} \mathbf{P}^k \ \mathbf{Q}^k \end{bmatrix} = \begin{bmatrix} \mathbf{I} \ \mathbf{0} \\ \mathbf{0} \ \mathbf{I} \end{bmatrix} . \tag{3.1}$$

Simply having a matrix $\mathbf{A}^k$ that satisfies (3.1) does not necessarily produce coarse data C^{k-1} that is a good approximation of C^k. Consequently, this condition should also be taken to account in our construction of $\mathbf{A^k}$.

Wavelet transform. We can repeatedly decompose a signal C^k to $C^\ell, C^{\ell+1}, \ldots, C^{k-1}$ and details $D^\ell, D^{\ell+1}, \ldots, D^{k-1}$ where $\ell < k$. The original signal C^k can be recovered from the sequence $C^\ell, D^\ell, D^{\ell+1}, \ldots,$ D^{k-1}; this sequence is known as a *wavelet transform*. Based on the properties mentioned above the total size of the transform $C^\ell, D^\ell, D^{\ell+1}, \ldots,$ D^{k-1} is the same as that of the original signal C^k. In addition, the time required to transform C^k to $C^\ell, D^\ell, D^{\ell+1}, \ldots, D^{k-1}$, and vice versa, is a linear function of the size C^k.

Details interpretation. If C^k represents a high-resolution approximation of a curve, then C^ℓ is a very coarse approximation of the curve showing the main outline, and D^i consist of vectors which perturb the curve into its original path. As Fig. 3.8(b) demonstrates, if we eliminate D^i, the reconstructed curve becomes much smoother but without any of the curve's individual finer structure. In fact, D^i can be considered as *characteristic* of the curves. It is possible to apply D^i to a new coarse curve to obtain a new curve but with the same character (see Fig. 3.8(d)). Consequently, D^i at different levels are important features for synthesizing techniques.

B-spline multiresolution. B-splines are often chosen as scaling functions [6]. The first order (zero degree) B-splines form a set of step functions and Haar functions are their associated wavelets [18,19]. The resulting matrix filters are very simple and efficient. However, these scaling functions and wavelets are non-continuous. This is a problem when we have discrete data that is a sample of smooth signals and objects. Higher order B-splines and their wavelets can be considered for smooth signals [6,8,16].

A common knot arrangement, the standard arrangement, for B-splines of order k is to have knots of single multiplicity uniformly spaced everywhere except at the ends of the domain where knots have multiplicity k [1,15]; this arrangement produces endpoint-interpolation. Conventionally, B-spline wavelets are constructed with a goal of semiorthogonality, which results in full analysis matrices.

An alternative approach to generating multiresolution matrices is *reverse subdivision*, originally introduced by Bartels and Samavati [2]. Based on this approach, it is possible to obtain banded matrices for biorthogonal B-spline wavelets whose bands are narrower than the ones conventionally produced. In this work, we construct and report multiresolution filter matrices for quadratic and cubic B-splines, which are important practical cases. Because of the similarity in the constructions, we just describe the process in detail for cubic B-splines.

Notation. For clarity of notation, the remainder of the chapter will forgo the superscript k for denoting the k-th level of subdivision. Let $C = C^k$ and $F = C^{k+1}$, such that

$$C = \{c_1, \ldots, c_n\} \ ,$$
$$F = \{f_1, \ldots, f_m\} \ .$$

Further, let $\mathbf{P} = \mathbf{P}^k$, $\mathbf{Q} = \mathbf{Q}^k$, $\mathbf{A} = \mathbf{A}^k$, and $\mathbf{A} = \mathbf{A}^k$. These matrices are assumed to be of the proper size so that the following equations hold

$$C = \mathbf{A}F \qquad\qquad (3.2)$$
$$D = \mathbf{B}F \qquad\qquad (3.3)$$
$$F = \mathbf{P}C + \mathbf{Q}D \ . \qquad\qquad (3.4)$$

3.3. Review of Construction

We construct multiresolution of B-splines by reversing their *subdivision* schemes. In general, a subdivision process takes some coarse data as input. To this is applied a set of rules that replace the coarse data with a finer (smoother) representation. The set of rules could again be applied to this finer data. In the limit of repeated application, the rules yield data with provable continuity. The standard midpoint knot insertion process results in a subdivision scheme for B-splines [1,15].

Though subdivision is usually discussed in the context of curves and surfaces, it is a general process that can operate on any data type upon which linear combinations are defined. Thus we will consider subdivision to operate on some "coarse set" C of samples, and the process of subdivision is expressed in matrix form as $F = \mathbf{P}C$, whereby C is converted into a larger "fine set" F by the subdivision matrix $\mathbf{P}$.

The construction of multiresolution assumes that F is not the result of subdivision; that is, $F \neq \mathbf{P}C$ for any vector C. In this case we wish to find a vector C so that $F \approx \widetilde{F} \equiv \mathbf{P}\,C$, so that the residuals $F - \widetilde{F}$ are small, and so that complete information about these residuals can be stored in the space used for $\{f\} \setminus \{c\}$ (or one of an equivalent size). Informally, this describes the features of a biorthogonal multiresolution built upon $\mathbf{P}$, which is the goal of the construction.

If the components of C and F are arranged in sequence, the subdivision matrices will be banded, repetitive, and slanted. That is, each column j of $\mathbf{P}$ has only a finite number of nonzero entries, located from some row r_j through a lower row $r_j + \ell$; these nonzero numbers appear in all columns save for a few exceptions (corresponding to the boundaries of the data $\{c\}$), and the entries of each succeeding or preceding column are shifted down or up by some fixed number of rows (which is determined by the *dilation scale* of the nested function spaces underlying the subdivision).

$$\mathbf{P} \; = \; \begin{bmatrix} \frac{1}{2} & \frac{1}{2} & 0 & 0 & 0 \\[4pt] \frac{1}{8} & \frac{3}{4} & \frac{1}{8} & 0 & 0 \\[4pt] 0 & \frac{1}{2} & \frac{1}{2} & 0 & 0 \\[4pt] 0 & \frac{1}{8} & \frac{3}{4} & \frac{1}{8} & 0 \\[4pt] 0 & 0 & \frac{1}{2} & \frac{1}{2} & 0 \\[4pt] 0 & 0 & \frac{1}{8} & \frac{3}{4} & \frac{1}{8} \\[4pt] 0 & 0 & 0 & \frac{1}{2} & \frac{1}{2} \end{bmatrix} \begin{matrix} \\ \\ c_{i-1} \\ c_i \\ c_{i+1} \\ \\ \\ \end{matrix} \qquad (3.5)$$

$$c_{j-1} \; c_j \; c_{j+1}$$

We demonstrate the construction of multiresolution by reversing cubic B-spline subdivision. For this construction, we take the subdivision provided by midpoint knot insertion for uniform cubic B-splines (which provides a 2-scale dilation for a nesting of uniform-knot spline spaces). A finite (7-row, 5-column) portion of the interior of the $\mathbf{P}$ matrix for this subdivision is given in (3.5).

A biorthogonal multiresolution based upon $\mathbf{P}$ consists of the matrix $\mathbf{P}$ together with matrices $\mathbf{A}$, $\mathbf{B}$, and $\mathbf{Q}$ that satisfy (3.1). The construction method of Bartels and Samavati [2] is directed toward finding examples of $\mathbf{A}$, $\mathbf{B}$, and $\mathbf{Q}$ that are also banded, repetitive, and slanted; specifically, these characteristics should be true of the columns of $\mathbf{Q}$ and of the rows of $\mathbf{A}$ and $\mathbf{B}$.

The construction is staged as follows:

(1) a matrix $\mathbf{A}$ is produced that satisfies $\mathbf{AP} = \mathbf{I}$;
(2) trial versions of $\mathbf{B}$ and $\mathbf{Q}$ are produced, containing partially constrained symbolic entries, that satisfy $\mathbf{BP} = \mathbf{0}$ and $\mathbf{AQ} = \mathbf{0}$;
(3) the final step to fix $\mathbf{B}$ and $\mathbf{Q}$ by solving $\mathbf{BQ} = \mathbf{I}$.

In each stage, we can take advantage of the fact that the matrices are banded, repetitive, and slanted. This means that any scalar equation that forms a part of the matrix equation (3.1) is entirely characterized by the interaction of a row of the left-hand matrix with only one of a small number of adjacent columns of the right-hand matrix. (Alternatively, the scalar equations can be studied by looking at the interaction of a column of the right-hand matrix with only a small number of adjacent rows in the left-hand matrix.) The repetitiveness offers us the benefit of being able to characterize the entire matrix-matrix product (or at least, all of it except

for a few special cases at the boundary) by studying how one representative
row (or column) interacts with a small number of columns (or rows).

The construction is, of course, carried out only once for each choice of
regular subdivision and connectivity. The rows of $\mathbf{A}$, $\mathbf{B}$, $\mathbf{P}$, and $\mathbf{Q}$ are
treated as *filters* that *decompose* the fine data F as in (3.2) and (3.3)

$$c_j = \sum_\lambda a_\lambda f_\lambda \qquad d_\ell = \sum_\mu b_\mu f_\lambda$$

and to *reconstruct* it as in (3.4)

$$f_i = \sum_\rho p_\rho c_\rho + \sum_\sigma q_\sigma d_\sigma \ .$$

As an example, the following illustrates the complete setup of equations
to specify the elements of any general, interior (regular) row of $\mathbf{A}$ for
cubic B-spline subdivision under the assumption that there are 7 nonzero
elements in the row, and that in the row defining the value of c_j they are
centered on the position corresponding to f_i

$$[\, a_{i-3} \ a_{i-2} \ a_{i-1} \ a_i \ a_{i+1} \ a_{i+2} \ a_{i+3} \,] \begin{bmatrix} \frac{1}{2} & \frac{1}{2} & 0 & 0 & 0 \\ \frac{1}{8} & \frac{3}{4} & \frac{1}{8} & 0 & 0 \\ 0 & \frac{1}{2} & \frac{1}{2} & 0 & 0 \\ 0 & \frac{1}{8} & \frac{3}{4} & \frac{1}{8} & 0 \\ 0 & 0 & \frac{1}{2} & \frac{1}{2} & 0 \\ 0 & 0 & \frac{1}{8} & \frac{3}{4} & \frac{1}{8} \\ 0 & 0 & 0 & \frac{1}{2} & \frac{1}{2} \end{bmatrix} = [\, 0 \ 0 \ 1 \ 0 \ 0 \,] \quad (3.6)$$

These are the only nontrivial scalar equations obtainable from the interior
rows of $\mathbf{A}$ and interior columns of $\mathbf{P}$, assuming this width and positioning
for the elements in each row of $\mathbf{A}$. The interaction of this row of $\mathbf{A}$ with any
other interior column of $\mathbf{P}$ involves only sums of products with one factor
in each product equal to zero. Interactions coming from the boundary will
produce a small number of scalar equations distinct from the ones in (3.6).
These distinct equations have no effect on the ones shown in (3.6). They
will be solved separately to yield $\mathbf{A}$ values that are to be applied only
to specific samples at the boundary. An example of this will be given in
Section 3.5.

By solving the equations implied by (3.6) for the elements a_i which yield
a minimum Euclidean norm (to have a good coarse approximation), we find

that $[a_{i-3} \ \cdots \ a_{i+3}]$ is

$$\left[\ \frac{23}{196} \ -\frac{23}{49} \ \frac{9}{28} \ \frac{52}{49} \ \frac{9}{28} \ -\frac{23}{49} \ \frac{23}{196} \ \right]$$

The sample $c_j = a_{i-3}f_{i-3} + \cdots + a_{i+3}f_{i+3}$ represents a local least squares estimate based upon the 7 consecutive fine samples $f_{i-3}, \ldots, f_{i+3}$ [2]. Figure 3.1 illustrates that in a curve, these 7 consecutive fine samples are those that are physically nearest to and symmetrically placed about c_j. This is arguably the configuration of choice for estimating c_j in a least squares sense from a local neighborhood about f_i. With the same motivation, 1, 3, 5, 9, or more consecutive samples $f_{i\pm\lambda}$ could be chosen for the estimate, producing other options for $\mathbf{A}$, and then correspondingly for $\mathbf{B}$ and $\mathbf{Q}$.

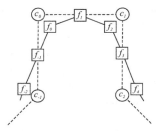

Fig. 3.1. The filters are based on a local indexing centered about a representative sample c_0.

To handle the second matrix equation, $\mathbf{BP} = \mathbf{0}$, scalar equations corresponding to the nontrivial interactions of one row of $\mathbf{B}$ with the columns of $\mathbf{P}$ are set up in a similar way, assuming a number of nonzeros in a row of $\mathbf{B}$ and a position for those nonzeros in the row defining the generic element d_λ. These scalar equations (along with any additional ones desired to enforce, for example, symmetry in the values of the row elements) are solved using a symbolic algebra system. Enough elements should be assumed in a row of $\mathbf{B}$ so that the solution is not fully defined and has free variables.

To handle the third matrix equation, $\mathbf{QA} = \mathbf{0}$, scalar equations corresponding to the nontrivial interactions of one column of $\mathbf{Q}$ with the rows of $\mathbf{A}$ are set up in a similar way, with assumptions about number and position of nonzeros being made. Additional conditions of symmetry are also possible. The equations are solved in symbolic algebra, and the result must also contain free variables.

The final matrix equation, $\mathbf{BQ} = \mathbf{I}$, is handled by using the symbolic results of the preceding two steps to generate the scalar equations representing the nontrivial interactions of a single row of $\mathbf{B}$ with the columns of $\mathbf{Q}$ (or a single column of $\mathbf{Q}$ with the rows of $\mathbf{B}$), and the resulting (bilinear) equations are solved. Any remaining free variables may be fixed at will (our preference being to establish a norm of approximately unity for any column of $\mathbf{Q}$).

A consistent set of solutions for cubic B-spline subdivision is given in Appendix A.

The construction of $\mathbf{B}$ and $\mathbf{Q}$ ends with the selection of leftover free parameters, and it is our custom to use these so that the maximum magnitude element (the infinity vector norm) of each column of $\mathbf{Q}$ is comparable to 1 in magnitude, expecting that this means the contribution of each d_ℓ in any residual will be comparable to the magnitude of d_ℓ itself. The residuals $f_i - \tilde{f}_i$ to which d_ℓ contributes are those corresponding to the nonzero elements of the ℓth column of $\mathbf{Q}$. The number of elements in $\{f\}$ corresponds to the number of rows in $\mathbf{P}$. The number of columns of $\mathbf{P}$ corresponds to the number of elements in $\{c\}$ and the number of columns of $\mathbf{Q}$ corresponds to the number of elements in $\{d\}$. If the columns of $\mathbf{P}$ and $\mathbf{Q}$ are adjoined, the result is a square matrix (whose inverse is the matrix with the rows of $\mathbf{A}$ adjoined below by the rows of $\mathbf{B}$). The number of columns of $[\mathbf{P}\ \mathbf{Q}]$, being the number of elements in $\{c\}\bigcup\{d\}$, is also the number of elements in $\{f\}$.

Throughout the remainder of this chapter we shall be using the term **matrix** to refer to the *decomposition* information, $\mathbf{A}$ and $\mathbf{B}$, and the *reconstruction* information, $\mathbf{P}$ and $\mathbf{Q}$, in its entire matrix format; i.e., capable of acting simultaneously on all the information $\{f\}$, $\{c\}$, and $\{d\}$, as laid out in vectors, in the manner of (3.2), (3.3), and (3.4). We shall be using the term **filter** to refer to the nonzero entries in a representative row of $\mathbf{A}$ and $\mathbf{B}$ and a representative column of $\mathbf{P}$ and $\mathbf{Q}$. We simply denote theses filters by $\mathbf{a}, \mathbf{b}, \mathbf{p}$ and $\mathbf{q}$. This helps us to compactly represents all involving filters of cubic B-spline as

$$\begin{aligned}
\mathbf{a} &= \begin{bmatrix} \frac{23}{196} & -\frac{23}{49} & \frac{9}{28} & \frac{52}{49} & \frac{9}{28} & -\frac{23}{49} & \frac{23}{196} \end{bmatrix} \\
\mathbf{b} &= \begin{bmatrix} \frac{13}{98} & -\frac{26}{49} & \frac{39}{49} & -\frac{26}{49} & \frac{13}{98} \end{bmatrix} \\
\mathbf{p} &= \begin{bmatrix} \frac{1}{8} & \frac{1}{2} & \frac{3}{4} & \frac{1}{2} & \frac{1}{8} \end{bmatrix} \\
\mathbf{q} &= \begin{bmatrix} -\frac{23}{208} & -\frac{23}{52} & -\frac{63}{208} & 1 & -\frac{63}{208} & -\frac{23}{52} & -\frac{23}{208} \end{bmatrix} .
\end{aligned} \qquad (3.7)$$

An important caveat to the filter vector notation in (3.7) is that $\mathbf{a}$ and $\mathbf{b}$ represent regular *rows* of $\mathbf{A}$ and $\mathbf{B}$, while $\mathbf{p}$ and $\mathbf{q}$ represent regular *columns* of $\mathbf{P}$ and $\mathbf{Q}$. Thus the application of $\mathbf{a}$ (or $\mathbf{b}$) to a sample vector is similar to convolution, as the filter vector is slid along the sample vector. Conversely, the application of $\mathbf{p}$ (or $\mathbf{q}$) is two convolutions, with one kernel consisting of the even entries and another filled with the odd entries (due to the column shift required by a 2-scale dilation). Section 3.6 illustrates how to interpret these filters algorithmically.

3.4. Other B-Spline Multiresolution Filters

The construction in Sec. 3.3 is general enough to be used for other subdivision methods, particularly B-spline subdivisions. Due to the fact that having two levels of smoothness is enough for most applications in imaging and graphics, only quadratic and cubic B-spline subdivisions are considered here. However, the multiresolution filters obtained from this method of construction are not unique and there are variety of options. For example, the filters in (3.7) are result of starting with the width seven for **A**. Different filters can be derived by changing the width of **A**. Wider filters result in a better coarse approximation of the fine samples but they are harder to implement and require more computations. In addition, it is possible to add constraints to the construction to obtain better filter values, such as inverse powers of two. Here we report some other alternative filter sets that may be useful in different applications.

3.4.1. *Short Filters for Cubic B-Spline*

If we start with a width of three for **A** in the construction, we obtain

$$
\begin{aligned}
\mathbf{a} &= \left[\ -\tfrac{1}{2}\ 1\ -\tfrac{1}{2}\ \right] \\
\mathbf{b} &= \left[\tfrac{1}{4}\ -1\ \tfrac{3}{2}\ -1\ \tfrac{1}{4}\right] \\
\mathbf{p} &= \left[\tfrac{1}{8}\ \tfrac{1}{2}\ \tfrac{3}{4}\ \tfrac{1}{2}\ \tfrac{1}{8}\right] \\
\mathbf{q} &= \left[\ \tfrac{1}{4}\ 1\tfrac{1}{4}\ \right]\ .
\end{aligned}
\tag{3.8}
$$

Although these filters are very compact and easy to implement, they often fail to generate a good coarse approximation.

3.4.2. *Cubic B-Spline Filters: Inverse Powers of Two*

Having powers and inverse powers of two is desirable for implementing multiresolution in hardware. It is possible to add constraints to the construction that is described in Sec. 3.3 to obtain filter values as inverse powers of two. This is not always successful, but it is certainly gratifying when it is. The construction is nicely successful for cubic B-spline with a width of seven for **a**

$$\mathbf{a} = \left[\begin{array}{ccccccc} \frac{1}{8} & -\frac{1}{2} & \frac{3}{8} & 1 & \frac{3}{8} & -\frac{1}{2} & \frac{1}{8} \end{array}\right]$$

$$\mathbf{b} = \left[\begin{array}{cccccc} -\frac{1}{8} & \frac{1}{2} & -\frac{3}{4} & \frac{1}{2} & -\frac{1}{8} \end{array}\right]$$

$$\mathbf{p} = \left[\begin{array}{ccccc} \frac{1}{8} & \frac{1}{2} & \frac{3}{4} & \frac{1}{2} & \frac{1}{8} \end{array}\right]$$

$$\mathbf{q} = \left[\begin{array}{ccccccc} \frac{1}{8} & -\frac{1}{2} & \frac{3}{8} & 1 & \frac{3}{8} & -\frac{1}{2} & \frac{1}{8} \end{array}\right] .$$

The quality of the coarse approximation in this case is near to optimal filters in (3.7).

3.4.3.   *Short Filters for Quadratic B-Spline*

The local filters of quadratic B-spline can be constructed based on reversing Chaikin subdivision [5], for which the underlying scale functions are the quadratic B-splines. The smallest width that can generate a non-trivial filters is four, which results in the following filters

$$\mathbf{a} = \left[\begin{array}{cccc} -\frac{1}{4} & \frac{3}{4} & \frac{3}{4} & -\frac{1}{4} \end{array}\right]$$

$$\mathbf{b} = \left[\begin{array}{cccc} -\frac{1}{4} & \frac{3}{4} & -\frac{3}{4} & \frac{1}{4} \end{array}\right] \tag{3.9}$$

$$\mathbf{p} = \left[\begin{array}{cccc} \frac{1}{4} & \frac{3}{4} & \frac{3}{4} & \frac{1}{4} \end{array}\right]$$

$$\mathbf{q} = \left[\begin{array}{cccc} -\frac{1}{4} & -\frac{3}{4} & \frac{3}{4} & \frac{1}{4} \end{array}\right] .$$

These filters are appealingly simple, yet their quality is reasonably good (see Sec. 3.8).

3.4.4.   *Wide Filters for Quadratic B-Spline*

Starting with a width of eight for **a**, the following filters are obtained

$$\mathbf{a} = \left[\begin{array}{cccccccc} \frac{3}{40} & -\frac{9}{40} & -\frac{1}{40} & \frac{27}{40} & \frac{27}{40} & -\frac{1}{40} & -\frac{9}{40} & \frac{3}{40} \end{array}\right]$$

$$\mathbf{b} = \left[\begin{array}{cccc} -\frac{27}{160} & \frac{81}{160} & -\frac{81}{160} & \frac{27}{160} \end{array}\right] \tag{3.10}$$

$$\mathbf{p} = \left[\begin{array}{cccc} \frac{1}{4} & \frac{3}{4} & \frac{3}{4} & \frac{1}{4} \end{array}\right]$$

$$\mathbf{q} = \left[\begin{array}{ccccccc} -\frac{1}{9} & -\frac{1}{3} & \frac{1}{27} & 1 & -1 & -\frac{1}{27} & \frac{1}{3} & \frac{1}{9} \end{array}\right] .$$

As shown in Section 3.8.1, these filters generate very high compression rates for images compression.

3.5.   **Extraordinary (Boundary) Filters**

All multiresolution filters in Section 3.3 and 3.4 are *regular*, meaning they are applicable only to data with full neighborhoods. Using symmetric extension (Section 3.7.3.1), we can apply such regular filters to curves, surfaces, and images with boundaries. However, boundary interpolation is often strongly desired. To have this property, we must sacrifice the regularity of the filters near to the boundary.

To fulfill the interpolation condition for B-spline representations, multiple knots are used at the ends of the knot sequence, corresponding to the beginning and ending portions of any data that the filters might operate on. This knot multiplicity creates irregular or *extraordinary* parts in the subdivision matrix. We use a block matrix notation to separate the boundary filters from the regular filters. For example, the $\mathbf{P}$ matrix for cubic B-Splines with the interpolation condition is shown in (3.11). In this notation $\mathbf{P}_s$ shows the extraordinary parts of the subdivision matrix near to the start of the sample vector. Similarly, $\mathbf{P}_e$ refers to the extraordinary parts near to the end. And finally, $\mathbf{P}_r$ shows the regular portion of this matrix.

$$\mathbf{P} = \begin{bmatrix} \mathbf{P}_s \\ \mathbf{P}_r \\ \mathbf{P}_e \end{bmatrix} , \tag{3.11}$$

where

$$\mathbf{P}_s = \begin{bmatrix} 1 & 0 & 0 & 0 & 0 & 0 & \cdots \\ \frac{1}{2} & \frac{1}{2} & 0 & 0 & 0 & 0 & \cdots \\ 0 & \frac{3}{4} & \frac{1}{4} & 0 & 0 & 0 & \cdots \\ 0 & \frac{3}{16} & \frac{11}{16} & \frac{1}{8} & 0 & 0 & \cdots \end{bmatrix} ,$$

$$\mathbf{P}_r = \begin{bmatrix} 0 & 0 & \frac{1}{2} & \frac{1}{2} & 0 & 0 & 0 & 0 & 0 & \cdots \\ 0 & 0 & \frac{1}{8} & \frac{3}{4} & \frac{1}{8} & 0 & 0 & 0 & 0 & \cdots \\ & & & \vdots & & & & & \end{bmatrix} ,$$

$$\mathbf{P}_e = \begin{bmatrix} \cdots & 0 & 0 & \frac{1}{8} & \frac{11}{16} & \frac{3}{16} & 0 \\ \cdots & 0 & 0 & 0 & \frac{1}{4} & \frac{3}{4} & 0 \\ \cdots & 0 & 0 & 0 & 0 & \frac{1}{2} & \frac{1}{2} \\ \cdots & 0 & 0 & 0 & 0 & 0 & 1 \end{bmatrix} .$$

3.5.1. *Boundary Filters for Cubic B-Spline*

Again we present our construction in the context of cubic B-Spline subdivision.

3.5.1.1. *Construction of* $\mathbf{A}$

Having extraordinary filters at the boundary of $\mathbf{P}$ affects our construction method and usually causes extraordinary filters for $\mathbf{A}$, $\mathbf{B}$ and $\mathbf{Q}$ too. In the first step, we would like to find $\mathbf{A}$ such that $\mathbf{AP} = \mathbf{I}$. Any width of $\mathbf{A}$ filter can be investigated, but we have tried to find a width consistent with the regular filters' widths. For the first row of $\mathbf{A}$ consisting of only a_0, the only interaction with $\mathbf{P}$ corresponds to the first $\mathbf{P}$ column.

By the way the subdivision is defined at the boundary, investigating a minimal $\mathbf{A}$ filter is the obvious thing to do, since the subdivision simply reproduces the extreme samples $c_{\min} = f_{\min}$. Nevertheless, for completeness, the setup for determining that fact is:

$$c_{\min} = a_0 \, f_{\min} \;\Rightarrow\; a_0 = 1.$$

The second row of $\mathbf{A}$, for estimating the second coarse sample, is more interesting. The $\mathbf{A}$ filter investigated is five elements long, which is the closest possible to the seven-element filter being used for the interior samples. For the five-element $\mathbf{A}$ filter being investigated, there are only four relevant $\mathbf{P}$ columns, and the equations that are generated by forming the relevant section of $\mathbf{AP} = \mathbf{I}$ are

$$1\,a_0 + \frac{1}{2}\,a_1 = 0; \qquad \frac{1}{2}\,a_1 + \frac{3}{4}\,a_2 + \frac{3}{16}\,a_3 = 1;$$

$$\frac{1}{4}\,a_2 + \frac{11}{16}\,a_3 + \frac{1}{2}\,a_4 = 0; \qquad \frac{1}{8}\,a_3 + \frac{1}{2}\,a_4 = 0 \,.$$

This creates the second row of $\mathbf{A}$

$$\mathbf{a}_2 = \left[-\frac{49}{139} \quad \frac{98}{139} \quad \frac{135}{139} \quad -\frac{60}{139} \quad \frac{15}{139} \right] \,.$$

The third row $\mathbf{A}$ can be found with the same method; for this row a seven element filter $[a_0, a_1, a_2, a_3, a_4, a_5, a_6]$ can be considered. This row has non-zero interaction with the first five columns of $\mathbf{P}$, resulting in

$$\mathbf{a}_3 = \left[\frac{9}{50} \quad -\frac{9}{25} \quad -\frac{2}{25} \quad \frac{32}{25} \quad \frac{43}{100} \quad -\frac{3}{5} \quad \frac{3}{20} \right] \,.$$

If we use the same kind of the blocked matrix notation for $\mathbf{A}$, where $\mathbf{A}_s$, $\mathbf{A}_r$ and $\mathbf{A}_e$ respectively refer to the extraordinary block near to the start, the regular block and the extraordinary block near to the end, then the result of the boundary analysis is:

$$\mathbf{A}_s = \begin{bmatrix} 1 & 0 & 0 & 0 & 0 & 0 & 0 & 0 & \cdots \\ -\frac{49}{139} & \frac{98}{139} & \frac{135}{139} & -\frac{60}{139} & \frac{15}{139} & 0 & 0 & 0 & \cdots \\ \frac{9}{50} & -\frac{9}{25} & -\frac{2}{25} & \frac{32}{25} & \frac{43}{100} & -\frac{3}{5} & \frac{3}{20} & 0 & \cdots \end{bmatrix},$$

$$\mathbf{A_r} = \begin{bmatrix} 0 & 0 & \frac{23}{196} & -\frac{23}{49} & \frac{9}{28} & \frac{52}{49} & \frac{9}{28} & -\frac{23}{49} & \frac{23}{196} & 0 & 0 & \cdots \\ 0 & 0 & 0 & 0 & \frac{23}{196} & -\frac{23}{49} & \frac{9}{28} & \frac{52}{49} & \frac{9}{28} & -\frac{23}{49} & \frac{23}{196} & \cdots \\ & & & & \vdots & & & & & & & \end{bmatrix},$$

$$\mathbf{A_e} = \begin{bmatrix} \cdots & 0 & 0 & \frac{3}{20} & -\frac{3}{5} & \frac{43}{100} & \frac{32}{25} & -\frac{2}{25} & -\frac{9}{25} & \frac{9}{50} \\ \cdots & 0 & 0 & 0 & 0 & \frac{15}{139} & -\frac{60}{139} & \frac{135}{139} & \frac{98}{139} & -\frac{49}{139} \\ \cdots & 0 & 0 & 0 & 0 & 0 & 0 & 0 & 0 & 1 \end{bmatrix}.$$

3.5.1.2. B *and* Q

To establish **B** filters near the boundary, we proceed in the way that we did for **A**. To begin, we would try solving for the fist row of **B** making it as near to the size of interior **B** filters as possible in that position. The first **B** filter configuration that yields nontrivial elements has the width five. The interactions of such a filter with the boundary **P** filter contribute to the equations **BP** = **0** as

$$1\,b_0 + \frac{1}{2}\,b_1 = 0; \quad \frac{1}{2}\,b_1 + \frac{3}{4}\,b_2 + \frac{3}{16}\,b_3 = 0;$$

$$\frac{1}{4}\,b_2 + \frac{11}{16}\,b_3 + \frac{1}{2}\,b_4 = 0; \quad \frac{1}{8}\,b_3 + \frac{1}{2}\,b_4 = 0\,.$$

For the second boundary row of **B** filter, we have:

$$\frac{3}{4}\,b_0 + \frac{3}{16}\,b_1 = 0; \quad \frac{1}{4}\,b_0 + \frac{11}{16}\,b_1 + \frac{1}{2}\,b_2 + \frac{1}{8}\,b_3 = 0;$$

$$\frac{1}{8}\,b_1 + \frac{1}{2}\,b_2 + \frac{3}{4}\,b_3 + \frac{1}{2}\,b_4 = 0; \quad \frac{1}{8}\,b_3 + \frac{1}{2}\,b_4 = 0\,.$$

Similar steps will be set up to generate the equations **AQ** = **0**. In addition, we need to the set up for contributing to the equations **BQ** = **I** from a **Q** columns. When all equations have been generated and solved, proceeding as we have done in terms of the lengths chosen for the boundary filters, we obtain the boundary **B** and **Q**. The resulting filter for **B** is

$$\mathbf{B_s} = \begin{bmatrix} -\frac{45}{139} & \frac{90}{139} & -\frac{135}{278} & \frac{30}{139} & -\frac{15}{278} & 0 & 0 & 0 & 0 & \cdots \\ 0 & 0 & \frac{57}{490} & -\frac{114}{245} & \frac{171}{245} & -\frac{114}{245} & \frac{57}{490} & 0 & 0 & \cdots \end{bmatrix},$$

$$\mathbf{B_r} = \begin{bmatrix} 0 & 0 & 0 & 0 & \frac{13}{98} & -\frac{26}{49} & \frac{39}{49} & -\frac{26}{49} & \frac{13}{98} & 0 & 0 & 0 & \cdots \\ 0 & 0 & 0 & 0 & 0 & 0 & \frac{13}{98} & -\frac{26}{49} & \frac{39}{49} & -\frac{26}{49} & \frac{13}{98} & 0 & \cdots \\ & & & & & \vdots & & & & & & & \end{bmatrix},$$

$$\mathbf{B}_e = \begin{bmatrix} \cdots & 0 & 0 & 0 & \frac{57}{490} & -\frac{114}{245} & \frac{171}{245} & -\frac{114}{245} & \frac{57}{490} & 0 & 0 \\ \cdots & 0 & 0 & 0 & 0 & 0 & -\frac{15}{278} & \frac{30}{139} & -\frac{135}{278} & \frac{90}{139} & -\frac{45}{139} \end{bmatrix}.$$

The resulting filters for $\mathbf{Q}$ are given in Appendix B.

3.5.2. *Boundary Filters for Short Cubic B-Spline*

Using the same kind of construction as Sec. 3.5.1, we can construct extraordinary filters for the narrow cubic B-spline filters of Sec. 3.4.1 (see Appendix C).

3.5.3. *Boundary Filters for Short Quadratic B-Spline*

By the same method, we can generate a full set of multiresolution matrices for quadratic B-spline subdivision (commonly referred to as Chaikin subdivision). The $\mathbf{P}$ filter for quadratic B-spline subdivision is $\mathbf{p} = \begin{bmatrix} \frac{1}{4} & \frac{3}{4} & \frac{3}{4} & \frac{1}{4} \end{bmatrix}$. The blocked matrix notation for the synthesis filter $\mathbf{P}$ is given in Appendix D.

3.5.4. *Boundary Filters for Wide Quadratic B-Spline*

In the case of boundary filters for wide quadratic B-spline, $\mathbf{P}$ is the same as Sec. 3.5.3. The $\mathbf{A}$ matrix is given in Appendix E.

3.6. Efficient Algorithm

We show how an efficient algorithm can be made based on the multiresolution filters for quadratic B-spline subdivision, according to the matrix forms presented in Section 3.5.3.

For all algorithms, we have focused on doing just one step of decomposition or reconstruction. Each algorithm can be used multiple times to construct a hierarchical wavelet transform. In all cases F represents the vector of high-resolution data, C represent low-resolution data and D represents the detail vector.

Conceptually, a multiresolution algorithm performs the matrix-vector operations specified in (3.2), (3.3), and (3.4). However, $\mathbf{A}$, $\mathbf{B}$, $\mathbf{P}$, and

$\mathbf{Q}$ are regular banded matrices, so using matrix-vector operations is not efficient. By using the blocked and banded structure of these matrices, more efficient ($O(n)$ versus $O(n^2)$) algorithms can obtained.

The first algorithm is REDUCE-RESOLUTION. In this algorithm, $F[1..m]$ is the input fine data and the vector $C[1..n]$ is the output coarse approximation. The index i traverses F and j traverses C.

REDUCE-RESOLUTION($F[1..m]$)
1   $C_1 = F_1$
2   $C_2 = -\frac{1}{2}F_1 + F_2 + \frac{3}{4}F_3 - \frac{1}{4}F_4$
3   $j = 3$
4   **for** $i = 2$ **to** $m - 5$ **step** 2
5   **do** $C_j = -\frac{1}{4}F_i + \frac{3}{4}F_{i+1} + \frac{3}{4}F_{i+2} - \frac{1}{4}F_{i+3}$
6        $j = j + 1$
7   $C_j = -\frac{1}{4}F_{m-3} + \frac{3}{4}F_{m-2} + F_{m-1} - \frac{1}{2}F_m$
8   $C_{j+1} = F_m$
9   **return** $C[1..j + 1]$

Lines 1–2 in REDUCE-RESOLUTION correspond to the $\mathbf{A}_s$ matrix, while lines 8–9 correspond to the $\mathbf{A}_e$ matrix. The `for` loop represents the application of the regular $\mathbf{A}_r$ block.

The second algorithm is FIND-DETAILS. We can again identify blocks corresponding to $\mathbf{B}_s$, $\mathbf{B}_r$, and $\mathbf{B}_e$.

FIND-DETAILS($F[1..m]$)
1   $D_1 = -\frac{1}{2}F_1 + F_2 - \frac{3}{4}F_3 + \frac{1}{4}F_4$
2   $D_2 = -\frac{1}{4}F_3 + \frac{3}{4}F_4 - \frac{3}{4}F_5 + \frac{1}{4}F_6$
3   $j = 3$
4   **for** $i = 5$ **to** $m - 5$ **step** 2
5   **do** $D_j = \frac{1}{4}F_i - \frac{3}{4}F_{i+1} + \frac{3}{4}F_{i+2} - \frac{1}{4}F_{i+3}$
6        $j = j + 1$
7   $D_j = \frac{1}{4}F_{m-3} - \frac{3}{4}F_{m-2} + F_{m-1} - \frac{1}{2}F_m$
8   **return** $D[1..j]$

For reconstruction, we need to compute $\mathbf{P}C + \mathbf{Q}D$. The 2-scale column shift property causes to have two kinds of regular rows(odd and even) for $\mathbf{P}$ and $\mathbf{Q}$. This only requires a simple odd/even regular rules as demonstrated by the algorithm RECONSTRUCTION.

RECONSTRUCTION($C[1..n], D[1..s]$)
1   $E_1 = 0D_1$
2   $E_2 = \frac{1}{2}D_1$
3   $E_3 = -\frac{3}{4}D_1 + \frac{1}{4}D_2$
4   $E_4 = -\frac{1}{4}D_1 + \frac{3}{4}D_2$
5   $E_5 = -\frac{3}{4}D_2 - \frac{1}{4}D_3$
6   $E_6 = -\frac{1}{4}D_2 - \frac{3}{4}D_3$

```
 7   j = 7
 8   for i = 3 to s − 1
 9       do E_j = 3/4 D_i − 1/4 D_{i+1}
10          E_{j+1} = 1/4 D_i − 3/4 D_{i+1}
11          j = j + 2
12   E_j = 1/2 D_s
13   E_{j+1} = 0 D_s
14
15   F_1 = C_1 + E_1
16   F_2 = (1/2 C_1 + 1/2 C_2) + E_2
17   j = 3
18   for i = 2 to n − 2
19       do F_j = (3/4 C_i + 1/4 C_{i+1}) + E_j
20          F_{j+1} = (1/4 C_i + 3/4 C_{i+1}) + E_{j+1}
21          j = j + 2
22   F_j = (1/2 C_{n-1} + 1/2 C_n) + E_j
23   F_{j+1} = C_n + E_{j+1}
24   return F[1..j + 1]
```

Lines 1–14 in RECONSTRUCTION construct the $E = \mathbf{Q}D$ term. Lines 1 through 6 correspond to $\mathbf{Q}_s$, and lines 13–14 apply $\mathbf{Q}_e$. The `for` loop at line 8 is for the regular block $\mathbf{Q}_r$. In line 1, E_1 has been set to $0D_1$ instead of 0 to have general algorithm that can work for the data with any dimension induced by D. Lines 24 trough 37 make $F = \mathbf{P}C + E$. Again the terms $\mathbf{P}_e$, $\mathbf{P}_r$ and $\mathbf{P}_e$ are distinguishable in the algorithm.

Note that m, the size of the high-resolution data F, is equal to $n + s$; it is clear that the running time of all three algorithms is linear in m.

3.7. Extensions

The regular and extraordinary filters presented thus far are intended for use on non-periodic data sets, such as open-ended curves. For many applications, we may have data that does not fit this definition; for example, periodic curves, tensor-product surfaces, or 2D and 3D images. In this section, we show how to use the local filters for these kinds of objects.

3.7.1. *Periodic (Closed) Curves*

For many applications, boundary-interpolating filters are not desirable. For closed curves (isomorphic to a circle), we can instead use periodic filters. In periodic (equivalently *closed*) curves, the regular filter values are applied to all samples; there is no concept of a boundary, as the signal F is assumed to wrap around on itself. For $F = \{f_1, \ldots, f_m\}$, we implicitly set $f_{m+x} = f_x$ for $x \geq 1$ and $f_x = f_{m-x}$ for $x < 1$. Figure 3.2 illustrates this wrapping.

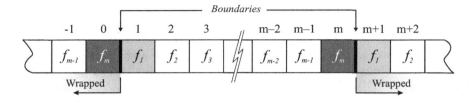

Fig. 3.2. For periodic curves, the left and right boundaries are implicitly connected, wrapping the sample vector back on itself.

In the matrix form of periodic subdivision, the regular columns of $\mathbf{P}$ will wrap around the top and bottom of the matrix, rather than being terminated with the special boundary filters in the non-periodic (open) case. Consider the matrix corresponding to periodic cubic B-spline subdivision

$$
\mathbf{P}_{periodic} =
\begin{bmatrix}
\frac{3}{4} & \frac{1}{8} & 0 & \cdots & 0 & \frac{1}{8} \\
\frac{1}{2} & \frac{1}{2} & 0 & \cdots & 0 & 0 \\
& \vdots & & & & \\
\cdots & \frac{1}{8} & \frac{3}{4} & \frac{1}{8} & 0 & \cdots \\
\cdots & 0 & \frac{1}{2} & \frac{1}{2} & 0 & \cdots \\
& \vdots & & & & \\
\frac{1}{8} & 0 & \cdots & 0 & \frac{1}{8} & \frac{3}{4} \\
\frac{1}{2} & 0 & \cdots & 0 & 0 & \frac{1}{2}
\end{bmatrix}
$$

The remaining matrices $\mathbf{A}$, $\mathbf{B}$, and $\mathbf{Q}$ are formed by a similar wrapping of the rows or columns.

From an implementation perspective, periodic data is easier to work with because there are no special boundary cases. We need only modify the indexing of the samples to ensure that the wrapping is done correctly.

3.7.2.   *Tensor Product Surfaces*

Multiresolution schemes for 1D data, such as the cubic or quadratic B-splines schemes developed earlier, can be applied to surface patches by a straightforward extension.

A surface patch is defined by a regular 2-dimensional grid of vertices. The regularity allows the patch to be split into two arbitrary dimensions, usually denoted as the u and v directions. Each row aligned along the u direction is referred to as a v-curve (because the v value is constant), and vice versa.

To apply a multiresolution filter to the patch, all u and v curves can be considered as independent curves to which the ordinary multiresolution algorithms can be applied. For instance, to decompose a grid of vertices, the REDUCE-RESOLUTION algorithm could be called for all rows in the grid, and then for all columns in the smaller grid that results from reducing the resolution of all rows.

As discussed in the previous section, we can interpret a set of point samples as defining an open (non-periodic) or closed (periodic) curve. With a tensor product surface, there are three unique ways to interpret the point grid.

3.7.2.1.   *Open-Open Surfaces*

In an open-open tensor product surface, both the u and v curves are considered to be open curves. Open-open surfaces are isomorphic to a bounded plane (sheet). See Fig. 3.3(a).

3.7.2.2.   *Open-Closed Surfaces*

We can treat a tensor-product surface as a set of open curves in one direction, and a set of closed curves in the other. In this case, the surface is isomorphic to an uncapped cylinder. See Fig. 3.3(b).

3.7.2.3.   *Closed-Closed Surfaces*

The final configuration of the u and v curves in a tensor product surface is when both dimensions are closed or periodic. In this configuration, the surface will be isomorphic to a torus, as shown in Fig. 3.3(c).

3.7.3.   *2D Images*

Conceptually, there is no need to distinguish between tensor product surfaces and 2D images. Each is a collection of samples (nD points

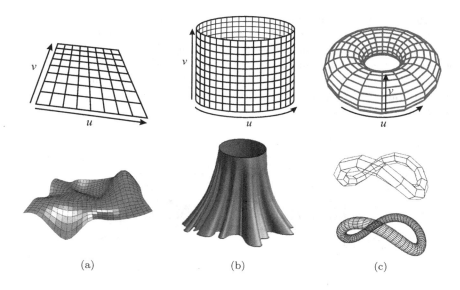

Fig. 3.3. There are three unique isomorphs for a tensor product surface, depending on how the u and v curves are interpreted: (a) open-open (bounded plane); (b) open-closed (uncapped cylinder); (c) closed-closed (torus).

and intensity values, respectively) arranged in a regular grid, and linear combinations are valid operations on each sample type. Multiresolution filters can be applied to 2D images just as with tensor product surfaces: by treating all rows, and then all columns, of the image as independent 1D sample vectors.

In practice, however, there are some subtle but important differences that should be accounted for. In an image, positionality is implied by the location of a pixel, rather than the content of the pixel. As well, multiresolution operations on images are usually employed for filtering purposes, so having boundary interpolation gives incongruous importance to the image boundary. Thus our typical approaches to handling boundaries — interpolation or periodicity — make little sense in the image domain.

3.7.3.1. *Symmetric Extension*

The wrapping of samples done in the periodic case is a particular case of a more general approach called *symmetric extension*. The goal of symmetric extension is to avoid special boundary case evaluations by filling in sensible values for samples outside of the bounds of the real samples.

While wrapping may make sense for tileable images, in general it will not give a logical result because mixing the intensity values of the left and the right boundaries of the image is not reasonable. Due to the implied positionality of samples in images, the most natural "neighbor" sample when none exists would be the mirrored neighbor from the other side [12]. More formally, if $F = \{f_1, \ldots, f_m\}$, then we could set $f_{x<1} = f_{1+(1-x)}$ to mirror about the lower boundary, and similarly set $f_{x>m} = f_{m-(x-m)}$. See Fig. 3.4(top) for a diagrammatic representation of this type of symmetric extension, referred to as Type A.

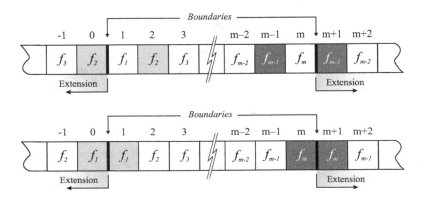

Fig. 3.4. Symmetric extension mirrors samples near the boundary. *Top:* mirrored about the first and last samples (Type A); *Bottom:* mirrored about the boundary (Type B).

An alternative approach is to mirror exactly about the boundary, which would produce duplicate entries of the first and last samples f_1 and f_m. In particular, we set $f_{x<1} = f_{1-x}$ to mirror about the lower boundary, and similarly set $f_{x>m} = f_{m-(x-m)+1}$. This is known as Type B symmetric extension; see the bottom image of Fig. 3.4.

The appropriate choice of symmetric extension depends on the multiresolution scheme. Consider cubic B-spline, with the filters given in (3.7). Cubic B-spline is known as a *primal* or *edge-split* scheme, meaning that each coarse point and coarse edge has a corresponding fine point. For such schemes, Type A symmetric extension can be used for decomposing F. For reconstruction, the correct interpretation is achieved by using Type A for extending C and Type B for extending D (see the shaded entries in Fig. 3.5).

The Chaikin scheme is classified as a *dual* or *vertex-split* scheme, because each coarse sample is split into two fine samples, and there is no

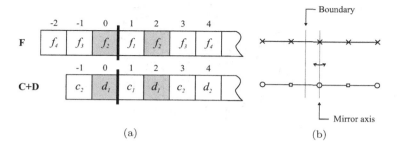

Fig. 3.5. Symettric extension for cubic B-spline. *Left:* Type A symmetric extension can be used when interpreting the samples in both decomposition and reconstruction, but Type B is required to interpret the details. *Right:* this is because the desired mirror axis is the same in both cases.

unambiguous relationship between points at each level. To use symmetric extension on such a scheme, we need both Type A and Type B. During decomposition, it is more natural to use Type B because the coarse vertex corresponding to the first fine vertex is split into a left and right component; see Fig. 3.6(b). During reconstruction, however, Type A extension for C and Type B extension for D provide the proper relationships.

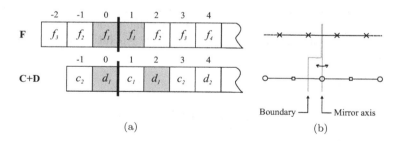

Fig. 3.6. Symmetric extension for Chaikin multiresolution. *Left:* Type A and Type B symmetric extension are used in reconstruction (for the samples C and the details D, respectively), while Type B is used to decompose F. *Right:* this is because the desired mirror axis is different in each process.

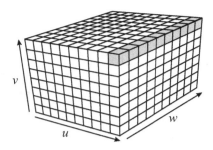

Fig. 3.7. Multiresolution techniques for 3D images are similar to those for 2D images: the appropriate filter is applied along each dimension independently.

3.7.4. *3D Images*

Multiresolution techniques for 3D images, such as volumetric data, naturally follow from 2D images. Consider the 3D image to have three axes: u, v, and w. Each w "curve" is actually an ordered set of samples (Fig. 3.7).

To apply multiresolution techniques to 3D image data, we can use the 2D image approach for each "slice" along the w direction, followed by each u-v curve aligned with w (highlighted in grey in Fig. 3.7).

As an example, consider a 3D image to be a set of still images from a video sequence. If the w axis quantifies time, then applying subdivision to each "slice" would smooth the image and subdividing each curve aligned with w would compute intermediate frames in the sequence. Conversely, decomposing a set of still images would reduce the resolution of each frame, while also removing every other frame.

3.8. Results, Examples and Applications

Multiresolution provides a tool for decomposing data into a hierarchy of components with different scales or resolutions. This hierarchy can be used for noise removal, compression, synthesizing and recognition of the objects [13,18,21]. In particular, multiresolution of smooth scalings and wavelets can nicely be used for smooth objects. Here we show some examples and applications of the quadratic and cubic B-spline multiresolution filters. We discuss iris synthesis and real-time visualization of volumetric data in detail.

3.8.1. *Example Applications*

Curves by example. We can use multiresolution filters on curves, such as those representing artistic silhouettes and line hand-gesture styles. Using analysis filters, we can extract styles (based on the characteristic details) from the curves, and these styles can then be applied to new base curves. Figure 3.8 shows an application of this, based on the interpolating quadratic B-spline wavelets from Section 3.5.3. This example illustrates the use of multiresolution for the common biometric task of feature extraction.

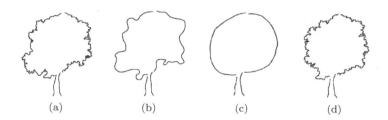

(a) (b) (c) (d)

Fig. 3.8. Capturing stroke styles: (a) the original curve. (b) the curve from (a), reconstructed without D^i. (c) A new base curve. (d) The reconstructed curve with the (c) as the base curve and details D^i from (a).

Removing noise from Curves. If we reconstruct a data set without using any (or using only a small portion) of the details, a simple de-noising is achieved as demonstrated in Fig. 3.9. In this example the cubic B-Spline filters of Sec. 3.5.1 have been used.

Fig. 3.9. Left: original silhouette from an inner-ear mesh. Right: the silhouette is de-noised after decomposing twice with cubic B-spline filters and then reconstructing with partial details.

Image Compression. After decomposing an image to a low-resolution approximation and corresponding details D^i, we can lossily compress the image by removing small magnitude details. This is one major step in current image compression techniques such as JPEG. We have compared our filters with Haar filters. All of our filters reported in Section 3.4 and (3.7) have better compression rates. We have also compared our filters with more successful image compression filters, D9/7 and D4, that have been used in JPEG2000[11]. Although our local filters are based on smooth scalings and wavelets (in contrast to D9/7 and D4), the resulting compression rate is comparable; see Fig. 3.10. In this comparison, we have removed the same amount of the details and compared the PSNR of the reconstructed images; higher PSNR indicates better detail retention. In particular, the wide quadratic B-spline filters given in (3.10) perform very close to the D9/7 filter.

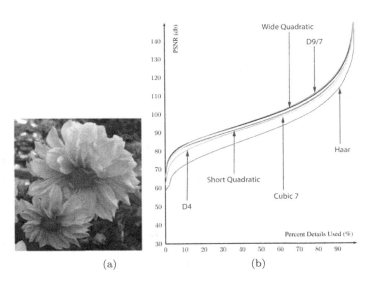

(a)   (b)

Fig. 3.10. Comparison of quadratic B-spline wavelets image compression with established techniques: (a) a sample image containing high-frequency data; (b) the resulting compression rates for various filters.

Terrain by Example. Terrain is typically represented with a height map (a regular grid of elevation data), which directly maps to an open-open tensor-product surface. Multiresolution helps us to use an existing terrain to synthesize new terrain by transferring the characteristic details [4]. For

Fig. 3.11. Left: A smooth base terrain. Middle: a model terrain with high-frequency details. Right: the synthesized terrain has low-frequency characteristics of the base terrain and high-frequency traits of the model terrain.

terrain synthesis, we can capture the details of a high-resolution target terrain and use them for a smooth base terrain to add realistic and more predictable noise. Figure 3.11 shows the application of the quadratic filters of (3.9) to terrain synthesis.

3.8.2. *Iris Synthesis*

Biometrics conventionally involves the analysis of biometric data for identification purposes. Due to logistical and privacy issues with collecting and organizing large amounts of biometric data, a new direction of biometric research concentrates on the synthesis of biometric information. One of the primary goals of the synthesis of biometric data is to provide databases upon which biometric algorithms can be tested [25]. Along this direction, Wecker et al. employ the quadratic B-spline filters of (3.9) to augment existing iris databases with synthesized iris images [22]. Looking at any iris image, it is apparent that most of its characteristics are made of high frequency data. In addition, underlying sweep structures seems to be smooth (circular curves). Therefore, using smooth multiresolution filters for capturing these details is a promising technique. In this method, extracted details from different irises are combined to generate new irises. In fact, because we just combine portions of real irises, realistic iris are obtained. Because of the circular structure of iris, it is better to transform the iris images into polar coordinates, as shown in Fig. 3.12. For a 256×256 iris image, four levels of decomposition was found to produce good experimental results. Given a database of N input images, we decompose each of these images into their five components (four levels of details, and the very coarse approximation). Therefore, when synthesizing an iris image, we have N choices available for each of the necessary components which allows us to create a total of N^5

possible combinations. The original N iris images are included in these new combinations, as they are recreated when each of the selected components is from the same iris image. Clearly, given even a reasonable small database to start with, this method can generate an exponential increase in size. Some of these combinations may not result in good irises and extra conditions should also be taken to account [22].

Figure 3.13 shows a sample iris database, and Fig. 3.14 presents some of the iris images synthesized from the originals.

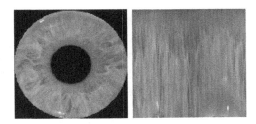

Fig. 3.12.   A polar transform is used to unwrap the iris image.

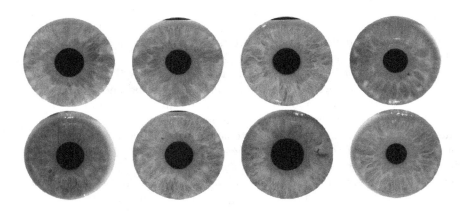

Fig. 3.13.   A sample database of iris images.

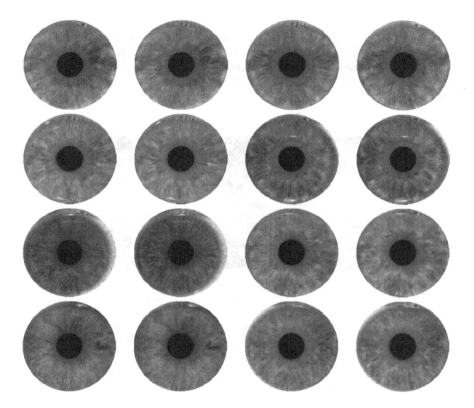

Fig. 3.14. Some synthetic iris images generated by combining elements of the original iris images from Fig. 3.13.

3.8.3. *Real-Time Contextual Close-up of Volumetric Data*

Medical image data sets can be very large. For instance, a CT scan of the head can have spatial dimensions up to $512 \times 512 \times 256$ voxels, while an MRI dataset can be as large as 256^3 voxels. Although many graphics display devices are now equipped with enough memory to store the entire volume, few commodity devices are actually capable of rendering the full resolution volume with the desired quality at interactive rates. Taerum et al. [20] propose a method based on B-spline filters that allow the data to be viewed and manipulated in real-time, while still allowing for full resolution viewing and detailed high resolution exploration of the data. In this method, the three different resolutions of the data-set are used. The lowest resolution is used during user interaction to maintain real-time feedback (Fig. 3.15). The medium resolution is considered for presenting the overall context, while a "super-resolution" is used when rendering the contents of a user-defined lens, as demonstrated in Fig. 3.16. The method is based on the representation of volumetric data set as described in Section 3.7.4. Both quadratic (3.9) and cubic B-spline filters (3.7) have been used in this work.

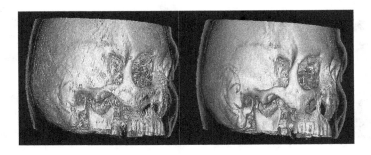

Fig. 3.15. Rendering CT scan data at multiple resolutions allows for real-time interactions while preserving visual clarity. *Left*: the lowest resolution is used for real-time interaction. *Right*: a medium resolution is shown when the data is not being manipulated.

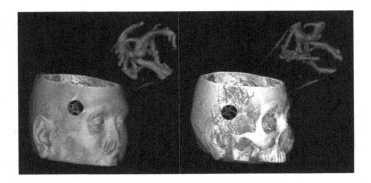

Fig. 3.16. The highest resolution of the CT scan data is used for generating contextual close-up views of the data.

3.9. Conclusion

We have presented several local multiresolution filters that can be employed in biometric applications involving discrete samplings of smooth data. Unlike Haar wavelets, these filters satisfy either the first or second level of smoothness. The biorthogonal construction also produces compact filters than semiorthogonal constructions, leading to more efficient algorithms.

We also consider several situations for the boundary, including interpolating and symmetric extension. With these considerations, the filters can be employed on many types of data, such as open and closed curves and surface patches, images, and volumes.

Finally, we discussed several applications of these filters to biometric-related problems, such as de-noising, feature (characteristic) extraction and transfer, data synthesis, and visualization.

Acknowledgement

This work has been supported by the National Science and Engineering Research Council (NSERC) of Canada. The authors would like to thank Reza Pakdel, Torin Taerum, Lakin Wecker, Katayoon Etemad, Meru Brunn and John Brosz for their help and input to this work.

Bibliography

1. R. H. Bartels, J. Beatty, and B. Barsky (1987). *An Introduction to Splines for Use in Computer Graphics and Geometric Modeling.*
2. R. H. Bartels and F. F. Samavati (2000). *Reversing subdivision rules: Local linear conditions and observations on inner products.* J. of Computational and Applied Mathematics, 119(1–2):29–67.
3. W. W. Boles and B. Boashash (1998). *A Human Identification Technique Using Images of the Iris and Wavelet Transform.* IEEE Trans. Signal Processing, 46(4), pp.1185–1188.
4. J. Brosz, F. F. Samavati and M. C. Sousa (2006). *Terrain Synthesis By-Example.* Proc. 1st Int. Conf. on Computer Graphics Theory and Applications.
5. G. Chaikin (1974). *An Algorithm for High Speed Curve Generation.* Comp. Graph. and Im. Proc. 3, 346–349.
6. C. Chui (1992). *An Introduction to Wavelets.* Academic Press.
7. J. Daugman (2004). How Iris Recognition Works. *IEEE Trans, CSVT* , Vol. 14, PP. 21–30.
8. A. Finkelstein and D. H. Salesin (1994). *Multiresolution curves.* A. Glassner, ed., Proc. SIGGRAPH '94, ACM Press, New York, 261–268.

9. Y. Huang, S. Luo and E. Chen (2002). *An Efficient Iris Recognition System.* Proc. IEEE in Machine Learning and Cybernetics, Vol. 1, PP. 450–454, 2002.

10. A. K. Jain, A. Prabhakar, L. Hong, and S. Pankanti (2000). *Filterbank-based Fingerprint Matching.* IEEE Trans. Image Processing, Vol. 9, No.5, pp. 846–859, May.

11. S. Lawson and J. Zhu (2002). *Image Compression using Wavelets and JPEG2000: A Tutorial.* Electronics and Communication Engineering J., Vol. 14, No.3, pp. 112–121, June.

12. S. Li and W. Li (2000). *Shape-Adaptive Discrete Wavelet Transforms for Arbitrarily Shaped Visual Object Coding.* IEEE Trans. Circuits and Systems for Video Technology, Vol. 10, No. 5.

13. T. Li, Q. Li, S. Zhu and M. Ogihara (2002). *A survey on Wavelets Applications in Data Mining.* ACM SIGKDDExplorations Newsletter archive, Vol. 4 , No. 2, PP. 49–68.

14. S. Lim, K. Lee, O. Byeon, and T. Kim (2001). *Recognition Through Improvement of Feature Vector and Classifier.* ETRI Journal, Vol. 23, Number 2.

15. L. Piegl and W. Tiller (1997). *The NURBS Book.* 2nd Edition. Springer.

16. F. F. Samavati and R. H. Bartels (1999). *Multiresolution Curve and Surface Representation by Reversing Subdivision Rules.* Computer Graphics Forum, 18(2), PP. 97–119, June.

17. F. F. Samavati and R. H. Bartels (2003). *Diagrammatic Tools for Generating Biorthogonal Multiresolutions.* Technical Report 2003-728-31, Computer Science Department, University of Calgary.

18. E. J. Stollnitz, T. D. DeRose, and D. H. Salesin (1996). *Wavelets for Computer Graphics.* Morgan Kaufmann Publishers.

19. G. Strang and T. Nguyen. (1996)*Wavelets and Filter Banks.* Wellesley-Cambridge Press.

20. T. Taerum, M. C. Sousa, F. F. Samavati, S. Chan, and R. Mitchell (2006). *Real-Time Super Resolution Contextual Close-up of Clinical Volumetric Data.* To appear in Eurographics/IEEE-VGTC Symposium on Visualization (EuroVIS 2006), Lisbon, Portugal, May.

21. M. Unser and A. Aldroubi (1996). *A Review of Wavelets in Biomedical Applications.* Proceedings of the IEEE, Vol. 84, No. 4, PP. 626–638.

22. L. Wecker, F. F. Samavati and M. Gavrilova (2005) *Iris Synthesis: A Reverse Subdivision Application.* Proc. Graphite 2005, in association with ACM SIGGRAPH, PP. 121–125, Dunedin, New Zealand, November.

23. R. Wildes and J. Asmuth (1994). *A system for Automated Iris Recognition.* Proc. 2nd IEEE Workshop on Application of Computer Vision, pp. 121–128.

24. M. Yang, D. J. Kriegman, and N. Ahuja (2002). *Detecting Faces in Images: A Survey* IEEE Trans. on Pattern Analysis and Machine Intelligence, Vol. 24, No. 1, PP. 34–58.

25. Yanushkevich, S. N., Stoica, A., Shmerko, V. P. and Popel, D. V. (2005). *Biometric Inverse Problems,* CRC Press/Taylor & Francis Group, Boca Raton, FL.

Appendixes

Appendix A: A consistent set of solutions for cubic B-spline subdivision

A consistent set of solutions for cubic B-spline subdivision yields **A** as follows (a 5 × 7 slice)

$$
\mathbf{A} =
\begin{bmatrix}
\frac{9}{28} & -\frac{23}{49} & \frac{23}{196} & 0 & 0 & 0 & 0 \\
\frac{9}{28} & \frac{52}{49} & \frac{9}{28} & -\frac{23}{49} & \frac{23}{196} & 0 & 0 \\
\frac{23}{196} & -\frac{23}{49} & \frac{9}{28} & \frac{52}{49} & \frac{9}{28} & -\frac{23}{49} & \frac{23}{196} \\
0 & 0 & \frac{23}{196} & -\frac{23}{49} & \frac{9}{28} & \frac{52}{49} & \frac{9}{28} \\
0 & 0 & 0 & 0 & \frac{23}{196} & -\frac{23}{49} & \frac{9}{28}
\end{bmatrix}
\begin{matrix}
\\ c^k_{j-1} \\ c^k_j \\ c^k_{j+1} \\ \\
\end{matrix}
$$

$$c^{k+1}_{i-1} \quad c^{k+1}_i \quad c^{k+1}_{i+1}$$

A corresponding (5 × 7) slice of one possible **B** matrix is

$$
\mathbf{B} =
\begin{bmatrix}
\frac{39}{49} & -\frac{26}{49} & \frac{13}{98} & 0 & 0 & 0 & 0 \\
\frac{13}{98} & -\frac{26}{49} & \frac{39}{49} & -\frac{26}{49} & \frac{13}{98} & 0 & 0 \\
0 & 0 & \frac{13}{98} & -\frac{26}{49} & \frac{39}{49} & -\frac{26}{49} & \frac{13}{98} \\
0 & 0 & 0 & 0 & \frac{13}{98} & -\frac{26}{49} & \frac{39}{49} \\
0 & 0 & 0 & 0 & 0 & 0 & \frac{13}{98}
\end{bmatrix}
\begin{matrix}
\\ d^k_{\ell-1} \\ d^k_\ell \\ d^k_{\ell+1} \\ \\
\end{matrix}
$$

$$c^{k+1}_{i-1} \quad c^{k+1}_i \quad c^{k+1}_{i+1}$$

And a corresponding (7 × 5) portion of a possible **Q** matrix is

$$
\mathbf{Q} =
\begin{bmatrix}
1 & -\frac{23}{52} & 0 & 0 & 0 \\
-\frac{63}{208} & -\frac{63}{208} & -\frac{23}{208} & 0 & 0 \\
-\frac{23}{52} & 1 & -\frac{23}{52} & 0 & 0 \\
-\frac{23}{208} & -\frac{63}{208} & -\frac{63}{208} & -\frac{23}{208} & 0 \\
0 & -\frac{23}{52} & 1 & -\frac{23}{52} & 0 \\
0 & -\frac{23}{208} & -\frac{63}{208} & -\frac{63}{208} & -\frac{23}{208} \\
0 & 0 & -\frac{23}{52} & 1 & -\frac{23}{52}
\end{bmatrix}
\begin{matrix}
\\ \\ c^{k+1}_{i-1} \\ c^{k+1}_i \\ c^{k+1}_{i+1} \\ \\ \\
\end{matrix}
$$

$$d^k_{\ell-1} \quad d^k_\ell \quad d^k_{\ell+1}$$

Appendix B

The resulting filters for $\mathbf{Q}$ are

$$\mathbf{Q}_s = \begin{bmatrix} 0 & 0 & 0 & 0 & 0 & 0 & \cdots \\ 1 & 0 & 0 & 0 & 0 & 0 & \cdots \\ -\frac{2033}{3000} & -\frac{49}{152} & 0 & 0 & 0 & 0 & \cdots \\ \frac{2137}{12000} & -\frac{289}{608} & -\frac{23}{208} & 0 & 0 & 0 & \cdots \\ \frac{139}{500} & 1 & -\frac{23}{52} & 0 & 0 & 0 & \cdots \\ \frac{139}{2000} & -\frac{347}{912} & -\frac{63}{208} & -\frac{23}{208} & 0 & 0 & \cdots \\ 0 & -\frac{115}{228} & 1 & -\frac{23}{52} & 0 & 0 & \cdots \\ 0 & -\frac{115}{912} & -\frac{63}{208} & -\frac{63}{208} & -\frac{23}{208} & 0 & \cdots \end{bmatrix}$$

$$\mathbf{Q}_e = \begin{bmatrix} \cdots & 0 & -\frac{23}{208} & -\frac{63}{208} & -\frac{63}{208} & -\frac{115}{912} & 0 \\ \cdots & 0 & 0 & -\frac{23}{52} & 1 & -\frac{115}{228} & 0 \\ \cdots & 0 & 0 & -\frac{23}{208} & -\frac{63}{208} & -\frac{347}{912} & \frac{139}{2000} \\ \cdots & 0 & 0 & 0 & -\frac{23}{52} & 1 & \frac{139}{500} \\ \cdots & 0 & 0 & 0 & -\frac{23}{208} & -\frac{289}{608} & \frac{2137}{12000} \\ \cdots & 0 & 0 & 0 & 0 & -\frac{49}{152} & -\frac{2033}{3000} \\ \cdots & 0 & 0 & 0 & 0 & 0 & 1 \\ \cdots & 0 & 0 & 0 & 0 & 0 & 0 \end{bmatrix}$$

$$\mathbf{Q_r} = \begin{bmatrix} 0 & 0 & -\frac{23}{52} & 1 & -\frac{23}{52} & 0 & 0 & \cdots \\ 0 & 0 & -\frac{23}{208} & -\frac{63}{208} & -\frac{63}{208} & -\frac{23}{208} & 0 & \cdots \\ & & & \vdots & & & & \end{bmatrix}$$

Appendix C: The Filters for the Narrow Cubic B-Spline

In this case, $\mathbf{A}$ becomes

$$\mathbf{A}_s = \begin{bmatrix} 1 & 0 & 0 & 0 & 0 & 0 & 0 & \cdots \\ -1 & 2 & 0 & 0 & 0 & 0 & 0 & \cdots \end{bmatrix}$$

$$\mathbf{A_r} = \begin{bmatrix} 0 & 0 & -\frac{1}{2} & 2 & -\frac{1}{2} & 0 & 0 & 0 & 0 & 0 & \cdots \\ 0 & 0 & 0 & 0 & -\frac{1}{2} & 2 & -\frac{1}{2} & 0 & 0 & 0 & \cdots \\ & & & & \vdots & & & & & & \end{bmatrix}$$

$$\mathbf{A}_e = \begin{bmatrix} \cdots & 0 & 0 & 0 & 2 & -1 \\ \cdots & 0 & 0 & 0 & 0 & 1 \end{bmatrix}.$$

The corresponding **B** matrix is $\mathbf{B}_s = \left[\begin{array}{ccccccc} \frac{3}{4} & -\frac{3}{2} & \frac{9}{8} & -\frac{1}{2} & \frac{1}{8} & 0 & \cdots \end{array} \right]$,

$$\mathbf{B_r} = \left[\begin{array}{cccccccccccc} 0 & 0 & \frac{1}{4} & -1 & \frac{3}{2} & -1 & \frac{1}{4} & 0 & 0 & 0 & 0 & 0 & \cdots \\ 0 & 0 & 0 & 0 & \frac{1}{4} & -1 & \frac{3}{2} & -1 & \frac{1}{4} & 0 & 0 & 0 & \cdots \\ & & & & & \vdots \end{array} \right],$$

$$\mathbf{B}_e = \left[\begin{array}{ccccccccc} \cdots & 0 & 0 & 0 & 0 & \frac{1}{8} & -\frac{1}{2} & \frac{9}{8} & -\frac{3}{2} & \frac{3}{4} \end{array} \right].$$

And finally, the **Q** matrix is $\mathbf{Q}_s = \left[\begin{array}{cccccc} 0 & 0 & 0 & 0 & 0 & \cdots \\ 0 & 0 & 0 & 0 & 0 & \cdots \end{array} \right]$,

$$\mathbf{Q_r} = \left[\begin{array}{cccccc} 1 & 0 & 0 & 0 & 0 & \cdots \\ \frac{1}{4} & \frac{1}{4} & 0 & 0 & 0 & \cdots \\ 0 & 1 & 0 & 0 & 0 & \cdots \\ 0 & \frac{1}{4} & \frac{1}{4} & 0 & 0 & \cdots \\ & & \vdots \end{array} \right] \quad \mathbf{Q}_e = \left[\begin{array}{cccccc} 0 & 0 & 0 & 0 & 0 & \cdots \\ 0 & 0 & 0 & 0 & 0 & \cdots \end{array} \right].$$

Appendix D

$$\mathbf{P}_s = \left[\begin{array}{cccccc} 1 & 0 & 0 & 0 & 0 & \cdots \\ \frac{1}{2} & \frac{1}{2} & 0 & 0 & 0 & \cdots \end{array} \right], \quad \mathbf{P}_r = \left[\begin{array}{cccccc} 0 & \frac{3}{4} & \frac{1}{4} & 0 & 0 & \cdots \\ 0 & \frac{1}{4} & \frac{3}{4} & 0 & 0 & \cdots \\ 0 & 0 & \frac{3}{4} & \frac{1}{4} & 0 & \cdots \\ 0 & 0 & \frac{1}{4} & \frac{3}{4} & 0 & \cdots \\ & & \vdots \end{array} \right],$$

$$\mathbf{P}_e = \left[\begin{array}{cccccc} \cdots & 0 & 0 & 0 & \frac{1}{2} & \frac{1}{2} \\ \cdots & 0 & 0 & 0 & 0 & 1 \end{array} \right].$$

Similarly, the **A** matrix becomes

$$\mathbf{A}_s = \left[\begin{array}{ccccccc} 1 & 0 & 0 & 0 & 0 & 0 & \cdots \\ -\frac{1}{2} & 1 & \frac{3}{4} & -\frac{1}{4} & 0 & 0 & \cdots \end{array} \right],$$

$$\mathbf{A}_r = \left[\begin{array}{ccccccccc} 0 & 0 & -\frac{1}{4} & \frac{3}{4} & \frac{3}{4} & -\frac{1}{4} & 0 & 0 & \cdots \\ 0 & 0 & 0 & 0 & -\frac{1}{4} & \frac{3}{4} & \frac{3}{4} & -\frac{1}{4} & \cdots \\ & & & & \vdots \end{array} \right],$$

$$\mathbf{A}_e = \begin{bmatrix} \cdots & 0 & 0 & -\frac{1}{4} & \frac{3}{4} & 1 & -\frac{1}{2} \\ \cdots & 0 & 0 & 0 & 0 & 0 & 1 \end{bmatrix}.$$

And **B** is

$$\mathbf{B}_s = \begin{bmatrix} -\frac{1}{2} & 1 & -\frac{3}{4} & \frac{1}{4} & 0 & 0 & 0 \cdots \\ 0 & 0 & -\frac{1}{4} & \frac{3}{4} & -\frac{3}{4} & \frac{1}{4} & 0 \cdots \end{bmatrix},$$

$$\mathbf{B}_r = \begin{bmatrix} 0 & 0 & 0 & 0 & \frac{1}{4} & -\frac{3}{4} & \frac{3}{4} & -\frac{1}{4} & 0 & 0 & \cdots \\ 0 & 0 & 0 & 0 & 0 & 0 & \frac{1}{4} & -\frac{3}{4} & \frac{3}{4} & -\frac{1}{4} & \cdots \\ & & & & & \vdots \end{bmatrix},$$

$$\mathbf{B}_e = \begin{bmatrix} \cdots 0 & 0 & \frac{1}{4} & -\frac{3}{4} & 1 & -\frac{1}{2} \end{bmatrix}.$$ Finally, for **Q** we have

$$\mathbf{Q}_s = \begin{bmatrix} 0 & 0 & 0 & 0 & \cdots \\ \frac{1}{2} & 0 & 0 & 0 & \cdots \\ -\frac{3}{4} & \frac{1}{4} & 0 & 0 & \cdots \\ -\frac{1}{4} & \frac{3}{4} & 0 & 0 & \cdots \\ 0 & -\frac{3}{4} & -\frac{1}{4} & 0 & \cdots \\ 0 & -\frac{1}{4} & -\frac{3}{4} & 0 & \cdots \end{bmatrix} \quad \mathbf{Q}_r = \begin{bmatrix} 0 & 0 & \frac{3}{4} & -\frac{1}{4} & 0 & 0 & \cdots \\ 0 & 0 & \frac{1}{4} & -\frac{3}{4} & 0 & 0 & \cdots \\ 0 & 0 & 0 & \frac{3}{4} & -\frac{1}{4} & 0 & \cdots \\ 0 & 0 & 0 & \frac{1}{4} & -\frac{3}{4} & 0 & \cdots \\ & & & \vdots \end{bmatrix}$$

$$\mathbf{Q}_e = \begin{bmatrix} \cdots 0 & 0 & 0 & \frac{1}{2} \\ \cdots 0 & 0 & 0 & 0 \end{bmatrix}$$

Appendix E

$$\mathbf{A}_s = \begin{bmatrix} 1 & 0 & 0 & 0 & 0 & 0 & 0 & 0 & 0 & \cdots \\ -\frac{41}{141} & \frac{82}{141} & \frac{45}{47} & -\frac{5}{141} & -\frac{15}{47} & \frac{5}{47} & 0 & 0 & 0 & \cdots \\ \frac{41}{425} & -\frac{82}{425} & -\frac{41}{425} & \frac{287}{425} & \frac{297}{425} & -\frac{11}{425} & -\frac{99}{425} & \frac{33}{425} & 0 & \cdots \end{bmatrix},$$

$$\mathbf{A}_r = \begin{bmatrix} 0 & 0 & \frac{3}{40} & -\frac{9}{40} & -\frac{1}{40} & \frac{27}{40} & \frac{27}{40} & -\frac{1}{40} & -\frac{9}{40} & \frac{3}{40} & 0 & 0 & 0 & \cdots \\ 0 & 0 & 0 & 0 & \frac{3}{40} & -\frac{9}{40} & -\frac{1}{40} & \frac{27}{40} & \frac{27}{40} & -\frac{1}{40} & -\frac{9}{40} & \frac{3}{40} & 0 & \cdots \\ & & & & & \vdots \end{bmatrix},$$

$$\mathbf{A}_e = \begin{bmatrix} \cdots & 0 & \frac{33}{425} & -\frac{99}{425} & -\frac{11}{425} & \frac{297}{425} & \frac{287}{425} & -\frac{41}{425} & -\frac{82}{425} & \frac{41}{425} \\ \cdots & 0 & 0 & 0 & \frac{5}{47} & -\frac{15}{47} & -\frac{5}{141} & \frac{45}{47} & \frac{82}{141} & -\frac{41}{141} \\ \cdots & 0 & 0 & 0 & 0 & 0 & 0 & 0 & 0 & 1 \end{bmatrix}.$$

The second decomposition matrix, **B**, is

$$\mathbf{B}_s = \begin{bmatrix} -\frac{27}{80} & \frac{27}{40} & -\frac{81}{160} & \frac{27}{160} & 0 & 0 & \cdots \end{bmatrix},$$

$$\mathbf{B}_r = \begin{bmatrix} 0 & 0 & -\frac{27}{160} & \frac{81}{160} & -\frac{81}{160} & \frac{27}{160} & 0 & 0 & 0 & 0 & \cdots \\ 0 & 0 & 0 & 0 & -\frac{27}{160} & \frac{81}{160} & -\frac{81}{160} & \frac{27}{160} & 0 & 0 & \cdots \\ & & & & \vdots & & & & & & \end{bmatrix},$$

$$\mathbf{B}_e = \begin{bmatrix} \cdots & 0 & 0 & \frac{27}{160} & -\frac{81}{160} & \frac{27}{40} & -\frac{27}{80} \end{bmatrix}. \text{ And } \mathbf{Q} \text{ is}$$

$$\mathbf{Q}_s = \begin{bmatrix} 0 & 0 & 0 & 0 & 0 & \cdots \\ \frac{4000}{3807} & -\frac{400}{1269} & 0 & 0 & 0 & \cdots \\ -\frac{2296}{3995} & -\frac{1928}{21573} & -\frac{88}{765} & 0 & 0 & \cdots \\ -\frac{328}{323595} & \frac{21416}{21573} & -\frac{88}{255} & 0 & 0 & \cdots \\ \frac{164}{765} & -\frac{49}{51} & \frac{58}{2295} & -\frac{1}{9} & 0 & \cdots \\ \frac{164}{2295} & -\frac{11}{459} & \frac{254}{255} & -\frac{1}{3} & 0 & \cdots \end{bmatrix},$$

$$\mathbf{Q}_r = \begin{bmatrix} 0 & \frac{1}{3} & -1 & \frac{1}{27} & -\frac{1}{9} & 0 & 0 & \cdots \\ 0 & \frac{1}{9} & -\frac{1}{27} & 1 & -\frac{1}{3} & 0 & 0 & \cdots \\ 0 & 0 & \frac{1}{3} & -1 & \frac{1}{27} & -\frac{1}{9} & 0 & \cdots \\ 0 & 0 & \frac{1}{9} & -\frac{1}{27} & 1 & -\frac{1}{3} & 0 & \cdots \\ & & & & \vdots & & & \end{bmatrix},$$

$$\mathbf{Q}_e = \begin{bmatrix} \cdots & 0 & \frac{1}{3} & \frac{254}{255} & -\frac{11}{459} & \frac{164}{2295} \\ \cdots & 0 & \frac{1}{9} & \frac{58}{2295} & -\frac{49}{51} & \frac{164}{765} \\ \cdots & 0 & 0 & -\frac{88}{255} & \frac{21416}{21573} & -\frac{328}{323595} \\ \cdots & 0 & 0 & -\frac{88}{765} & -\frac{1928}{21573} & -\frac{2296}{3995} \\ \cdots & 0 & 0 & 0 & -\frac{400}{1269} & \frac{4000}{3807} \\ \cdots & 0 & 0 & 0 & 0 & 0 \end{bmatrix}.$$

Chapter 4

Computational Geometry and Image Processing in Biometrics: On the Path to Convergence

Marina L. Gavrilova

Biometric Technology Laboratory: Modeling and Simulation,
SPARCS Laboratory, Department of Computer Science,
University of Calgary, Calgary, AB, Canada
marina@cpsc.ucalgary.ca

The rapid development of biometric technologies is one of the modern world's phenomena, which can be justified by the strong need for the increased security by the society and the spur in the new technological developments driven by the industries. This chapter examines a unique aspect of the problem — the development of new approaches and methodologies for biometric identification, verification and synthesis utilizing the notion of proximity and topological properties of biometric identifiers. The use of recently developed advanced techniques in computational geometry and image processing is examined with the purpose of finding the common denominator between the different biometric problems, and identifying the most promising methodologies. The material of the chapter is enhanced with recently obtained experimental results for fingerprint identification, facial expression modeling, iris synthesis, and hand boundary tracing.

Contents

Glossary

FAR — False Acceptance Rate

FRR — False Refused Rate

RBF — Radial Basis Function

NPR — Non-Photorealistic Rendering

VD   — Voronoi Diagram

DT   — Delaunay Triangulation

4.1. Introduction

The term "biometric authentication" refers to the automatic identification, or identity verification using physiological and behavioural characteristics. According to J. L. Wayman [1], there are two distinct functions for biometric devices: to prove that you are who you say you are or to prove that you are not who you say you are not.

A negative claim to identity can only be accomplished through biometrics.   For positive identification, however, there are multiple alternative technologies, such as passwords, PINs (Personal Identification Numbers), cryptographic keys, and various tokes, including identification cards. However, the use of passwords, PINs, keys and tokens carries the security problem of verifying that the presenter is the authorized user, and not an unauthorized holder.   Consequently, passwords and tokens can be used in conjunction with biometric identification to mitigate their vulnerability to unauthorized use.   Most importantly, properly designed biometric systems can be faster and more convenient for the user, and cheaper for the administrator, than the alternatives.

Security systems traditionally use biometrics for two basic purposes: *to verify* and *to identify* users [2]. Identification tends to be the more difficult task as the system must perform the search on a database of enrolled users to find a match (a one-to-many search). The biometric that the security system employs depends in part on what the system is protecting and what it is trying to protect against.

For decades, many highly secure environments have used biometric technology for monitoring the entry access. Today, the primary application of biometrics is shifting from the *physical security*, where the access to the specific locations is usually monitored using standard security identification mechanisms such as signature, access cards, passwords, fingerprint scanning, palm prints and others, to *remote security*. By this we mean the advanced methods of crowd monitoring at banks, car registries or in the mall using video cameras, infrared and ultrared scanners, temperature measuring devices, gait and behavioural pattern recording made at the distance from the studied subject. The popularity of such approaches has increased dramatically as the new technological devices are coming on the market every week, capability to process massive amount of data is doubling every few month, and the algorithm development by the leading IT companies and the University research centers is tripled in the last year.

> *Biometric modeling or synthesis of the biometrics is the new area of the research at the frontier of biometric science. It is driven by two major stimulating factors: the need to verify the effectiveness and correctness of the newly developed methodologies on the extensive biometric data sets and the desire to study dynamically evolving situations from the different perspectives.*

In order to satisfy the first demand, the data sets containing fingerprint, facial and other biometric data are generated using specifically developed methods. For the second need, dynamical modeling of processes, situations, and subject behavior is utilized. Such computations are usually complex and require either distributed environment, parallel algorithms or complex 3D visualization system.

Finally, as the demand for fast remote access and distributed processing of biometric data had been increasingly growing, and the need to develop more precise and reliable ways of person identification is ever pressing, the combination of biometrics and data processing methods are gaining high popularity as it undoubtedly increases the quality of the results and the level of security protection.

Driven by the above motivations, the rest of this Chapter is devoted to displaying the variety of computational geometry techniques in biometric synthesis and identification, as well as the use of multiple algorithms to enhance the solution to a specific problem. Our goal is not to provide a comprehensive analysis of all the methods used in modern biometrics (even this book with all its chapters would not be enough for this purpose), but rather to concentrate on unique aspects of geometric computing and exploring the topological relationship among objects in the biometric world with the purpose of showcasing the strength and robustness of such approaches. All algorithms presented in the Chapter have been implemented by students of Biometric Technologies Laboratory under supervision of *Prof. Marina Gavrilova.*

4.2. Biometric Methods

There seems to be virtually no limit to personal characteristics and methods that have been suggested and used for biometric identification: fingers, hands, feet, faces, eyes, ears, teeth, voices, signatures, typing patterns, gaits etc. Among the recently emerging biometric characteristics are temperature maps, obtained by analyzing the amount of heat emitted by different body parts, infrared scanning and bio-magnetic fields.

Typically, biometrics can be separated into two distinctive classes: *physiological* characteristics and *behavioural* characteristics. Common physiological biometric identifiers include fingerprints, hand geometry, eye patterns (iris and retina), facial features and other physical characteristics. Typical behavioural identifiers are the voice, signature and typing patterns. Analyzers, which are based on behavioural identifiers, are often less conclusive because they are subject to limitations and can exhibit complex patterns.

However, there has been no agreement on which biometric measure is the best, since they vary greatly in their availability, ease of accessing, storing and transmitting, and even in their ability to identify a person uniquely (fingerprint seems to be a unique measure, while hand geometry is not). Some are also much easier to digitize and classify than others — thus making their use dependent on a variety of factors, such as availability of technological base and specifics of the goals of the applications.

4.3. Chapter Overview

The interest of general public as well as research community to the area of biometric can be illustrated by many examples, perhaps one of the most appealing is the number of manuscripts being written in the recent years and devoted exclusively to biometric. After a very successful book on fingerprint identification [10], a book on face recognition as well as the book devoted to various biometrics recently saw the light of publishing [33,22,23]. While these books concentrate on the fundamental techniques developed for addressing biometric identification, classification and recognition problems (and doing an exceptional job at it), the new area of biometrics devoted to continuously emerging new techniques as well as new applications (such as biometric synthesis), remain outside the scope of these fine publications.

Within two years since the manuscript "Computational Geometry and Biometrics: on the Path to Convergence" has found the audience as the part of *International Workshop on Biometrics Technologies* 2004 [19], the number of attempts to extract geometric information and apply topology to solve biometric problems has increased significantly. There has been research on application of topological methods, including Voronoi diagram, Delaunay triangulation and distance transform, for hand geometry detection, iris synthesis, signature recognition, face modeling and fingerprint matching [14,18,24]. As the methodology is new, many questions on what is the best way to utilize the complex geometric data structures, which topological information to use and in which context, how to make implementation decisions and whether the performance is superior to other techniques remain largely unanswered. This chapter attempts to fill this gap and introduces a number of methodologies and techniques, illustrated by concrete examples from different biometric areas.

4.4. Biometric System Architecture

We start with the basics of introducing the generic biometric system components and then identifying the appropriate placement of the topological methods in the above content.

Although biometric devices rely on a scope of different technologies, much can be said about them in general. In Fig. 4.1, we depict a generic biometric authentication system, divided into five sub-systems: data collection, transmission, processing, decision and data storage. Data Collection subsystem constitute the set of electronic devices and collection methods, intended to measure accurately some biometric characteristics. This information is then either sent directly to Processing module or compressed and stored in a Biometric Database (located locally or at a remote location). Processing module includes preprocessing of the biometric data (a scanned fingerprint image, for instance), with the purpose of enhancing its quality, accommodating for variations in the conditions under which the data was obtained. Then the extracting important features (such as minutiae of a fingerprint) are extracted. The decision making module uses obtained information in order to search in the database and compare new biometric data with the existing one with the purpose of identification or verification. It then reports the results after which the new query can be originated.

It is important to note that the process in non-linear. In the case of insufficient quality of an image, multiple conflicting feature points or erroneous matching, data can be re-sampled, preprocessed again or another matching technique could be used.

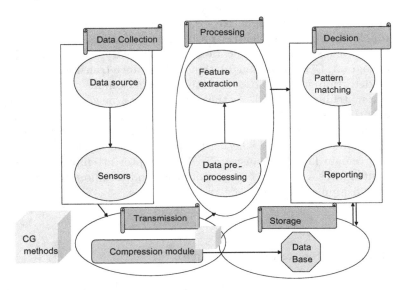

Fig. 4.1. Computation Geometry Methods (identified by yellow cubes) in Generic Biometric System.

The above process is not necessary error-proof. Various factors might influence false identification, or in contrast not finding any matches. Therefore, a long-term goal in biometric testing is to establish acceptable error rates for biometric devices in as many different application categories as possible, and to continue to improve methodology in order to decrease a possibility of an error and make implementation more robust.

4.5. Computational Geometry Methods in Biometrics

As can be seen from Fig. 4.1, the most important steps in the processing and identification of biometric data are pattern matching, extraction of important features and pre-processing. The techniques for optimization of these steps are crucial for successful design and functioning of the biometric system. These steps can be made more efficient and implemented using advanced research in the area of computational geometry (see Fig. 4.1), based specifically on concepts of proximity of geometric sets and extracting and utilizing topological information on the data studies. The methods include:

- Utilization of the Voronoi diagram and Delaunay triangulation computational geometry data structures for data processing and matching
- Development of distance distribution methods using weighted metric functions and distance transform for studying image properties and pattern matching
- Constructing medial-axis transform for boundary and skeleton extraction and topological properties identification
- Introducing topology-based approach for generation and synthesis of new biometric data

These methods and numerous applications are described below according to their classification.

4.5.1. *Voronoi Diagram Techniques in Biometrics*

On the perpetual human quest for perfection, techniques for increasing the efficiency and reliability of biometric identification and matching are continuosly being developed; computational geometry and image processing inspiring many of those techniques.

As one of the most fundamental data structures of computational geometry, *Voronoi diagram* and it's dual *Delaunay triangulation* [32] is being perhaps one of the most popular data structures being utilized in various areas of applied sciences and recently making its way into the area of biometrics. Applications of Voronoi diagrams for proximity studies and material structure analysis has gained considerable momentum in the past few decades; traditional application areas being Molecular Biology, Physics, Chemistry, Astronomy, Mechanical Engineering, Material Sciences, and more recently Visualization, Geographical Information Systems and Bioinformatics.

In the area of biometrics, Biometric Technologies Laboratory at the University of Calgary has pioneered some of the important applications of Voronoi diagrams in fingerprint matching, iris synthesis, hand geometry identification and face modeling [36,37,35,34]. For instance, in 3D facial expression modeling, mesh is used as a medium to convey facial expression transformation through a set of control areas. Delaunay triangulation technique is often utilized to represent and control mesh deformations, which is useful for facial structure analysis, face synthesis and identification [27,36,28]. Among other studies on the subject, Voronoi diagrams were utilized for face partitioning onto segments and facial feature extraction in [24]. In fingerprint identification area, Bebis et. al. [14] used the Delaunay triangle as the comparing index in fingerprint matching in 1999. Recently, a new method for binary fingerprint image denoising based on Distance Transform realized through Voronoi method was introduced in [18]. In related areas of geographical information systems and navigation planning, Voronoi diagrams have been applied for 3D object reconstruction from point clouds and in clustering analysis. Examples can be found in recent works by [12,13,26]. The above is just a small fraction of the research undertaken in this area. There are other application areas of biometrics where Voronoi diagram methodologies are being utilized and developed. The concentration of the following sections is on the example of Voronoi technique application in fingerprint matching.

4.5.1.1. *Voronoi Diagram Preliminaries*

Let us start with the brief introduction of Voronoi Diagram and Delaunay Triangulation to the reader. Typically,

> The Euclidean Voronoi diagram is used to store the topological information about the system of objects (points, spheres, polygons) in the plane or in the higher dimension, as well as to assist in performing various nearest-neighbors and point-location queries.

We start with the most general definition of the Voronoi Diagram.

Definition 1. The Voronoi diagram for a set of objects S in d-dimensional space is a partitioning of the space into regions, such that each region is the locus of points from S closer to the object $P \in S$ than to any other object $Q \in S$, $Q \neq P$.

The above general definition can be specialized to the set of spheres in the Euclidean metric [32]:

Definition 2. A generalized Euclidean Voronoi diagram for a set of spheres S in R^d is the set of Voronoi regions $\{\mathbf{x} \in R^d | d(\mathbf{x}, P) \leq d(\mathbf{x}, Q), \forall Q \in S - \{P\}\}$, where $d(\mathbf{x}, P)$ is the Euclideandistance function between a point $\mathbf{x}$ and a sphere $P \in S$.

The above definition is often used in material sciences for representation of circular objects (balls, particles, molecules) as well as in dynamic behavior modeling (of growing plants, in game design, etc). In the simplest case, the set of spheres can be replaced by the set of points and the distance between the sphere and the point will reduce to the simple distance between two points in Euclidean space. Following the classification of Voronoi diagrams based on the way of how to compute such distance, presented in [32], the Euclidean weighted Voronoi diagram is an instance of the class of additively weighted Voronoi diagrams. Thus, the distance between a point $\mathbf{x}$ and a sphere P with center at $\mathbf{p}$ and radius r_p is defined as $d(\mathbf{x}, P) = d(\mathbf{x}, \mathbf{p}) - r_p)$ (see Fig. 4.2). The distance $d(\mathbf{x}, \mathbf{p})$ between points $\mathbf{x}(x_1, x_2, ..., x_d)$ and $\mathbf{p}(p_1, p_2, ..., p_d)$ in the Euclidean metric is computed as $d(\mathbf{x}, \mathbf{p}) = \sqrt{\sum_{i=1}^{d} (x_i - p_i)^2}$. According to the definition:

> The generalized Voronoi region of an additively weighted Voronoi diagram of n spheres is obtained as the intersection of $n - 1$ quasi-halfspaces with hyperbolic boundaries.

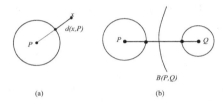

(a)   (b)

Fig. 4.2.   The Euclidean distance and the Euclidean bisector in the plane.

The boundary of the generalized Voronoi region consists of up to $n - 1$ curvilinear facets, where each facet is a $d - 1$ curvilinear dimensional face. The boundary of the $(d-1)$-dimensional face consists of $(d-2)$-dimensional faces and so on. An 1-dimensional face is called a generalized Voronoi edge (a straight-line segment or a hyperbolic arc), and a 0 dimensional face is called a *generalized Voronoi vertex*. It was shown in [32] that the weighted Euclidean Voronoi diagram for a set of spheres is the set of singly-connected Voronoi regions with the hyperbolic boundaries, and each Voronoi region is star-shaped relative to sphere P.

The example of the Euclidean Voronoi diagram in R^3 is given in Fig. 4.3.

> The straight-line dual to the Voronoi diagram, called a Delaunay tessellation, is often used instead of the Voronoi diagram to store topological information for a set of spheres.

Definition 3.   A generalized Delaunay tessellation corresponding to a generalized Voronoi diagram for a set of spheres S in d-dimensional space

Fig. 4.3.   The Euclidean Voronoi diagram in R^3.

is a collection of d-dimensional simplices such that for each generalized Voronoi vertex $\mathbf{v} = EVor\,(P_1) \cap EVor\,(P_2) \cap ... \cap EVor\,(P_{d+1})$ there exists a simplex $(p_1, p_2, , p_{d+1})$ in the generalized Delaunay tessellation.

The example of the Delaunay triangulation in the plane is given in the figure below (see Fig. 4.4).

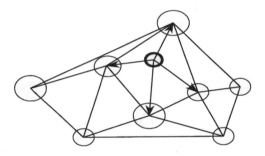

Fig. 4.4.   Neighbors identified by the edges of the Delaunay triangulation.

The concept is illustrated further by Fig. 4.5: left image displays the Voronoi diagram in supremum metric, right image — its dual, the Delaunay triangulation. A number of useful properties of the above data structure, such as "empty circle" and "nearest-neighbor property, as well as regularity of corresponding Delaunay triangulation, make this data structure highly popular in a variety of applications [7,6,8,32,9,3]. The methodology is making its way to the core methods of biometrics, such as fingerprint identification, iris and retina matching, face analysis, ear geometry and others [24,17,18,2,14,1,11]. The methods are using Voronoi diagram to partition the area of a studied image and compute some important features (such as areas of Voronoi region, boundary representation, etc.) and compare with similarly obtained characteristics of other biometric data.

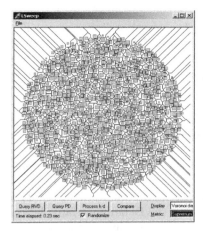 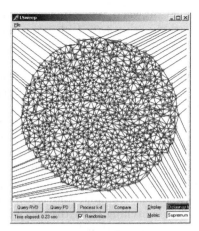

Fig. 4.5.   Voronoi diagram and corresponding Delaunay triangulation of 1,000 weighted sites in supremum (L_{inf}) metric (left to right).

The Delaunay triangulation in the plane possesses important properties that make its application in biometric and computer graphics, among other areas, very appealing. These properties are described on the example of fingerprint matching in the section below.

4.5.1.2. *Voronoi Diagrams in Fingerprint Matching*

Let us now discuss how Delaunay triangulation can be utilized in fingerprint matching problem. The large number of approaches to fingerprint matching can be coarsely classified into three classes: correlation-based matching, minutiae-based matching and ridge-based matching.

In correlation-based matching, two fingerprint images are superimposed and the correlation between corresponding pixels is computed for different alignments. During minutiae-based matching, the sets of minutiae are extracted from the two fingerprints and stored as sets of points in the two dimensional plane [15,1]. Ridge-based matching is based on such features as orientation map, ridge lines and ridge geometry [15]. Among the above techniques, minutiae-matching is most widely used approach that yields best matching results. Traditional algorithms (such as [14]) work under the restricting assumption that at least one corresponding triangle pair can be found between the input and template fingerprint images. Unfortunately, this assumption might not hold due to low quality of fingerprint images, unsatisfactory performance of the feature extraction algorithm or distorted images. Another problem in fingerprint matching is noticeable when the distance between two points in template image increases. It becomes more challenging to match point pairs in the corresponding images. For instance, Kovacs-Vajna [17] have shown that local deformation of less than 10% can cause global deformation reaching 45% in edge length. In the following, we discuss how to deal with shortcoming of the above methods using Delaunay triangulation technique and applying radial functions for deformation modeling.

Some of the outlined above problems can be alleviated by the use of a special data structure for data representation and processing, such as Delaunay triangulation. We propose to use this data structure for minutiae matching and singular-point comparison in the fingerprints. The method was implemented as part of the global fingerprint recognition system [20] and is sustainable under presence of elastic deformations [21]. The minutiae-matching algorithm, in which context the Delaunay triangulation approach is utilized, is based on classic fingerprint matching technique [22]. Using Delaunay triangulation brings unique challenges and advantages:

(1) Delaunay edges rather than minutiae or whole minutiae triangles are selected as matching index which provides an easier way to compare two fingerprints.
(2) The method is combined with deformation model, which helps to preserve consistency of the results under elastic finger surface deformations.
(3) To improve matching performance, we introduce feature based on spatial relationship and geometric attribute of ridges, and combine it with information from both singular point sets and minutiae sets to increase matching precision.

4.5.1.3. *Delaunay Triangulation for Fingerprint Matching and Deformation Modeling*

The purpose of fingerprint identification is to determine whether two fingerprints are from the same finger or not. In order to complete this task, the input fingerprint needs to be aligned with the template fingerprint represented by its minutia pattern. We divide our algorithm into three stages. During the first stage, identification of feature patterns utilizing Delaunay triangulation takes place. During the second step, Radial Basis Function (RBF) is applied to model the finger deformation and align images. Finally, the global matching algorithm is employed to compute the combined matching score, using additional topological information extracted from ridge geometry (see Fig. 4.6).

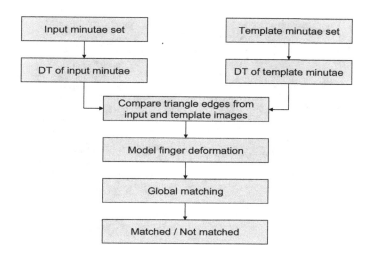

Fig. 4.6.   Flow chart of generic fingerprint identification system.

Using triangle edge as comparing index has many advantages. For local matching, we first compute the Delaunay triangulation of minutiae sets Q and P. Second, we use triangle edge as our comparing index. To compare two edges, parameters such as $Length$, $Type1$, $Type2$, $Ridgecount$ are used. Note that these parameters are invariant of the translation and rotation. Conditions for determining whether two edges match identified as the set of linear inequalities depending on the above parameters and the specified thresholds (usually derived empirically depending on image size and quality). A sample set of such conditions can be found in [21,37]. If the threshold is selected successfully, then the transformation to align the input and template images is obtained. Matching using Delaunay triangulation edge is realized as follows. If one edge from an input image matches two edges from the template image, we need to consider the triangulation to which this triangle edge belongs to and compare the triangle pair. For a certain range of translation and rotation dispersion, we detect the peak in the transformation space, and record transformations that are neighbors of the peak in the transformation space. Note that those recorded transformations are close to each other, but not identical. Also note that elastic deformations should be dealt with separately as such deformations can not be ignored. Figure 4.7 shows the successfully matched Delaunay triangle edge pairs. Detailed description of the matching algorithm can be found in [21,37].

The deformation problems arise due to the inherent flexibility of the finger. In [21], we proposed a framework aimed at quantifying and modeling

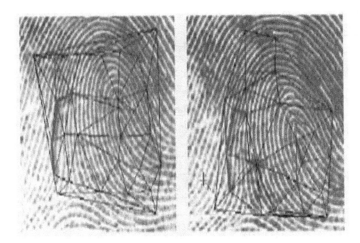

Fig. 4.7. Matched Delaunay triangle edges are shown in red. On contrary, no single triangle is matched.

the local, regional and global deformation of the fingerprint. The method is based on the use of RBFs, that represent a practical solution to the problem of modeling of a deformable behavior. The application of RBF has been explored in medical image matching as well as in image morphing. For our fingerprint matching algorithm, deformation problem can be described as knowing the consistent transformations of specific control points from the minutiae set, and knowing how to interpolate the transformation of other minutiae which are not control points. We do not consider all the transformations obtained by the local matching. We rank them in the transformation space and pick those consistent transformations which form large clusters.

We then consider the transformation of input image as a rigid transformation. Experimentally, we established that the maximum number of matching minutiae pairs between input image and template image is 6 (labeled by the big dashed circle in Fig. 4.8, left). Circles denote minutiae of the input image after transformation, while squares denote minutiae of the template image. Knowing transformation of the five minutiae (control points) of the input image, we apply the RBF to model the non-rigid deformation, visualized by the deformed grid in Fig. 4.8 (right). In the example below, the number of matching minutiae pairs is 10, which greatly increased the matching scores of these two corresponding images.

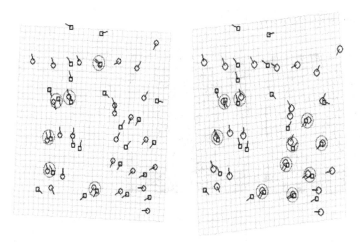

Fig. 4.8.   Comparison of rigid transformation and non-rigid transformation.

4.5.1.4. *System Implementation and Experiments*

The described above fingerprint identification system (including both local and global matching) was implemented in C++ and tested on a Pentium 4 2.8 GHZ CPU, 512 RAM PC. Experiments have been performed on two fingerprint databases: from Biometric System Laboratory at University of Bologna, that consists of 21 × 8 fingerprint images of 256 × 256 size; and on the publicly available database used in Fingerprint Verification Competition FVC2000. The captured fingerprint images vary in quality. The interface of the implemented system is presented below (Fig. 4.9).

Extensive experimentations were performed with the system, that confirmed its high efficiency, low FRR (False Refused Rate), low FAR (False Acceptance Rate) and good overall comparison with many other algorithms. We further conducted experiments to show resistance of the proposed matching algorithm to fingerprint deformations, which impact negatively most of the other fingerprint matching algorithms. For detailed experimental results, consult [21,37].

Note, that an additional improvement in matching rate was achieved by combining the Delaunay Triangulation approach with the distance transform method for ridge geometry computation in addition to minutiae matching, that will be described in the next section.

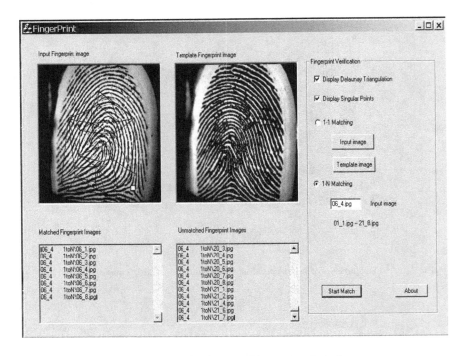

Fig. 4.9.   Fingerprint matching system interface.

4.5.2. Distance Distribution Computation in Biometrics

In this section, we first look at the new ways of establishing thresholds for biometric matching using weighted distance functions. Next, we describe how the concept of distance transform (in Euclidean or other Minkowski metric) can be used for ridge geometry computation and facial expression morphing. We conclude this chapter with links to the next topic, boundary and skeleton extraction for biometric image comparison.

4.5.2.1. Weighted Distance Metrics for Establishing the Biometric Threshold

The most basic technical measures which can be used to determine the distinctiveness and repeatability of the biometric patterns are the distance measures output by the processing module. Through testing, one can establish three application-dependent distributions based on these measures:

The first distribution, called the *genuine distribution,* is created from

distance measures resulting from comparison of samples to similar templates. It shows the repeatability of measures from the same person.
The second distribution is created from the distance measures resulting from the comparison of templates obtained from different individuals. It is called the *inter-template distribution*.
The third distribution is created from the distance between samples to non-like templates, called the *impostor distribution*. It shows the distinctiveness of measures from different individuals.

One common way to establish a decision policy is by picking a *threshold* distance, and declaring distances less than the threshold as a "match" and those greater to indicate "non-match". However, under less than perfect conditions, experiments shown that there is an overlap between the genuine and impostor distributions. This means that the method is not an error proof and in some instances no threshold could clearly distinguish the genuine from the impostor distances.

While the experiments are normally conducted using the traditional Euclideandistance, one possible way to approach the problem might be by using other distance functions and weighted metrics. The decision on which metric to use for studying the proximity of biometric sets can be based on the type of the biometric data as well as the task. For instance, the regularity of templates (in a case of fingerprint, or retina), might be a feature that will allow us to use distance measure more efficiently by applying a weighted Minkowski metric, such as L_1 (*Manhattan*) or L_{inf} (*supremum*) metrics.

To illustrate the idea, a number of possible metrics originated by the group of *additively weighted* metrics, is suggested. The distance is described by the general form

$$d_e(x, P) = d(x, p) - w_p \qquad (4.1)$$

where x is a sample point, P is a specific object or feature point and is a weight function w_p which represents any individual or cumulative parameter of the studies phenomena. Under traditional Euclidean metric, the above expression transforms into:

$$d(x, P) = \sqrt{\sum_{i=1}^{d}(x_i - p_i)^2} - w_p \qquad (4.2)$$

Using Manhattan and supremum metric (respectively), the formula is expressed as (Fig. 4.10):

$$d(x, P) = \sum_{i=1}^{d}|x_i - pi| - w_p \qquad (4.3)$$

$$d(x, P) = \max_{i=1\dots d} |x_i - pi| - w_p \qquad (4.4)$$

Our proposition is that the regularity of the metric will allow to measure the distances from some distinct features of the template more precisely, and ignore minor discrepancies originated from noise and imprecise measurement while obtaining the data. We presume that the behavioural identifiers, such as typing pattern, voice and handwriting styles will be less susceptible to improvement using the proposed weighted distance methodology than the physiological identifiers.

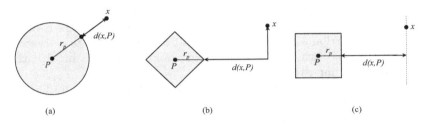

Fig. 4.10.   Distance computation using additively weighted metrics: (a) Euclidean (b) Manhattan (c) supremum (left to right).

4.5.2.2.   *Distance Transform Preliminaries*

Let us now take a look at one of the examples of using distance between pixels of the input image for the purpose of image matching and related problem of image morphing.

Given an $n \times n$ binary image $\mathcal{I}$ of white and black pixels, the distance transform of $\mathcal{I}$ is a map that assigns to each pixel the distance to the nearest black pixel, referred to as *feature*. The feature transform of $\mathcal{I}$ is a map that assigns to each pixel the feature that is nearest to it. The distance transform was first introduced by Rosenfeld and Pfaltz [31], and it has a wide range of applications in image processing, robotics, pattern recognition and pattern matching. The distance metrics used to compute the distance transform include the L_1, L_2 and L_∞ metrics, with the L_2 (Euclidean) metric being the most natural, and rotational invariant.

One approach is to grow clusters or neighborhoods around each feature p consisting of those pixels whose nearest feature is p. This approach has been taken in [30] and [29] to obtain sequential and parallel algorithms, respectively. Daniellson [30] describes a sequential nearest-neighbor transform algorithm that is nearly error-free. An alternative approach, pioneered by Rosenford and Pfaltz [31], is based on the idea of dimension reduction. The transform is first computed using the distance function in one lower dimension and then the two-dimensional distance transform is computed.

The approach we utilize in our research is based on simple and optimal algorithm for computing the Euclidean distance transform and the nearest feature transform based on image sweeping technique developed in [4]. The algorithm processes the rows of the image twice: in a top-down scan and a bottom-up scan. The polygonal chain C, containing all the necessary information to compute the nearest feature for each pixel of the currently processed row, is maintained. A marking characteristic of this algorithm is updating the polygonal chain dynamically as the image is swept row by row. The information gathered while processing the previous row is utilized to compute the nearest features for the next row. The algorithm is one of the most efficient algorithms of its sort in terms of the implementation efficiency when compared with other linear time algorithms, and thus it was selected for facial expression modeling problem. The full description of this method can be found in [4].

4.5.2.3.   *Face Modeling using Distance Transform*

We now illustrate the application of the Distance Transform algorithm to facial expression modeling problem.

Facial expression modeling and animation, as an identifiable area of computer graphics and biometrics, has long fascinated computer graphics researchers. Historically, the earliest attempts to model and animate realistic human faces date back to the early 1970s [38]. Since then, numerous research papers have been published on this topic.

NPR animation of faces offer several advantages over attempts to work with photorealistic images, including compact image storage, inclusion of selected representable lines for image displaying and manipulation, as well as additional freedom to transform image according to artist's desire. In BT Lab, we developed a new method for automatic animation of NPR faces from sample images. Our system takes two human facial images as input and outputs a particular NPR style facial animation. First, a particular NPR style portrait is created from a frontal facial photograph. While many NPR techniques have been proposed to generate digital artwork, the detail information, such as expressive wrinkles and creases, is usually missed in the generated pictures. As an extension, we propose a segmentation and tracking method to map those expressive lines, which makes the portrait more expressive. Then, a metamorphing algorithm using distance transform is utilized to produce the resulted animation. The system flowchart is show in the graph below (see Fig. 4.11).

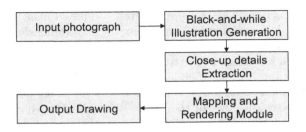

Fig. 4.11. The distance transform method.

After edge detection for feature lines and feature line tracking, mapping and rendering steps are performed (for detailed description, see [36]), the distance transform based morphing is applied to transform two input images. The process is outlined below.

Given two input photographs of different facial expressions, two NPR style portraits are to be produced by Portrait Generator module. The task is to create a smooth animation between these two NPR portraits. Given two images A and B, this smooth transition is computed as a sequence of frames S_0, S_1, ... S_k, where $S_0 = A$ and $S_k = B$. Traditionally, to achieve a good morphing sequence, a number of equal control points are selected in both images (often manually), and then a one-to-one mapping is defined. It can be difficult both to select the control points and compute the mapping. Thus, we developed a novel approach to computing morphing sequences, that relies on distance transforms rather than control points.

Let A, B be two bi-level $m \times n$ images, we treat the image matrix A as a set, that is $A = \{(i, j) \mid a_{ij} = 1\}$. B is defined in the same manner. In addition, the intermediate slides can be viewed either as bi-level images or as grey-scale images.

During the morphing sequence, we would like each pixel $a \in A$ to move to its new location in B, so that it travels the shortest distance. We can achieve this by sending each pixel to its nearest neighbor in B. In this way, the morphing problem between images A and B is reduced to the problem of animation of the motion of $N = |A| + |B|$ pixels along straight line segments. To do this, we define two functions, $f : A \mapsto B$, and $g : B \mapsto A$. Each function maps pixels to a nearest neighbor in the other image. Each $a \in A$ is really an ordered pair and represents the location of a foreground pixel in A: $a = (x_a, y_a)$ and $b = (x_b, y_b)$. We define distance between points a and b in the Euclidean sense $d(a, b) = \sqrt{(x_a - x_b)^2 + (y_a - y_b)^2}$. Next, we define f and g: $f(a) = \arg\min_{b \in B}\{d(a, b) \mid b \in B\}$, $g(b) = \arg\min_{a \in A}\{d(a, b) \mid a \in A\}$. Function f is called a *feature transform* of B and similarly g is a feature transform of A. $f'(a) = \min_{b \in B}\{d(a, b) \mid b \in B\}$ is a *distance transform*.

In the morphing sequence, each a will travel to its nearest neighbor in B, i.e. $a \mapsto f(a)$. This may cause some visual artifacts, as this mapping may not be one to one. Although all pixels in A will move, not all pixels in B may be involved in the transformation. Formally, domain$(f) = A$, but range(g) need not be equal to B. Thus, a set of straight line paths in the reverse direction are created from B to A, using g as a guide.

Formally, define a set of line segments L as follows

$$L = \{(a, f(a)) \mid a \in A\} \cup \{(g(b), b) \mid b \in B\}$$
$$= \{(u, v)\}, \text{ for notational simplicity}$$

Thus, L is a set of line segments, each line represented by its endpoints u and v (u and v are each ordered pairs).

Now we are ready to compute the frames $S_0, S_1, \ldots S_k$. We define $S_i = \{u + \frac{i}{k}(u - v) \mid (u, v) \in L\}$, where $L = \{(u, v)\} = \{(a_1, b_1), (a_2, b_2), (a_2, b_3)\}$.

Then, the maps f and g can be computed in linear time $O(mn)$, where A is an $m \times n$ image, using the algorithm in [4] using the L_2 norm.

Thus, the distance transform provides out a fast, smooth and visually continuous morphing process of facial expressions. A morphing sequence between a happiness and surprise facial expressions using this method is shown in Fig. 4.12.

Fig. 4.12.   Slices morphing from happiness to surprise in line drawing style, 15 frames as the interval.

4.5.2.4.   *Distance Transform for Ridge Extraction*

Another application of the described above methodology can be found in ridge extraction problem for fingerprints. It is known that the use of additional features in conjunction with the minutiae assists in increasing the system accuracy and robustness. In this respect, we propose to use a new feature based on the spatial relationship and geometry attributes of ridge lines. A two-dimensional binary image I of M by M pixels is a matrix of size M by M whose entries are 0 or 1. The pixel in a row i and column j is associated with the Cartesian coordinate (i, j). For a given distance function, the binary image is the thinned fingerprint image that can be transformed to a distance map to the feature points.

Let $m00$ represent the minutiae (end or bifurcation). Then it is possible to construct a vertical line perpendicular to its orientation field. Assume that the intersection with the binary image is $m10$, $m20$, $m30$ and $m40$. To add the curvature information of a associated ith ridge of minutiae $m00$, we use $m1$, $m2$, $m3$ and $m4$, the sampled points on the ridge close the $mi0$ and equally spaced from near to far, to represent the ridge. The local comprehensive fingerprint information contains the five associated ridges, each ridge include the four distance from sampled points to the intersection. Experiments show that the proposed approach increases the matching accuracy at the cost of some additional storage space for ridge. Application of such new feature (orientation field and ridge shape) in matching algorithm implementation decreases the false refused rate from 8 percent to 5 percent, according to [21].

4.5.2.5. *Pattern Matching*

Aside from a problem of measuring the distance, pattern matching between the template and the measured biometric characteristic is a very serious problem on its own. Some preliminary research, mainly in the area of image processing, should be utilized in order to approach the problem from the right angle. The advanced research in the area of pattern matching is being conducted in the recent years for the purpose of image comparison, database searches in the collection of images or historical texts, performing handwriting recognition, and integrating the technology into the cellular phones, notepads and electronic organizers. The most common methods are based on bit-map comparison techniques, scaling, rotating and modifying image to fit the template through the use of linear operators, and extracting template boundaries or skeleton (also called medial axis) for the comparison purposes.

In addition, template comparison methods also differ, being based on either pixel to pixel, important features (such as minutiae) positions, or boundary/skeleton comparison. The illustration of bit-based methods for image matching is presented in Fig. 4.13. While these methods are well studies and prove them to produce reliable results, the use of technique based on the *distance transform* method can provide necessary improvement in speed while preserving the methods simplicity and practicality.

4.5.2.6. *Distance Transform for Pattern Matching*

Definition 4. Given an $n \times m$ binary image I of white and black pixels, the *distance transform* of I is a map that assigns to each pixel the distance to the nearest black pixel (a *feature*).

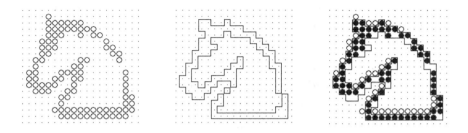

Fig. 4.13. Bit-map representation of an image; template of an image; pixel matching (from left to right) (images are courtesy of Prof. Jim Parker, University of Calgary).

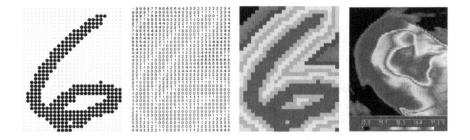

Fig. 4.14. The distance transform method. Top row from left to right: binary image of black and white pixels; distance transform of a binary image. Bottom row, left to right: "temperature map" based on the distance transform; thermogram of an ear (Brent Griffith, Infrared Thermography Laboratory, Lawrence Berkeley National Laboratory).

The distance transform method introduced in [4] is based on fast scans of image in the top-bottom and left-right directions using a fast polygonal chain maintenance algorithm for optimization. After the distance transform is build (Fig. 4.14), it can be used to visualize information on closeness in a form of so-called *temperature map*. As the distance from the black pixels (*features*) increases, the color intensity changes.

It would be interesting and highly beneficial, in our view, to extract the temperature map (and the original distance transform) from the thermograms that need to be compared against each other. Our experience with the method indicates that this process is deterministic and fast, thus will assure a high reliability and robustness when used for biometric identification and verification purposes.

4.5.3.   *Medial Axis Transform for Boundary and Skeleton Extraction and Topological Properties Identification*

Another way to approach the above biometric problems is through extraction of a boundary or a skeleton of a studied biometric samples.

Definition 5. The *medial axis*, or *skeleton* of the set D, denoted $M(D)$, is defined as the locus of points inside D which lie at the centers of all closed discs (or spheres) which are maximal in D, together with the limit points of this locus.

The medial axis transform was first introduced by Blum to describe biological shapes. It can be viewed as the locus of the center of a maximal disc as it rolls inside an object. Since its introduction, the concept was used in numerous applications that involved studying the topology and geometry of some objects. In the area of image processing and biological modeling, the algorithms for investigating connectivity problems and background skeletonization include: peeling, distance-based, Voronoi diagram based methods [32]. While the first two techniques has been traditionally used for skeleton extraction, the Voronoi diagram based method (the concept is described in detail in the next Section) provides a robust and efficient alternative, reducing significantly the computation time and also allowing for computation of additional features of the shape. The illustration of the skeleton (blue line) constructed for the randomly drawn shape using software we have developed in presented in Fig. 4.15. The yellow lines represent dual to Voronoi diagram, the Delaunay triangulation edges.

4.5.3.1.   *Topology-Based Approximation of Image for Feature Extraction along the Boundary*

In mane biometric problems, such as detecting *singular points* in fingerprint images, the quality of the result and false detection rates depend directly on the quality of the data (image, print, recording etc). To improve the result, pre-processing can be used. In some cases, however, it is not enough to simply enhance the image properties (such as valleys and ridges in a fingerprint image). Many cases of false detection happen at the boundary of an image or at place where lines are of irregular shape. To remedy the first situation, we propose a method based on extending the lines of the image beyond the boundary in the projected direction so that the singular point can be computed more precisely. For the second case, topology-based

Fig. 4.15.   Skeleton extraction using Delaunay triangulation.

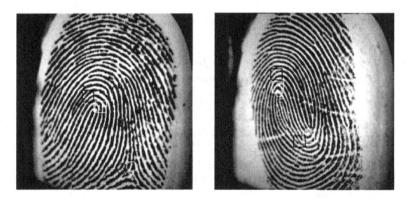

Fig. 4.16.   Singular point detection (top to bottom): singular point close to boundary (lower); regular pattern.

methods are traditionally used to smooth the irregularity (including the interpolation techniques) [5,28].

4.5.3.2.   *Edge Detection for Feature Lines*

One example of a particular application of the above methodology is edge detection for feature lines and feature line thinning.   All existing edge detection methods can be classified into the following three categories: first-order differential operators, zero-crossing detection of second-order derivatives and more complex heuristic algorithms (such as Canny operator). As the feature lines are not uniform in all directions, all the results extracted by first-order derivatives operators (Sobel, Prewitt, Roberts) are not of sufficient quality. Many detailed feature lines might be lost. Compared with first-order derivatives, the results of LoG and Canny are much better: LoG covers all the possible features, including minor features such as curved lines and dots; Canny connects some adjacent feature lines, so the resulting edges are smoother. In order to keep the continuity of feature lines and cover all the possible detail features, we combined the results of LoG and Canny for edge detection.

After the binary edge results are obtained, the morphological thinning algorithm is applied to get a unitary image, where the maximal width of the edge is one pixel. By thinning edges into skeletal pixels, which is an accurate representation of feature edges, we can also easily record the position of those feature lines.

After the tracking process is performed for all starting points, information on the total number of feature lines extracted, the number of points in each feature line, and x,y coordinates of every point, is recorded for subsequent use in mapping and rendering.

4.5.4.   *Topology-Based Approach for Generation and Synthesis of New Biometric Information*

Finally, one of the most challenging areas is a recently emerged problem of generating biometric information, or so-called inverse problem in biometrics. In order to verify the validity of algorithms being developed, and to ensure that the methods work efficiently and with low error rates in real-life applications, a number of biometric data can be artificially created, resembling samples taken from live subjects. In order to perform this procedure, a variety of methods should be used, but the idea that we explore is based on the extraction of important topological information from the relatively small set of samples (such as boundary, skeleton, important features etc.), applying variety of computational geometry methods, and then using these geometric samples to generate the adequate set of test data.

One example of such technique can be the generation of synthetic iris patterns for testing of the iris identification algorithms and for medical purposes. A multiresolution technique known as *reverse subdivision* is used to capture the necessary characteristics from existing irises, which are then combined to form a new iris image. In order to improve the results, a set of heuristics to classify iris images is developed in collaboration with Prof. *Faramarz Samavati*, University of Calgary. We propose to generate iris images as unique combinations of existing irises, and we also introduce heuristics used to classify iris images. The flowchart of the iris synthesis system is given in Fig. 4.17.

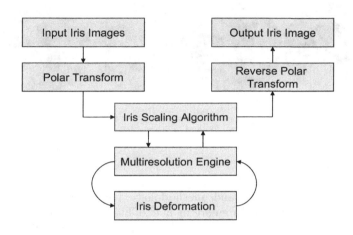

Fig. 4.17.   Iris generation system flowchart.

A more detailed description of the algorithm is now provided. First, all of the images are pre-processed in such a way that the center of the image corresponds to the center of the pupil, which significantly reduced the complexity of the following algorithms. Then we transform the iris into an alternate coordinate system which unwraps the circular iris information into columns and rows which is better suited to computer processing. Next we remove the pupil to guarantee that we are working with the iris information needed for the synthesis components. Finally, once an image has been synthesized, we reinsert the pupil information and return the iris to its natural coordinate system.

At the iris synthesis stage, we first irises are classified according to their patterns (number of concentric rings) and their frequency characteristics. Then low and high level details from the selected irises are combined according to a specific algorithm (for more information, consult [39]). The resulting set of synthetic irises is shown in Fig. 4.18.

The methodology has high potential for applications in other areas of biometric synthesis.

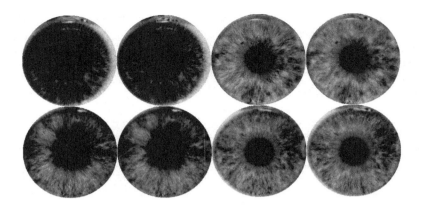

Fig. 4.18. Set of synthetic irises.

4.6. Conclusions

Geometric data structures and methodology based on proximity and topology prove to be useful for emerging field of biometric technologies. The chapter presents a detailed overview of existing computational geometry methods, their recently developed applications in biometrics and their implementation, and suggests a number of new approaches for investigation of specific biometric problems.

Acknowledgments

Author would like to thank Prof. Vlad P. Shmerko for his encouragement, advice and expertise that he generously shared during the process of preparation of this Chapter. Author also would like to acknowledge Prof. Jon Rokne and Prof. Jim Parker for their valuable suggestions, support and contributions to this Chapter. Last, but not the least, author would like to thank all graduate students and research associates of the Biometric Technologies Laboratory at the University of Calgary for their dedicated work on the project and enthusiasm for research.

Bibliography

1. Wayman, J.L. (1999). Technical testing and evaluation of biometric identification devices, in Biometrics: Personal Identification in Networked

Society (A.Jain, R.Bolle, S.Pankanti,eds), Kluwer, Dordrecht, pp. 345–352.

2. Burge, M. and Burger, W. (2000). Ear Biometrics in Computer Vision, *Intern. Conf. on Pattern Recognition Proceeding* **Vol.02**, (Barcelona, Spain, 2000), pp. 826–831.

3. Gavrilova, M. L. (2001). Robust algorithm for finding nearest-neighbors under L-1, L-inf and power metrics in the plane, LNCS 2073, **1**, Springer-Verlag, pp. 663–672.

4. Gavrilova, M.L. and Alsuwaiyel,M. (2001). Two Algorithms for Computing the Euclidean Distance Transform, *Intern. J. of Image φ Graphics*, World Scientific, **1**, 4, pp. 635–646.

5. Karasick, D.L., Nackman, L.R, Rajan,V.T., Mitchell,J.B.A. (1997). Visualization of Three-Dimensional Delaunay Meshes, Algorithmica, **19(1-2)**, pp. 114–128.

6. Luchnikov, V.A., Gavrilova,M.L., Medvedev,N.N., Voloshin,V.P. (2002). The Voronoi-Delaunay Approach for the Free Volume Analysis of a Packing of Balls in a Cylindrical Container, *J. Future Generation Computer Systems*, **8**, pp. 673–679.

7. Medvedev, N.N. (2000). Voronoi-Delaunay method for non-crystalline structures, SB Russian Academy of Science, Novosibirsk.

8. Mucke, E, Saias, I. and Zhu, B. (1996). Fast Randomized Point Location Without Preprocessing in Two- and Three-dimensional Delaunay Triangulations, *12th Annual Symposium on Computational Geometry*, pp. 274–283.

9. Agarwal, P., Guibas, L., Murali, T. and Vitter,J. (1997). Cylindrical static and kinetic binary space partitions, *13th Annual Symposium on Computational Geometry*, pp. 39–48.

10. Maltoni, D., Maio, D. and Jain, A.K. (2003). Handbook of Fingerprint Recognition, Springer-Verlag, 2003.

11. Jain, A.K., Hong, L. and Bolle, R. (1997). On-line Fingerprint Verification, *IEEE Transaction on Pattern Analysis and Machine Intelligence*, **19**, (April,1997), pp. 302–314.

12. Abellanas, M., Hurtado, F. and Palop, B. (2004). Transportation networks and Voronoi Diagrams, *Proc. 1st Intern. Symposium on Voronoi Diagrams in Science and Engineering*, Tokyo, Japan.

13. Lee, D.T., Chung-Shou Liao and Wei-Bung Wang (2004). Time-Based Voronoi Diagram, *Proc. 1st Intern. Symposium on Voronoi Diagrams*, Tokyo, Japan, pp. 229–243.

14. Bebis, G., Deaconu, T. and Georiopoulous, M. (1999). Fingerprint identification using Delaunay triangulation, ICIIS 99, Maryland, pp. 452–459.

15. Cappelli R., Maio D. and Maltoni D., (2001). Modelling Plastic Distortion in Fingerprint Images, ICAPR, LNCS 2013, pp. 369–376.

16. Jiang, X. and Yau, W.-Y, (2000). Fingerprint minutiae matching based on the local and global structures. *Proc. 15th Intern. Conf. Pattern Recognition* (ICPR, 2000) **2** pp. 1042-1045.

17. Kovacs-Vajna, Zs. and Miklos A. (2000). Fingerprint Verification System

Based on Triangular Matching and Dynamic Time Warping, *IEEE Trans. on PAMI*, **22**, 11, pp. 1266–1276.

18. Liang, X. and Asano, T. (2006). A Linear Time Algorithm for Binary Fingerprint Image Denoising Using Distance Transform, *IEICE TRANSACTIONS on Information and Systems*, **E89-D**, 4, pp. 1534–1542.

19. Gavrilova, M. (2004). Computational Geometry and Biometrics: on the Path to Convergence, *Proceedings of the International Workshop on Biometric Technologies 2004*, Calgary, AB, Canada, pp. 131–138.

20. Wang, C. and Gavrilova, M. (2004). A Multi-Resolution Approach to Singular Point Detection in Fingerprint Images, *Intern. Conf. of Artificial Intelligence*, **I**, pp. 506–511.

21. Wang, C. and Gavrilova, M. (2005). A Novel Topology-based Matching Algorithm for Fingerprint Recognition in the Presence of Elastic Distortions, *Intern. Conf. on Computational Science and its Applications*, LNCS, Springer-Verlag, **1**, pp. 748–757.

22. Wayman, J, Jain. A, Maltoni, D. and Maio, D. (2006). Biometric Systems: Technology, Design and Performance Evaluation, Book, Springer.

23. Yanushkevich, S. N., Stoica, A., Shmerko, V. P. and Popel, D. V. (2005). *Biometric Inverse Problems*. CRC Press/Taylor & Francis Group, Boca Raton, FL.

24. Xiao, Y. and Yan, H. (2002). Facial Feature Location with Delaunay Triangulation/Voronoi Diagram Calculation, *Conferences in Research and Practice in Information Technology*, Feng, D. D., Jin, J., Eades, P. and Yan, H., Eds., ACS. **11** pp. 103–108.

25. Yanushkevich, S., Gavrilova, M. and Shmerko, V. (2004). Editors, Proceedings of the International Workshop on Biometric Technologies 2004, Special Forum on Modeling and Simulation in Biometric Technologies, University of Calgary, Calgary, AB, Canada, 120p.

26. Tse, R, Dakowicz, M, Gold, C.M. and Kidner,D.B. (2005). Building Reconstruction Using LIDAR Data, *In Proc. 4th ISPRS Workshop on Dynamic & Multi-dimensional GIS*, pp. 156–161.

27. Cohen-Or, Daniel and Levanoni Yishay. (1996). Temporal continuity of levels of detail in Delaunay triangulated terrain, *Visualization96*, IEEE. Press, pp. 37–42.

28. Apu, R. and Gavrilova, M. (2005). Adaptive Spatial Memory Representation for Real-Time Motion Planning, *3IA 2005 Intern. Conf. on Computer Graphics and Artificial Intelligence*, France, pp. 21 32.

29. Yamada, H. (1984). Complete Euclidean distance transformation by parallel operation, *Proc. 7th International Conference on Patter Recognition*, pp. 69–71.

30. Danielsson, P. (1980). Euclidean distance mapping, *Computer Graphics and Image Processing*, **14**, pp. 227–248.

31. Rosenfeld, A. and Pfalz, J.L. (1966). Sequential operations in digital picture processing, *J. of ACM*, **13**, pp. 471–494.

32. Okabe, A., Boots, B., Sugihara, K. and Sung Nok Chiu. (2000). Spatial tessellations: concepts and applications of Voronoi diagrams, 2nd Ed., John

Wiley and Sons, Chichester, West Sussex, England.

33. Li, S. and Jain ,A. (2004). Handbook of Face Recognition, Springer, 2004.

34. Apu, R. and Gavrilova, M. (2006). An Efficient Swarm Neighborhood Management for a 3D Tactical Simulator, IEEE-CS Proceedings, ISVD 2006, Banff, AB, Canada.

35. Bhattachariya, P. and Gavrilova, M. (2006). CRYSTAL - A new density-based fast and efficient clustering algorithm, IEEE-CS Proceedings, ISVD 2006, Banff, AB, Canada.

36. Luo, Y., Gavrilova, M. L. and Costa-Sousa, M. (2006). NPAR by Example: line drawing facial animation from photographs, IEEE CS Proceedings of CGIV06, Australia, August 2006 (upcoming).

37. Wang, H. and Gavrilova, M. (2006). An Efficient Algorithm for Fingerprint matching, *International Conference on Pattern Recognition*, 20 - 24 August 2006, Hong Kong, IEEE-CS publishers, 2006 (upcoming).

38. Parke, F.I. (1974). A Parameterized Model for Human Faces, PhD Thesis, University of Utah, Salt Lake City, UT, December 1974.

39. Wecker, L., Samavati, F. and Gavrilova M. (2005). Iris Synthesis: A Multi-Resolution Approach, *GRAPHITE 2005*, New Zealand, pp. 121–125.

PART 2
ANALYSIS IN BIOMETRICS

PART 2

BIOMETRIC VERIFICATION: STATISTICAL MODEL

- *Discriminating elements and similarity*
- *Strength of evidence*
- *Applications*

SIGNATURE RECOGNITION: COMPOSITE MODEL

- *Distances between signatures*
- *Experimental protocols*
- *Composite classifiers*
- *Signature synthesis*

EAR RECOGNITION: FORCE FIELD MODEL

- *Ear topology*
- *Force field transform*
- *Feature extraction*

FACIAL RECOGNITION: NONTENSOR-PRODUCT WAVELET MODEL

- *Formal design*
- *Filter banks*
- *Design examples*

PALMPRINT RECOGNITION: FUSED WAVELET MODEL

- *Palprint images preprocessing*
- *Wavelet-based feature extraction*
- *Palmprint feature matching*

KEYSTROKE AND MOUSE DYNAMICS ANALYSIS

- *Keystroke and free text detection*
- *Mouse analysis movement*

Chapter 5

A Statistical Model for Biometric Verification

Sargur Srihari*, Harish Srinivasan[†]

*Center of Excellence for Document Analysis and Recognition (CEDAR),
Department of Computer Science and Engineering, University at Buffalo,
State University of New York, Buffalo, NY, USA*
*srihari@cedar.buffalo.edu
[†]hs32@cedar.buffalo.edu

The biometric verification task is one of determining whether an input consisting of measurements from an unknown individual matches the corresponding measurements of a known individual. This chapter describes a statistical learning methodology for determining whether a pair of biometric samples belong to the same individual. The methodology involves four parts. First, discriminating elements or features, are extracted from each sample. Second, similarities between the elements of each sample are computed. Third, using conditional probability estimates of each difference, the log-likelihood ratio (LLR) is computed for the hypotheses that the samples correspond to the same individual and to different individuals; the conditional probability estimates are determined in a learning phase that involves estimating the parameters of various distributions such as Gaussian, gamma or mixture of gamma/Gaussian. Fourth, the LLR is analyzed with the Tippett plot to provide a measure of the strength of evidence. The methods are illustrated in two biometric modalities: friction ridge prints — which is a physical biometric, and handwriting — which is a behavioural biometric. The statistical methodology has two advantages over conventional methods such as thresholds based on receiver operating characteristics: improved accuracy and a natural provision for combining with other biometric modalities.

Contents

Glossary

AER — Average Error Rate

DE — Discriminating element

EER — Equal Error Rate

FAR — False Acceptance Rate

LLR — The log-likelihood ratio

LR — The likelihood ratio

p.d.f. — Probability density functions

ROC — Receiver Operating Characteristic

TAR — True Acceptance Rate

5.1. Introduction

The task of biometric verification is defined as that of determining whether two samples of biometrics correspond to (or were generated by) the same or by different individuals [1]. The biometric samples can be of any modality such as fingerprints, handwriting, signature, speech etc. For each modality different terminology is used for the verification task:

In *fingerprint* verification, one of the two samples is called the *template*, and the sample being tested for match is called the *input*.

In *writer* verification, they are called *known* and *questioned* respectively, which is derived from the domain of forensic document examination.

In *signature* verification they are referred to as *genuine* and *input*.

This chapter describes a statistical approach to the design of a verification system to automatically make a decision of *same*(match) or *different*(non-match) when two biometric samples are compared. The approach involves a statistical learning phase and an evaluation phase.

The learning phase begins with a training set of ensemble of pairs, which capture the distribution of *intra-class* (or same individual) variations and *inter-class* (or two different individuals) differences. It consists of four steps:

> **Step 1.** Discriminating element extraction,
>
> **Step 2.** Similarity computation,
>
> **Step 3.** Estimating conditional probability density estimates for the difference being from the same individual or from different individuals, and
>
> **Step 4.** Determining the log-likelihood ratio (LLR) function and from it the Tippett plot, which is a function of the distribution of the LLR.

The evaluation phase for a given pair of samples is as follows:

(i) Discriminating element extraction,
(ii) Similarity computation,
(iii) Determining the LLR from the estimated conditional probability density estimates, and
(iv) Determining the strength of evidence from the Tippet plot.

Each of the four steps in the model are described in the following sections.

5.2. Discriminating Elements and Similarity

Discriminating elements (DEs) are dependent upon the biometric modality. They are objects that are located within a sample that will be used for verification. For example, in the case of writer identification based on handwriting on paper, an allograph or character is a DE. The shape of the DE within a sample of handwriting is used to match against the shape of the same DE in another sample. There is a hierarchy of DEs involving 21 classes of DEs such as elements of execution, style, etc., which in turn consist of finer DEs [2]. For fingerprints, the most commonly used DEs are the location and orientation of minutiae(Ridge endings and Ridge bifurcations) present in the fingerprint image.

Once a DE is determined to be present in the sample we extract a feature or measurement from the DE. Some DEs are determined from the entire sample-the corresponding features are referred to as macro features. DEs and features that capture finer details at the individual level are called as *micro-features*. Some DEs capture the class characteristics of an individual such as gender and ethnicity.

In order to match DEs between two samples, the presence of the DE has to be first recognized in each sample. Matching is performed between the same elements in each sample. A formal statement of these considerations follow.

Assume that there are r different DEs that can possibly be used. The value of r can range from 1 to many. In the case of present-day automatic fingerprint identification systems (AFIS), only the minutiae are considered as the DEs, thus making $r = 1$. Since the features of a DE are usually fixed, the symbol for a DE, say x_i, is taken to denote the features of the DE. Denote by $\mathbf{x_i}(k_j)$, vector of all the occurrences of the i_{th} DE, from the j_{th} sample of the k_{th} individual, where $k = 1, \ldots, n$ is the index of the individuals whose samples are available for training and $j = 1 \ldots, m$ is the index of those multiple samples available for each individual.

Assume that there exists a suitable similarity or distance measure between two DEs of the same type. For the i_{th} DE that occurs in two particular samples A and B (A and B may or may not belong to the same individual), the distance between all the occurrences of that DE in A and the those in B is computed as as

$$D_i[\mathbf{x_i}(A), \mathbf{x_i}(B)].$$

The definition of the distance measure D_i is dependent upon the particular discriminating element i. It should be noted that the symbol D_i can also be used to mean a similarity measure, as opposed to a distance measure. The two are interchangeable depending on the definition of D_i.

5.3.   Statistical Formulation

The probability of a particular pair of sample A and B belonging to the same individual, conditioned on all the occurrences of a DE i, that occurs in both samples, is denoted as

$$P_{\text{same}}(D_i[A, B]).$$

This probability distribution P_{same} is determined from the set

$$\{D[\mathbf{x_i}(s_j), \mathbf{x_i}(s_m)],$$

where j, m vary over all the samples of each individual s, and $s = 1, \ldots, n\}$ and $j <> m$. The probability of their belonging to different individuals is denoted as

$$P_{\text{different}}(D_i[A, B]),$$

which is determined from the set

$$\{D[\mathbf{x_i}(s_j), \mathbf{x_i}(t_m)],$$

where j, m vary over all the samples of each pair of individuals s, t, $s = 1, \ldots, n$, $t = 1, \ldots, n$ and $s <> t$. It is useful to represent the two distributions as probability density functions (p.d.f.s) so that they can be used in the computation of likelihood functions. The p.d.f.s corresponding to the two similarity distributions will be denoted as

$$p_{\text{same}}(D_i[A, B]) \quad \text{and} \quad p_{\text{different}}(D_i[A, B]),$$

respectively.

Assume that the distances(or similarities) along different DEs are conditionally independent given whether or not the two samples belong to the same or different individuals. In cases, where the DEs are not conditionally independent, then the need to learn a joint p.d.f.,

$$p_{\text{same}}(\mathbf{D}[A, B]) \quad \text{and} \quad p_{\text{different}}(\mathbf{D}[A, B])$$

arises, where $\mathbf{D}[A, B]$ is a *vector* of distance measures, one for each DE. In most scenarios, the independence assumption holds good and leads to a model that can be learnt easily. When trying to match two samples A and B, if there are r DEs of available, then the probability of samples A and B belonging to the same or different individual can be written as Eqs. (5.1) and (5.2).

$$p_{\text{same}}(A, B) = \prod_{i=1}^{r} p_{\text{same}}(D_i[\mathbf{x_i}(A), \mathbf{y_i}(B)]) \tag{5.1}$$

$$p_{\text{different}}(A, B) = \prod_{i=1}^{r} p_{\text{different}}(D_i[\mathbf{x_i}(A), \mathbf{y_i}(B)]) \tag{5.2}$$

It is useful to determine the ratio of the two p.d.f.s to determine the relative strength of the two values. We refer to

$$\frac{p_{\text{same}}(A, B)}{p_{\text{different}}(A, B)}$$

as the likelihood ratio (LR). The log-likelihood ratio (LLR), obtained by taking the logarithm of LR is more useful since LR values tend to be very large (or small). Then the log-likelihood ratio has the form of Eq. (5.3)

$$LLR(A, B) = \sum_{i=1}^{r} \log[p_{\text{same}}(D_i(\mathbf{x_i}(A), \mathbf{x_i}(B))]$$
$$- \log[p_{\text{different}}(D_i(\mathbf{x_i}(A), \mathbf{x_i}(B))] \tag{5.3}$$

Similarity p.d.f.s p_{same} and $p_{\text{different}}$ for scanned signatures are shown in Fig. 5.1. In this case, the DE consists of the entire signature and hence the number of occurrences of the DE is 1. The feature vector of this

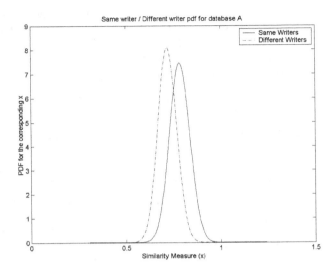

Fig. 5.1.   Probability density functions of similarities of same and different pairs of DEs.

single DE x, consists of 1024 binary-values which respectively capture the finest variations in the contour, intermediate stroke information and larger concavities and enclosed regions of a signature have been shown to be useful for writer recognition [1]. Since micro features are binary valued several binary string distance measures can be used, the most effective of which is the correlation measure [3]. The data is from pairs of signatures: genuine-genuine pairs and genuine-forgery pairs. The closeness of the two p.d.f.s is due to skilled forgeries. It is useful to represent the two distributions as p.d.f.s so that they can be used in the computation of likelihood functions. Figures 5.2(a) and 5.2(b) show the likelihood function and LLR values for normally distributed DEs.

5.3.1.   *Gaussian Case*

If the similarity data has both positive and negative scores, then the similarity data can be acceptably represented by Gaussian distributions-which can be expected since we are dealing with several factors such as the feature measurement, distance computation, etc. — we represent the probability density functions of distances conditioned upon the same- and different-individual categories for a single DE x as the parametric forms

$$p_{\text{same}}(x) \ \mathcal{N}(\mu_s, \sigma_s^2) \ \text{ and } \ p_{\text{different}}(x) \ \mathcal{N}(\mu_d, \sigma_d^2),$$

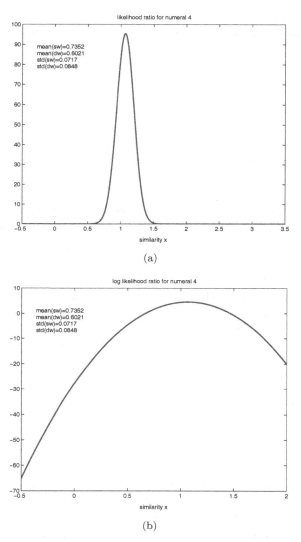

Fig. 5.2.   Likelihood ratio (a) and log-likelihood ratio of a DE whose similarity p.d.f.s are normally distributed (b).

whose parameters are estimated using maximum likelihood estimates. The LLR value can be expressed as Eq. (5.4).

$$LLR = \frac{(x - \mu_d)^2}{2\sigma_d^2} - \frac{(x - \mu_s)^2}{2\sigma_s^2} + \log \frac{\sigma_d}{\sigma_s} \qquad (5.4)$$

which has the quadratic form $a_1 x^2 + a_2 x + a_3$, where a_1, a_2 and a_3 are constants defined in terms of the means and variances of the two p.d.f.s as Eq. (5.5)

$$a_1 = \frac{(\sigma_s^2 - \sigma_d^2)}{2\sigma_s^2 \sigma_d^2} \qquad a_2 = \frac{\mu_s \sigma_d^2 - \mu_d \sigma_s^2)}{2\sigma_s^2 \sigma_d^2}$$

$$a_3 = \frac{1}{2\sigma_s^2 \sigma_d^2}(\mu_s \sigma_d^2 - \mu_d \sigma_s^2) + \log\frac{\sigma_d}{\sigma_s} \tag{5.5}$$

5.3.2. *Gamma Case*

In many cases, the distance measure $D_i(A, B)$ is always positive. This is because, the simplest distance measure, may be defined as an absolute difference between different measurements on the biometric sample. Hence, a Gaussian distribution is not ideal to model such a distribution. Here a more robust distribution, a Gamma distribution can be used to model the distributions of

$$P_{\text{same}}(D_i[A, B]) \quad \text{and} \quad P_{\text{different}}(D_i[A, B]).$$

The p.d.f. of a Gamma distribution is given in Eq. (5.6). The gamma p.d.f. has two sufficient statistics namely the scale parameter θ and the shape parameter α. These two parameters can be learnt from Eq. (5.7) where σ and μ are the standard deviation and mean of the distribution. The robustness of the Gamma distribution to model different kind of shapes is shown in Figs. 5.3(a), 5.3(b) and 5.3(c).

$$P(x|\alpha, \theta) = x^{\alpha-1}\frac{e^{-\frac{x}{\theta}}}{\theta^\alpha \cdot \Gamma(\alpha)} \qquad \text{for} \qquad x > 0 \tag{5.6}$$

$$\theta = \frac{\sigma^2}{\mu} \qquad \alpha = \frac{\mu}{\theta} \tag{5.7}$$

Thus analogous the Gaussian case, we represent the probability density functions of distances conditioned upon the same- and different-individual categories for a single DE x as the parametric forms

$$p_{\text{same}}(x) \; \mathbf{Gamma}(\alpha_s, \theta_s^2) \; \text{and} \; p_{\text{different}}(x) \; \mathbf{Gamma}(\alpha_d, \theta_d^2),$$

whose parameters are estimated using maximum likelihood estimates. The LLR value analogous to Eq. (5.4) for the Gamma case is given below in Eq. (5.8)

$$LLR = (\alpha_s - \alpha_d)\log x - \left[\frac{1}{\theta_s} - \frac{1}{\theta_d}\right]x$$

$$- \alpha_s \theta_d + \alpha_d \theta_s + \log\Gamma(\alpha_s) - \log\Gamma(\alpha_d) \tag{5.8}$$

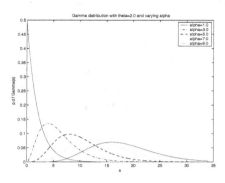 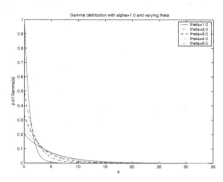

(a) Gamma distribution curves for fixed $\theta = 2.0$ and varying α.

(b) Gamma distribution curves for fixed $\alpha = 1.0$ and varying θ.

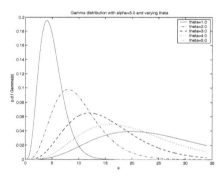

(c) Gamma distribution curves for fixed $\alpha = 5.0$ and varying θ.

Fig. 5.3. Gamma distribution curves. (a) Varying α, (b) Varying θ, $\alpha = 1.0$. Note how the θ parameter defined the rate of decay of the exponential. (c) Varying θ, $\alpha = 5.0$.

5.3.3. Mixture Model Case

In some cases, the distribution of the distance measures, is not unimodal. Hence, neither unimodal Gamma or Gaussians can effectively model the distributions. Hence, a mixture model is used to learn the distribution. A mixture model always has one extra parameter (π) known as the mixing proportion, in addition to the sufficient statistics of the underlying distribution used. For example, a mixture of Gamma distribution as three parameters namely α, θ (as defined previously in Eq. (5.7)) and π. The p.d.f. of a mixture of gamma is given in Eq. (5.9)

$$P(s|\alpha, \theta, \pi) = \sum_{k=1}^{K} \pi_k P(s|\alpha_k, \theta_k) \qquad (5.9)$$

where $P(s|\alpha_k, \theta_k)$ can be given by a unimodal Gamma distribution as in Eq. (5.6). The parameters of a mixture model can be learnt using the EM algorithm. One example of a mixture of gamma using two mixtures to model a complex distribution with two peaks is shown in Fig. 5.4.

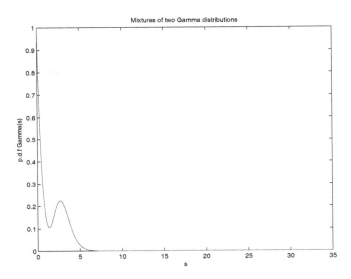

Fig. 5.4.   Mixture of two Gamma distributions, $\alpha_1 = 1, \theta_1 = 0.5$ and $\alpha_2 = 10, \theta_2 = 0.3$, Mixing proportion $\pi = \{0.5, 0.5\}$. A mixture of Gamma can model multi-modal distributions with irregular peaks such as one above.

5.3.4. *Strength of Evidence*

The LLR value provides a measure of the strength of the conclusion. Of greatest interest to the jurist is the reliability of the method. This is an issue that has arisen in several forensic examination fields. Paralleling the work in DNA and speaker verification we use the Tippett plot to provide this basis.

In work on DNA matching [4] the plot of probability versus likelihood ratio is known as the Tippett plot in reference to [5]. This approach has gained significant popularity in the speaker verification domain, e.g. [6].

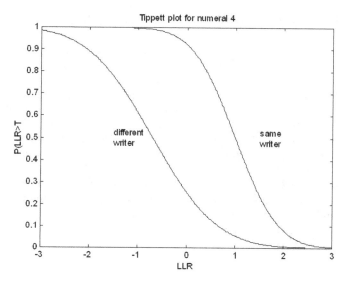

Fig. 5.5. The Tippet plot indicates for a given LLR score, the percentage of cases that are weaker than the present score when the pair of samples belonged to either the same individual or to different individuals.

The Tippett plot can be derived analytically in the case of normally distributed univariate p.d.f.s. The Tippet plot contains a representation of the conditional probability of the LLR being greater than a particular value. The conditioning is based on whether the underlying sample pair belongs to the same or different

5.4. Application of Model

The proposed approach to biometric verification has been applied to three domains namely fingerprint, writer and signature verification. The first two use handwriting as a biometric and the second fingerprints.

5.4.1. *Fingerprint Verification*

In fingerprint verification, the original model as in Eqs. (5.1) and (5.2) can be used as such. Most commonly only one DE is used, namely the minutiae. There are variable number of occurences of minutiae in each fingerprint. Hence the DE is a vector a minutiae each represented as triplets x, y, θ (x-location, y-location and orientation). A fingerprint matching algorithm is used to produce a similarity score between two pairs of fingerprint samples.

From our definitions, a fingerprint matching algorithm computes

$$D_i[\mathbf{x_i}(A), \mathbf{x_i}(B)],$$

between two samples A and B. Figure 5.6 shows two fingerprints that are compared and the corresponding score given by a standard Bozorth [7] algorithm.   To learn a model to automatically make a decision of match(same finger) or non-match(different finger), one assumes the availabliity of a training set. The FVC 2002 Db1 data set was used in the experiments below. The data set consists of 100 fingers, 8 samples for each. From these, 2880 same and 4950 different fingerprint pairs were taken, and half of these numbers were used for the purpose of learning and remaining for evaluation.

Fig. 5.6.   Fingerprint verification, Sscore obtained by Bozorth matcher was 85, with 45 minutiae available on the left image and 42 on the right.

5.4.1.1.   *ROC Method of Learning*

In the traditional method of learning, the distribution of scores obtained from the ensemble of pairs from the learning (training) set is used to plot an ROC curve. Figure 5.7 shows the histogram of the distribution of scores obtained on the training set using the Bozorth [7] algorithm. The ROC curve is plotted by, varying a threshold, and evaluating the *False Acceptance*

Rate (FAR)

FAR: Fraction of pairs decided as match when truly they were
non-matches

and *True Acceptance Rate* (TAR)

TAR: Fraction of pairs decided as match when truly there were
matches.

Thereby a graph of FAR vs TAR is plotted and Fig. 5.8 shows the
ROC curve for the score distribution in Fig. 5.7. From this ROC, an
operating point is chosen, as the threshold to make decisions on the test
data set. This operating point is usually the *Minimum Average Error*
operating point, or the *Equal Error Rate* (EER) operating point. Using
this threshold, a decision of match/non-match is made on the test data
set and the performance is evaluated on the basis of *Average Error Rate*
(AER)

AER: Fraction of incorrect decisions: both match and
non-match.

It was found to be 2.47%.

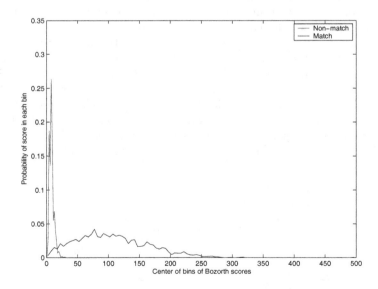

Fig. 5.7. Bozorth scores histogram on the FVC 2002 Db1 fingerprint data set.

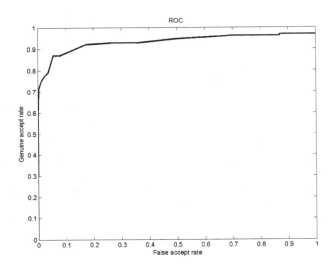

Fig. 5.8.   ROC curves for Bozorth on the FVC 2002 Db1 fingerprint data set.

5.4.1.2.   *Parametric Learning using Gamma Distributions*

The ROC method of learning is prone to noise(outliers data points) in the data set. An illustration on a toy data set is shown in Fig. 5.9, where there is ambiguity in choosing the operating point from the distribution of scores. The better way to learn from the training set, is to use a parametric model. Fingerprint matching scores are always positive and hence Gamma distributions/Mixture of Gamma distributions are used. Figure 5.10 shows the use Gamma p.d.f.s for modeling the scores from a standard matching algorithm Bozorth3 [7]. The parameters for the Gamma distributions were leart using Eq. (5.7). Now using likelihood methods for making decisions of match/non-match , an average error rate of 2.42% was obtained. Table 5.1 compares the ROC method against the statistical method. The parametric method is more robust to presence of noise in the training data set.

Table 5.1.   Comparison of Error rate between Likelihood and ROC method. Average Error Rate is the average error of False Positives and False Negatives.

Algorithm	Average Error Rate	
	Likelihood method	ROC method
Bozorth3 [7]	2.42%	2.47%

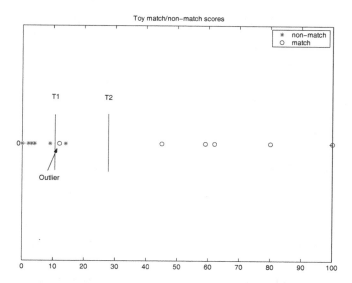

Fig. 5.9. Toy match/non-match scores. It is clear that at both operating points $T1$ and $T2$, the average error is the same. Ideally $T2$ is the better opearting point, but the outlier data point, introduces $T1$ also as a candidate. When modeling statistically, it is obvious that the parametric learning method will arrive at the correct decision boundary close to $T2$.

5.4.2. *Writer Verification*

In the writer verification application, the statistical parameters, viz., the means and variances of distances, for 13 DEs called as *macro features* of the entire sample of writing, were computed from writing samples of over 1000 writers. Figure 5.11 shows an example of a pair of documents being compared, listed along with the values and distances of the 13 macro features. From these, α and θ parameters for a Gamma distribution can be learnt. These DEs are available for any given sample of writing during the testing phase. On the other hand the 62 DEs corresponding to individual characters are available in varying quantities from each sample.

In this scenario, the distance measure $D_i[\mathbf{x_i}(A), \mathbf{x_i}(B)]$ between the vector occurrences of the i_{th} DE, from two samples A and B, can be further decomposed into product over the individual occurrences of that DE as in Eq. (5.10).

$$D_i[\mathbf{x_i}(A), \mathbf{x_i}(B)] = \prod_{j=1}^{n_i} \prod_{k=1}^{m_i} D_i[x_i(j, A), x_i(k, B)] \qquad (5.10)$$

where n_i and m_i are the total number of occurrences of the i_{th} element in the samples A and B respectively. $x_i(j, A)$ is the j_{th} occurrence of the i_{th}

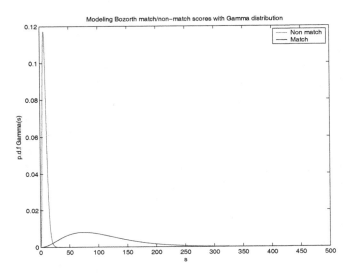

Fig. 5.10. Modeling of fingerprint matching scores of Bozorth using Gamma distribution. The FVC 2002 Db1 data set was used.

Fig. 5.11. Two pairs of handwritten documents compared. The feature values and the distance measure is given in the table.

DE from sample A and $x_i(k, A)$ is the k_{th} occurrence of the i_{th} DE from sample B. Hence Eqs. (5.1) and (5.2) are expanded as below Eqs. (5.11) and (5.12).

$$p_{\text{same}}(A, B) = \prod_{i=1}^{r} \prod_{j=1}^{n_i} \prod_{k=1}^{m_i} p_{\text{same}}(D_i[x_i(j, A), y_i(k, B)]) \qquad (5.11)$$

$$p_{\text{different}}(A, B) = \prod_{i=1}^{r} \prod_{j=1}^{n_i} \prod_{k=1}^{m_i} p_{\text{different}}(D_i[x_i(j, A), y_i(k, B)]). \qquad (5.12)$$

Table 5.2. Verifying two handwritten documents, The 13 macro features and the distances are listed. **f1**=Entropy, **f2**=Threshold, **f3**=No of black pixels, **f4**=No of exterior contours, **f5**=No of interior contours, **f6**=Horizontal slope, **f7**=Positive slope, **f8**=Vertical slope, **f9**=Negative slope, **f10**=Stroke width, **f11**=Average Slant, **f12**=Average Height, **f13**=Average Word gap.

Doc	f1	f2	f3	f4	f5	f6	f7	f8	f9	f10	f11	f12	f13
2a	0.03	226	8282	27	6	0.24	0.21	0.39	0.15	6	0.75	28	54.18
2b	0.03	226	7413	25	7	0.25	0.19	0.39	0.16	7	−0.44	27	48.76
D	0	0	869	2	1	0.01	0.02	0	0.01	7	1.2	1	0.07

Correspondingly the LLR is also expanded as below in Eq. (5.13)

$$LLR(A, B) = \sum_{i=1}^{r} \sum_{j=1}^{r} \sum_{k=1}^{r} \log[p_{\text{same}}(D_i(x_i(j, A), x_i(k, B))]$$

$$- \log[p_{\text{different}}(D_i(x_i(j, A), x_i(k, B))] \qquad (5.13)$$

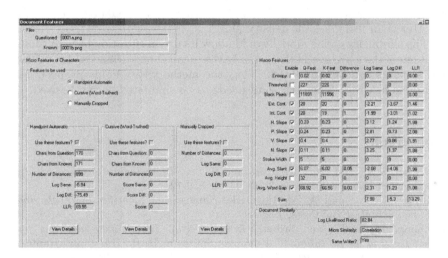

Fig. 5.12. Results of statistical model applied to writer verification. In this case there are 13 DEs computed from the entire sample, shown on the right and 62 DEs corresponding to upper, lower case characters and numerals.

The right hand side of Fig. 5.12 shows the LLRs for several DEs and corresponding features. Since each DE corresponds to the entire sample the features are called as macro features. The left hand side shows the results based on several DEs automatically located-called micro-features. Due to the independence assumption the scores can all be cumulatively added to determine the overall score-shown on the bottom of the right hand side.

A similar approach to signature verification with few samples available for estimating the parameters has given good results.

5.5. Concluding Remarks

A statistical model for biometric verification has been described. It involves computing the statistical distributions of the differences between same and different pairs of samples for each of several discriminating elements. The approach has been the basis for systems for signature verification, writer identification and lately fingerprint verification, which have yielded promising results. Other biometric modalities where significant ambiguities exist, e.g. speaker verification, are also likely to benefit from the methodology.

The approach also provides a natural means of combining multiple biometric modalities. It has in fact been shown that assuming statistical independence of modalities and combining them in a likelihood framework is better than using a joint density [9].

The simplicity of the model would allow its use in situations where small amounts of data are available and only the means and variances of the DEs and features can be computed. Thus the methodology holds promise for situations in which biometric data needs to be simulated or synthesized.

Bibliography

1. Srihari, S.N., S-H Cha, H. Arora and S. Lee, *Individuality of Handwriting*, Journal of Forensic Sciences, 2003, 44(4): 856-72.
2. Huber, R. A., and A. M. Headrick, *Handwriting Identification:Facts and Fundamentals*, CRC Press, 1999.
3. Zhang, B., and S. N. Srihari, *Binary vector dissimilarity measures for handwriting identification*, T. Kanungo, Smith, E. H. B., Hu, J. and Kantor, P.B., eds., Document Recognition and Retrieval X, Bellingham, WA: SPIE, 2003, 5010: 28-38
4. Evett, I., and J. Buckleton, (1996), *Statistical Analysis of STR data, Advances in Forensic Haemogenetics*, vol. 6, Springer-Verlag, Heidelberg, 1996, pp. 79-86.
5. Tippett, C. F., V. J. Emerson, M. J. Fereday, F. Lawton, and S. M. Lampert, *The Evidential Value of the Comparison of Paint Flakes from Sources other than Vehicules*, Journal of Forensic Sciences Society, vol. 8, 1968, pp. 61-65.
6. Drygajlo, D., D. Meuwly,, and A. Alexander, *Statistical Methods and Bayesian Interpretation of Evidence in Automatic Speaker Recognition*, Eurospeech 2003.

7. Tabassi, E., C. Wilson R. McCabe S. Janet C. Watson and M. Garris, *User's guide to NIST Fingerprint Image Software 2 (nfis2)*, 2004, pp. 80-86.

8. Srinivasan, H., S. N. Srihari, M. Beal, G. Fang and P. Phatak, *Comparison of ROC and likelihood methods for fingerprint verification*, Proc. Biometric Technology for Human Identification III: SPIE Defense and Security Symposium 2006, pp. 620209-1 to 620209-12.

9. Tulyakov, S., *A Complexity Framework for Combination of Classifiers in Verification and Identification Systems*, Ph. D. dissertation, Department of Computer Science and Engineering, University at Buffalo, 2006.

Chapter 6

Composite Systems for Handwritten Signature Recognition

Jim R. Parker

*Digital Media Laboratory, University of Calgary,
2500 University Dr. NW Calgary, Alberta, T2N 1N4, Canada
jparker@ucalgary.ca*

It has been a standard assumption that handwritten signatures possess significant within-class variation, and that feature extraction and pattern recognition should be used to perform automatic recognition and verification. Described here is a simple way to reliably compare signatures in a quite direct fashion. Reasonable speeds and very high success rates have been achieved. Comparisons are made to other methods, and a four algorithm voting scheme is used to achieve over 99% success.

Contents

Glossary

APBorda	—	Posteriori Borda Count
DWMV	—	Dissenting-Weighted Majority Vote
SMV	—	Simple Majority Vote
WMV	—	Weighted Majority Vote
wBorda	—	Weighted Borda count

6.1. Introduction

Handwritten signatures have been used to verify identity for centuries. It is widely believed that signatures are almost unique, or at least that identical signatures are rare enough that signatures can be used reliably as identification. It is also believed that an average person can verify a signature in a few moments, since bank tellers and cashiers do so with cheques and credit cards. In fact, experts are generally needed to properly verify signatures; only the fact that very few forgeries are attempted allows institutions to be casual about verification.

One problem with using computers for automatic signature verification is the fact that no two signatures are ever really the same. We must deal with degrees of similarity, and in the presence of noise two signatures could appear to be very different to a computer while being, in fact, largely the same. Signature verification is commonly done by applying a classifier, possibly a support vector machine or a neural network [1], [2], [11], to a collection of measurements on a digitized image [12], [8] of the signature. The multiple measurements, or features, are numerical values, and are collected into a feature vector. Thus, the signatures are not compared directly against each other [19], [22].

Features selected for signature recognition will allow for variations between instances of the same writer, while allowing the detection of differences that indicate a different writer. The hope is that the features will vary more for two different writers (between-class variance) than for the same writer (in-class variance). If a writer is constrained to use a certain form, or to write a signature within a box, then the signatures may display a great deal of consistency, especially with respect to orientation and scale. This is not sufficiently true in unconstrained cases to yield consistent sets of features for recognition, so scale and rotation invariance is clearly an issue.

Figure 6.1 shows a set of 50 signatures drawn freehand on a data tablet. The variation can be seen clearly, but is difficult to measure in any really meaningful way. For example, the maximum distance between temporally related points (points the same time from the time of the first pen down) is 76 pixels, or 19% of the signature width. The maximum distance between any two closest points is 22 pixels (5.5% of width). Its not clear what any of this really means. or how it could be used to verify the identity of the writer.

Fig. 6.1.   50 signatures overlaid, aligned on their centroids.

What is clear is that if, as is claimed, in-class signatures are in some sense more similar to each other than they are to out-class signatures, then there is some obvious (at least to the human visual system) measure that minimizes the small variations normal to signatures and uses larger ones to make distinctions.

6.2.   Simple Distances Between Signatures

When using feature vectors it is common practice to calculate a distance between an incoming exemplar and signatures of known persons. The distance used depends on the situation; it may simply be Euclidean distance, or something more complex. The known signature having smallest distance form the unknown can be used to classify it. It has often, for reasons mentioned above, been assumed that signatures cannot be compared directly in any meaningful way. The work described here is an effort to determine whether simple, relatively obvious distances can yield good results.

Many different things might be meant by 'relatively obvious'. Here it means that the distance will be computed in a Euclidean fashion, from points in two dimensional geometric space. It is not necessary that pixels actually be these points. All of the signatures used in this work were acquired using a camera attached to a computer. Software was used to follow the end of the pen while the signature is being drawn [12], [13]. The data are almost identical to those obtained using a data tablet — a series of (X, Y) pen positions collected at fixed time intervals. Linear interpolation can be used between sample points so as to permit arbitrary times, or to create a bitmap version of the signature. This would correspond to simply drawing lines between adjacent points.

6.2.1. *Simple Direct Comparison: Global Relative Distance*

Noise is variously described as "random variation" in a signal or as something unwanted that degrades the quality of signals and data. This could be describing slight variations in signatures, which could then be thought of as noise. One standard method for dealing with noisy data is to average multiple samples. Using multiple raster images, we can overlay all exemplars and create a single composite image, like the one in Fig. 6.1. Before superimposition, the signatures would have to be scaled to a standard size, and then a new image, the mask image, is created by adding all of these scaled signatures together, assuming that object pixels are 1 and background is 0.

We could simply look up each pixel in the target image to see if the corresponding pixel in the mask image was set. A simple similarity measure is the number of pixel matches, essentially an XOR. This produced very good results on a small number of images, but showed no better than 76% success on our larger data set. This was not good enough, but does show some promise.

Perhaps it would be better to use distances between corresponding signature pixels rather than simple overlap counts. One possible way do this on raw signatures would calculate the distance from a black pixel in one signature to the nearest black pixel in the other. A complete bipartite pixel match would probably be best, but the cost of doing this would be prohibitive, and the improvement may not be significant.

In order to compute a distance based on pixels in two images, a common coordinate system and scale have to be established. The centroid of the signature is a logical origin for such a coordinate system. Scaling is done using a bounding box for each signature, matching the box for the exemplar to the size of the unknown sample while maintaining the aspect ratio of the exemplar. Since the signatures are stored as line end-points, the scaling is a simple matter, but if the signatures were bitmaps then accurate scaling is much harder. In any case there will be some size variation remaining.

Next the distance is calculated using a *distance transform*. Consider the signature image $\mathbf{S}$, obtained by drawing the connections between the sampled pen positions into a bi-level image. The distance transform is an image $\mathbf{D}$ the same size as $\mathbf{S}$ in which the value of each pixel $\mathbf{D}_{ij}$ is the distance from the pixel $\mathbf{S}_{ij}$ to the nearest object pixel. The 8-distance was used, because it is easily possible to compute the entire transform in just two passes through the image [14]. The distance $\mathbf{E}_i - S$ for an exemplar image

$\mathbf{E}_i$ is found by computing the distance map $\mathbf{D}$ as above and accumulating the sum of all $\mathbf{D}_{ij}$ that correspond to object pixels. This is the distance from each object pixel in $\mathbf{E}_i$ to the nearest object pixel in $\mathbf{S}$.

We need to compute also the *reverse distance*, $\mathbf{S} - \mathbf{E}_i$, which calculates the distance transform of $\mathbf{E}_i$ in a symmetrical manner to the above. The net distance, used to characterize the difference between the two signatures, is the product of these two distances. Figure 6.2 illustrates the idea.

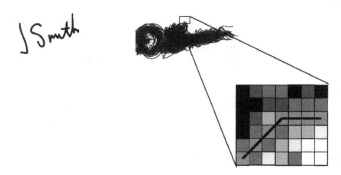

Fig. 6.2. Calculation of the distance from a simple distance map (Darker pixels are more distant from an object pixel). The map of the exemplar is overlaid with the signature, and a score is found by incrementing a counter by the distance found whenever the line passes over a pixel in the mask.

Most of the distance images can be computed in advance and stored instead using the raw images and recomputing the distances each time.

6.2.2. *Temporal Distance*

If two signatures are instances of the same class, it should be true that points on each that were sampled at the same time relative to the initial pen down time should be near to each other spatially [17]. To measure this, a standard coordinate system is needed; the centroid of the signature should serve very well again as the origin. It is also necessary to know at what relative time each pixel was drawn, and for this the signature must have been captured using a data tablet or similar device. These return pen coordinates at regular sampling intervals. Times associated with points between two sampled coordinates can be interpolated. There are issues around scale to be addressed as well.

The basic process involved in computing temporal distance is shown in Fig. 6.3, and is as follows: The signature being validated is $\mathbf{S}$; the exemplar being compared against currently is $\mathbf{E}_i$.

> **Step 1.** Rescale $\mathbf{E}_i$ to the same size as $\mathbf{S}$, but do not alter the aspect ratio of $\mathbf{E}_i$. Some size variation will therefore remain.
>
> **Step 2.** Because $\mathbf{E}_i$ has changed size, the time needed to draw it should change also. Based on the ratio of old arc length to new, compute the new value for the time needed to draw $\mathbf{E}_i$, $\mathbf{T}_e$.
>
> **Step 3.** Convert the coordinates for both $\mathbf{S}$ and $\mathbf{E}_i$ from those relative to the capture tablet to those relative to the centroid of each signature.
>
> **Step 4.** Rotate the signature $\mathbf{S}$ by its measured orientation angle; that is, attempt to make it horizontal.
>
> **Step 5.** Now divide the time interval needed to draw $\mathbf{S}$ into a fixed number of smaller intervals— we use 100 such subintervals. For each subinterval, calculate the (x, y) position of the pen for both $\mathbf{S}$ and $\mathbf{E}_i$. Determine the distance between these two points, and accumulate the sum of these distances in d.

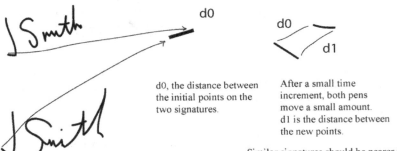

d0, the distance between the initial points on the two signatures.

After a small time increment, both pens move a small amount. d1 is the distance between the new points.

Similar signatures should be nearer to each other after each interval than should different signatures. Sum the distances over all time intervals to get a similarity measure.

Fig. 6.3.   How temporal distance is calculated. After the signatures are scaled, each is broken into 100 pieces by time taken to draw. The distance between the two signatures after each interval is summed.

The distance between $\mathbf{S}$ and $\mathbf{E}_i$ is $d/100$, or the average position error per subinterval. What we are doing in plain English is this: we start at the beginning of each signature, and step along each in units of constant time. In fact, we compute a distance in each signature that represents identical times and step in units of these respective distances. At each step we have a point on each signature where the pen was at the same time. We compute the Euclidean distance between these points, and sum this distance over all points sampled.

6.3. Experimental Protocol: Trial 1

A total of 1400 signatures were available, which consists of up to 25 samples of 56 individuals [14], [18]. It is difficult to collect signature data: University ethics committees become involved, and people are naturally hesitant to give aware such a valuable proof of their identity, in spite of the fact that they do so each day on cheques, letters, and credit card chits. As a result, we had a relatively small amount of data available, and a "leave-one-out" protocol was used initially. A database of signatures was constructed for each trial, omitting the target in each case. In other words, the set of exemplars never includes the signature being evaluated. However, every signature is compared against all others, both in -class and out-class, and the smallest single distance (or largest similarity value) represents the correct classification.

A confusion matrix was generated in each case, showing all of the information available for the trials: correct classifications, false positives, and false negatives are all present. These matrices are too large to present here, but the upper left corner of the matrices for the global relative distance method and the temporal distance method is given in Fig. 6.4.

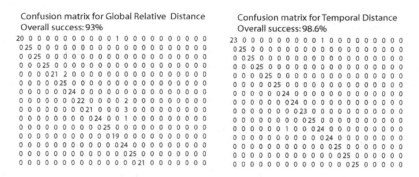

Fig. 6.4.   Results for the two signature recognition techniques discussed. Left and below — global relative distance. Right and below — temporal distance. This is only the first 15 rows/16-24 columns of the confusion matrices.

6.3.1. *Results: Trial 1*

The results from the two different methods for comparing signatures can be condensed into two numbers: the success rates. Global relative distance yielded a 93% success rate on the 1400 signatures in our test set, while the temporal distance algorithm yielded a success rate of 98.6%. This is very high given the relative simplicity of the algorithm, and so it was decided to compare in detail against other strategies.

Of course, a leave-one-out protocol implies that 24 out of the 25 instances of each signature would be used as training data. This is unrealistic, but does give a good idea of the best results one can achieve. We also performed trials using between 2-22 signatures as training data. Figure 6.4 shows a graph of the success rates of both global relative distance and temporal distance as a function of the number training items.

6.4. Two Other Methods — A Comparison

In the pursuit of a complete analysis, it was decided to compare the two procedures discussed so far against two others, using the same data set. The two methods to be used as a basis for comparison are relatively established ones, seen in the literature for some years. This was done so that all other individuals working on this problem could easily reproduce the programs needed, and would have some understanding of and instinct for the behavior of the methods.

6.4.1. *Slope Histograms*

The first of the comparison methods uses slope histograms [16], [26]. The slope histogram of an object in a bi-level image is the frequency histogram of the directions of the tangents (slopes) to the boundary of the object at all boundary pixels. The actual values in each bin are not as important as the relative relationship between bins, because this is merely a matter of scale. It is standard practice to reduce histograms to a standard scale by dividing each bin by the total number of boundary pixels or by the maximum value in any bin.

The tangents can be calculated using a small square region centred at each boundary pixel, say 5×5 pixels. The set of object pixels in this region that are 8-connected to the centre pixel are extracted, and a line is fit to those pixels passing through the centre one. The angle that this line makes to a standard axis is binned, and the relevant bin is incremented.

This method of calculating slopes from boundary pixels cannot distinguish between angles 180 degrees apart As a result, it is possible for two different objects to have identical slope histograms (Fig. 6.5). It is better to compute slope histograms using point data from a data tablet, as the points are in temporal order and all angles can be calculated.

As a classifier, we used nearest Euclideandistance to an exemplar, computed bin to bin on a normalized pair of histograms (by dividing all

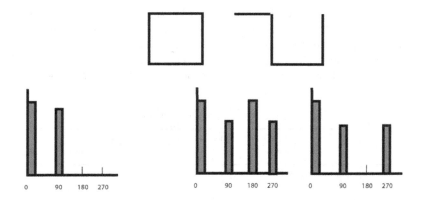

Fig. 6.5. Top: two figures having the same slope histogram. Left: slope histogram of the two objects. Right: the two distinct histograms found using chain code pixel order.

bins by the largest bin). The largest value in any bin will therefore be 1.0.

To determine whether the signature consistent with the samples that are known to belong to the target individual? To determine this, there must be a number of reliable signatures belonging to the target available for comparison. When a signature is being checked, its slope histogram is calculated and must be compared against these known exemplars. If it is 'too far away' in some measurable sense, then the signature is rejected.

This argues for some sort of distance measure, and for the use of a threshold to discriminate between acceptance and rejection. It is likely that for most practical systems, a more sophisticated scheme, such as use of a support vector machine, would be used. However, the issue here is the use of slope histogram variants, and not an evaluation of discrimination methods. It was decided to use a K nearest neighbor scheme for all evaluations, so that the comparison is fair to all methods tested. If Support Vector Machines were used, it may be found that different kernel functions yield better results for some kinds of slope histogram than others.

The distance between two slope histograms is easily calculated as

$$\sum_{i=1}^{36} \left(\frac{H1 - \mu_i}{\sigma_i} - \frac{H2 - \mu_i}{\sigma_i} \right).$$

Of course, geometric normalization needs to be performed. Each histogram is normalized for scale by dividing all bins by the largest one. The largest peak will therefore be 1.0. Rotation is normalized by measuring the orientation of the overall signature using moments. The histograms are circularly shifted by the orientation angle before the distance calculation is performed.

The result of verifying 650 available signatures using multiple component slope histograms appears in Table 6.1 below. The same experimental protocol was used as previously (Section 6.1) on the same data, so these results can be compared directly against those in Table 6.1. The use of multiple exemplars appears to be one of the best methods tried so far.

Table 6.1. Results using multiple components.

Method	Verification rate	False accept	False reject	Time per signature (sec)
Multiple exemplars (threshold)	96.8	0.0	3.2	0.010
Multiple exemplars (K nearest neighbors)	98.4	1.2	0.36	0.10

6.4.2. *Shadow Masks*

The second comparison method is the high success rate shadow mask technique [20], [21]. A shadow mask is a configuration of sampling squares which cover the entire signature image, each sampling a small area of the signature. A typical shadow mask configuration consists of a constant number of sampling squares, say $N \times M$ such boxes, to cover the entire image area. A feature vector is created for each image, each feature being a count of the number of pixels in the shadow that is cast onto each of the bars, cast from the center outward, as seen in Fig. 6.6. When all boxes and features are considered we have information about the shape of all parts of the signature.

Fig. 6.6. Definition of shadow masks: (a) An individual sampling square showing the six projection areas. (b) The signature is drawn over a grid of these squares, centered at the centroid. (c) Each calculation projects the signature in each square onto the projection areas, and a bitmap.

We use a shadow mask configuration consisting of a constant number of sampling squares, (where a sampling square consists of top, bottom, right, left, and two diagonal shadow mask bars at 45 and -45 degrees) arranged in n rows and m columns, where each square of size $P \times Q$ pixels. This configuration covers the entire image area. A feature vector is made for each image in which each "feature" of the feature vector is a count of the number of pixels representing the shadow that is cast onto each of the bars (cast from the center outward) that make up the sampling box. Because the sampling boxes are configured such that they touch one another at the borders, the feature vector stores only 4 of the 6 shadow mask bars making up a single sampling box. This means that most of the bars will be shared with another sampling box, and storing all 6 bars would be redundant.

In total, the feature vector contains $4 \times m \times n + m + n$ features. Once feature vectors have been computed for each image in the database, the usual nearest neighbor Euclideandistance calculation is made when classifying a feature vector.

While the algorithm was neither scale- nor rotation-invariant, it was made somewhat translation-invariant by computing the "hyper-centre of inertia", essentially the centroid, of each signature. This was then used to translate the signature to the center of the image prior to the casting of shadows. In this way the signatures of the same person would overlap each other more readily, assuming they were all placed one on top of another, even though they may not line up the same way with the boundaries of the sampling boxes.

The reported accuracy of the aforementioned signature verification method is extremely high, in the 98–99% range, and thus bears further investigation. Our implementation of the algorithm was tested on a database of genuine signatures (black and white in *pbm* format) and, using a leave-one-out method, keeps track of the percentage of signatures that are correctly classified. We should mention that our work has identified a minor flaw in the original shadow mask implementation in that it appears to use scale, or the signature size, as a feature. We corrected this in our own implementation, which may have affected the measured success rates [18].

6.4.3. *Intermediate Results*

We tested our version on 11 sets of signatures, for a total of 539 images. Each set, at one time, contained 50 signatures each. However, some had to be removed due to problematic data (e.g. an image consisting of all background). These signatures were relatively constant in terms of size and orientation. That is, signatures belonging to the same set were similar in size, although signatures belonging to different sets sometimes varied in size. Our measured success rate was 99.26%, which agrees with the published result.

It may be of interest to note that, when tested using a set of 50 traced forgeries, the program correctly classified them as belonging to a distinct class. Forgeries, however, have not been thoroughly tested.

Using a leave-one-out protocol, the slope histogram technique achieved a success rate of 89.1%, while the use of shadow masks gave 89.9%. This is, by the way, lower than the published rate of over 99%, but this reflects the method is neither scale nor rotation invariant. All the methods here are using the same data, and are therefore being compared on an equal basis.

The graph in Fig. 6.7 shows how the success rate of these two methods varies with the number of signatures used to train, and permits a direct comparison with the temporal distance, global relative distance, and the slope histogram methods discussed.

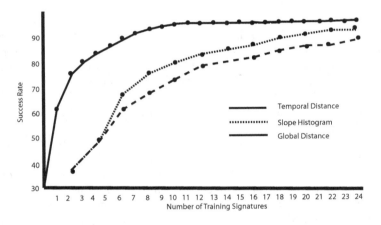

Fig. 6.7. Results for the signature recognition techniques discussed. Graphs of success rate VS number of training items. All four methods are shown.

6.5. Composite Classifiers

In the process of designing vision systems and pattern analysis algorithms, it is natural that a variety of techniques be tried, with the overall goal of improving the rate at which patterns can be correctly classified. Many diverse algorithms each have strengths and weaknesses, good ideas and bad. One way to take advantage of this variety is to apply many methods to the same recognition task, and have a scheme to merge the results [3], [9], [10]; this should be successful over a wider range of inputs than would any individual method. Indeed, it is what will be demanded of such a multiple classifier system: it must have a higher overall recognition rate than any of the individual classifiers that make it up, or there is no point to the extra effort involved.

Composite classifiers can be made up of similar classifier types (e.g. nearest K neighbor, geometric features) or of quite diverse types (non-homogeneous). In the latter case it is hoped to take advantage of the diversity in the classifier set to achieve an increase in robustness. Classifiers can differ in both the nature of the measurements used in the classification process and in the type of classification produced:

> Type I *classification is a simple statement of the class,*
> Type II *classification is a ranked list of probable classes, and*
> Type III *classification assigns probabilities to classes.*

A problem with the evaluation of composite classifiers is the availability of the needed large number of classifications for a valid characterization, along with the associated ground truth. A large data set is not enough; a correct implementation of a collection of classifiers is needed. Even then the nature of the classifications is pre-determined; there are many variables that simply cannot be controlled. Thus it is that many published combination algorithms have been tried only on relatively few data sets and in relatively few variations.

It is desirable to promise a set of predictable behaviours under a wide variety of circumstances. For instance, how does a composite classifier behave when more than half of the component classifiers have high correlations? When correlations are small? When one or more classifiers fail on a few classes? These questions need answers

6.5.1. *Merging Type 1 Responses*

Given that the output of each of the classifiers is a single, simple classification value (a type I response), the obvious way to combine them is by using a voting strategy. A majority voting scheme can be expressed as follows: let $C_i(x)$ be the result produced by classifier i for the image x, where there are k different classifiers in the system; then let $H(x,d)$ be the number of classifiers giving a classification of d for the image x, where d is one of 0,1,2,3,4,5,6,7,8,9. Then

$$E(x) = \begin{cases} j, & \text{if } MAX(H(x,i) = H(x,j)) \text{ and } H(x,j) > \frac{k}{2} \\ 10, & \text{otherwise.} \end{cases}$$

This is called a *simple majority vote* (SMV). An easy generalization of this scheme replaces the constant $k/2$ in the above expression with $k \times a$ for $0 \leq \alpha \leq 1$ 27. This permits a degree of flexibility in deciding what degree of majority will be sufficient, and will be called a weighted majority vote (WMV). This scheme can be expressed as:

$$E(x) = \begin{cases} j, & \text{if } MAX(H(x,i) = H(x,j) \text{ and } H(x,j) > \alpha k) \\ 10, & \text{otherwise.} \end{cases}$$

Neither of these takes into account the possibility that all of the dissenting classifiers agree with each other. Consider the following cases:

(a) In case A there are ten classifiers, with six of them supporting a classification of "6", one supporting "5", one supporting "2" and two classifiers rejecting the input digit.

(b) In case B, using the same ten classifiers, six of them support the classification "6" and the other four all agree that it is a "5".

Do cases A and B both support a classification of "6", and do they do so equally strongly?

One way to incorporate dissent into the decision is to let max1 be the number of classifiers that support the majority classification j ($MAX1 = H(x,j)$), and to let $MAX2$ be the number supporting the second most popular classification h ($MAX2 = H(x,h)$). The the classification becomes:

$$E(x) = \begin{cases} j, & \text{if } MAX(H(x,i) = H(x,j) \text{ and } MAX1 - MAX2 \geq \alpha k) \\ 10, & \text{otherwise.} \end{cases}$$

where α is between 0.0 and 1.0. This is called a Dissenting-Weighted Majority Vote (DWMV).

For the five classifier system being discussed, the SMV strategy gave a 99.4 success rate. Evaluation of WMV is a little more difficult, requiring an assessment of the effect of the value of a on the results. A small program was written that varied a from 0.05 to 0.95, classifying all samples on each iteration. This process was then repeated five more times, omitting one of the classifiers each time to again test the relative effect of each classifier on the overall success. With this much data a numerical value is needed that can be used to assess the quality of the results. The success rate could be used alone, but this does not take into account that a rejection is much better than a misclassification; both would count against the recognition rate. A measure of *reliability* can be computed as:

```
Reliability = recognition/((100%-rejection)
```

The reliability value will be low when few misclassifications occur. Unfortunately, it will be high if recognition is only 50%, with the other

50% being rejections. This would not normally be thought of as acceptable performance. A good classifier will combine high reliability with a high recognition rate; in that case, why not simply use the product

```
Reliability\index{subject}{Reliability}× Recognition
```

as a measure of performance? In the 50/50 example above this measure would have the value 0.5: reliability is 100% (1.0) and recognition is 50% (0.5). In a case where the recognition rate was 50%, with 25% rejections and 25% misclassifications, this measure will have the value 0.333, indicating that the performance is not as good. The value `Reliability×` `Recognition` will be called *acceptability*. The first thing that should be done is to determine which value of a gives the best results, and this is more accurately done when the data is presented in tabular form (Table 6.2).

Table 6.2. Acceptability of the Multiple Classifier Using a Weighted Majority Vote.

Alpha	Success
0.05	0.994
0.25	0.994
0.50	0.992
0.75	0.956

From this information it can be concluded that a should be between 0.45 and 0.5, for in this range the acceptability peaks without causing a drop in recognition rate. DWMV also uses the a parameter, and can be evaluated in a fashion identical to what has just been done for WMV. The optimal value of a, obtained from Table 6.3, was found to be 0.25.

6.5.2. *Merging Type 2 Responses*

The problem encountered when attempting to merge type 2 responses is as follows: given M rankings, each having N choices, which choice has the largest degree of support?

For example, consider the following 3 voter/4 choice problem [23]:

Voter 1: *a b c d* **Voter 2:** *c a b d* **Voter 3:** *b d c a*

This case has no majority winner; *a*, *b* and *c* each get one first place vote. Intuitively, it seems reasonable to use the second place votes in this case to see if the situation resolves itself. In this case b receives two second place

Table 6.3. Acceptability of
the Multiple Classifier Using a
Dissenting Weighted Majority
Vote.

Alpha	Success
0.05	0.993
0.25	0.985
0.30	0.985
0.45	0.960
0.55	0.960
0.65	0.932
0.80	0.932
0.85	0.984

votes to a's one, which would tend to support b as the overall choice. In
the general case there are a number of techniques for merging rank-ordered
votes, four of which will be discussed here.

The *Borda count* [5] is a well-known scheme for resolving this kind of
situation. Each alternative is given a number of points depending on where
in the ranking it has been placed. A selection is given no points for placing
last, one point for placing next to last, and so on up to $N - 1$ points for
placing first. In other words, the number of points given to a selection is
the number of classes below it in the ranking. For the 3 voter/4 choice
problem described above the situation is:

Voter 1: $a(3)$ $b(2)$ $c(1)$ $d(0)$,

Voter 2: $c(3)$ $a(2)$ $b(1)$ $d(0)$,

Voter 3: $b(3)$ $d(2)$ $c(1)$ $a(0)$,

where the points received by each selection appears in parentheses behind
the choice. The overall winner is the choice receiving the largest total
number of points:

$$a = 3 + 2 + 0 = 5$$
$$b = 2 + 1 + 3 = 6$$
$$c = 1 + 3 + 1 = 5$$
$$d = 0 + 0 + 2 = 2.$$

This gives choice b as the "Borda winner". However, the Borda count does
have a problem that might be considered serious. Consider the following 5

voter/3 choice problem: **Voter 1:** *a b c*, **Voter 2:** *a b c*, **Voter 3:** *a b c*, **Voter 4:** *b c a*, **Voter 5:** *b c a*.

The Borda counts are $a = 6, b = 7, c = 2$, which selects b as the winner. However, a simple majority of the first place votes would have selected *a*! This violates the so-called *majority criterion* [23]:

> If a majority of voters have an alternative X as their first choice, a voting rule should choose X.

This is a weaker version of the Condorcet Winner Criterion [6]:

> If there is an alternative X which could obtain a majority of votes in pair-wise contests against every other alternative, a voting rule should choose X as the winner.

This problem may have to be taken into account when assessing performance of the methods.

Finally, we have the monotonicity criterion: If X is a winner under a voting rule, and one or more voters change their preferences in a way favorable to X without changing to order in which they prefer any other alternative, then X should still be the winner. No rule that violates the monotonicity criterion will be considered as an option for the multiple classifier. This decision will not eliminate the Borda count. With the monotonicity criterion in mind, two relatively simple rank merging strategies become interesting. The first is by Black [4], and chooses the winner by the Condorcet criterion if such a winner exists; if not, the Borda winner is chosen. This is appealing in its simplicity, and can be shown to be monotonic. Another strategy is the so-called Copeland rule [23]: for each option compute the number of pair-wise wins of that option with all other options, and subtract from that the number of pair-wise losses. The overall winner is the class for which this difference is the greatest. In theory this rule is superior to the others discussed so far, but it has a drawback in that it tends to produce a relatively large number of tie votes in general.

The Borda, Black, and Copeland rules were implemented as described and applied to the five classifier problem, and the results are summarized in Table 6.4.

From Table 6.4, it would appear that the Borda scheme is tied with Black, followed by Copeland. It is important to temper this view with the fact that this result was obtained from basically one observation. Confirmation would come from applying these schemes to a large number of sets of characters.

Table 6.4. Results of the
Voting Rules for Rank
Ordering.

Rule	Success
Borda	99.9
Black	99.9
Copeland	99.6

6.5.3. *Merging Type 3 Responses*

The problem of merging type 3 responses was not pursued with as much vigor as were the type 1 and 2 problems, as very few systems yield probabilities as their output. Indeed, the solution may be quite simple. Xu decides that any set of type 3 classifiers can be combined using an averaging technique. That is,

$$P_E(x \in C_i | x) = \frac{1}{k} \sum_{j=1}^{k} P_j(x \in C_i | x), \ i = 1, 2, \ldots, M,$$

where P_E is the probability associated with a given classification for the multiple classifier, and P_k is the probability associated with a given classification for each individual classifier k. The overall classification is the value j for which is a maximum. There is little actual type 3 data, but it could be approximated by using the a posteriori method described previously, where it is used to convert type 1 responses to type 3 responses. Using this approximate data set, the result obtained by merging type 3 responses using averaging is given by:

Correct: 997 **Incorrect:** 3 **Rejected:** 0

Acceptability is 0.994.

6.5.4. *Results from the Multiple Classifier*

Using the acceptability measure to assess each of the merging methods discussed, we need to look only at the best method in each of the three groups; that is, the best multiple type 1 classifier, the best type 2, and the best type three. The best three are given in Table 6.5.

From the Table 6.5 above it can be seen that the best classifier explored uses the *Black* scheme for merging rank ordered responses.

Table 6.5.   Multiple classifier results.

Name	Type	Acceptability
SMV	1	0.994
Black	2	0.998
Average	3	0.994

6.5.5. *Further Classifier Combination Techniques*

Kittler [10] suggests five possible combination strategies, and compares these and a majority vote against each other on four actual classifiers and 2213 handprinted character samples. The best two of these combination strategies, the *median rule* and the *sum rule*, are of interest here. Both are based on the assumption, again, that the a *priori* probabilities associated with each classifier are very near the a *posteriori* probabilities. The sum rule, for **R** classifiers and **m** classes is:

Assign θ to class

$$(1 - R)P(w_j) + \sum_{i=1}^{k} P\langle w_j | x_i \rangle$$

$$= max_{k=1}^{m}[(1 - R)P(w_k) + \sum_{i=1}^{R} P\langle w_k | x_i \rangle].$$

The median rule can be expressed as:

Assign θ to class w_j

$$\text{MEDIAN}_{i=1}^{R} P\langle w_j | x_i \rangle = \text{MAX}_{k=1}^{m} \text{MEDIAN}_{i=1}^{R} P\langle w_k | x_i \rangle,$$

where the x_i represent measurements on the item: a feature vector. The value of **P** can be found in the confusion matrix for each classifier **i**.

Other rank-based schemes could be used as well [24], [25]. Behind the Borda count is the presumption that the second most likely classification is relatively near, in terms of likelihood or preference, to the best classification; its rank is only one away it. Consider a four-candidate vote and the result *A B C D*. The sum of the ranks is 6 (in general $N(N - 1)/2$ for N candidates). Treating these as scores, *A* gets 3 and *B* gets 2; the difference (1) is 1/6 of the total, the same as the difference between *B* and *C*, and the difference between *C* and *D*. In other words, a Borda count assumes that the distance between each candidate, once sorted, is the same; a presumption of uniformity.

This uniformity assumption is often flawed in the case of classifiers, although it may be the best that can be done for elections, the domain for which it was devised. Using prior information it is possible to more accurately estimate the relative distances between the ranked classifications and use these to calculate better weights for resolving the rankings in a Borda fashion [7].

What is being proposed is a variable weighting of the ranked items. One suggestion is to use the measured properties of the classifier directly to assign a value to each rank position. For the 5 classifier/3 class problem above the count computed for class a would be

$$k(a) = \sum_{i-1}^{5} P_i(a) f_a,$$

where $Pi(a)$ is the measured probability of classifier i issuing a classification of a, and fa is a scale value. This scale value could be a constant (e.g. $R-1$) or a function of the rank of a, as two examples. This will be called the a *posteriori Borda count*, or *APBorda* in the ensuing discussion [15].

A second idea is to ignore the specific probabilities and assign simple non-uniform values to the ranked items. These weights could still be based on the typical observed distances between classes in a given classification, or could be constructed to achieve a specific goal. For example, again using the 5 *classifier/3* class problem above, recall that the problem was that there was a conflict between the majority vote and the Borda count. Now use a weight of $R - 1$ on the first ranked class, $(R - 2) \times w$ on the second, and $(R - 3) \times w^2 (= 0)$ on the third. It is now possible to find a weight w such that the winner found by summing the weights over all classifiers is the majority winner. In this latter instance the weight would be $w = 0.67$; this could be thought of as a weighted Borda count, referred to as *wBorda* in further discussion.

In all of these cases the initial ranking is found by sorting the classes based on the measured probabilities for the classifier involved.

6.5.6. *Empirical Evaluation*

One of the problems with the empirical evaluation of classifier combination algorithms is the need for vast amounts of data. In this case, the data consists of classifier output from many different classifiers and a great many individual classification tasks. This means that the input data, with ground truth, must exist, and the classifiers must be correctly implemented. Even if sufficient data can be found, it has certain pre-defined properties that can't be altered. A classifier has a given recognition rate for a given type of data, and it may not be possible to provide all desired combinations of characteristics. For example, a classifier combination algorithm may work well when all individual classifiers operate at over 90% recognition, or when the classifiers have nearly the same rates. It is important to be able to determine benefits and limitations, either from a theoretical basis or empirically.

6.6.   Conclusions

We have presented some new signature features and classifiers, and have combined them with existing ones to create a composite signature recognition scheme having high reliability. We have also discussed classifier combination in the context of signature recognition in a general way, hoping to inspire further work in this area.

The simplest way to merge multiple methods is using a voting scheme. While it may be possible to improve things a little more using a more complex scheme, such as a rank based vote, a very large amount of data based on past performance is required to gain an accurate confusion matrix from which rankings can be extracted. So, a simple majority vote was constructed using the temporal distance, slope histogram, and global relative distance methods. The graph in Fig. 6.8 shows the result of this combining process for different numbers training signatures and shows the temporal distance graph overlaid on this. In no case is the result better than that achieved by simple temporal distance. The graph in Fig. 6.8 shows the result of this combining process for different numbers training signatures and shows the temporal distance graph overlaid on this. In no case is the result better than that achieved by simple temporal distance.

The graph does show, along with Fig. 6.7, how the success depends on training set size; a more robust algorithm will have a relatively flat graph, although it is clear that smaller training sets will invariably give poorer results. The temporal distance method is the most robust method presented, according to this definition, and robustness is unimproved by using multiple algorithms.

Adding a fourth method to the mix changes things a little. When the shadow mask algorithm is included in the vote the success rate overall can be made to reach 99.14%. Since there are four components in each vote a parliamentary majority vote is used, in which the target receiving the most votes, rather than the majority of votes, is selected. Thus, two votes are needed to select a winner; if there are two items receiving two votes each, we have a tie, and reject the signature. In actual practice this happened once in the 1400 signatures recognized, and there were fifty cases of a vote being determined by two agreeing votes. In no case did one of these two-votes misclassify a signature.

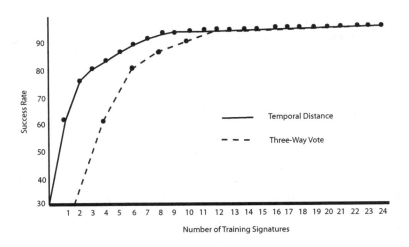

Fig. 6.8. Results for the voting technique. A three method vote, simple majority, compared against the simple temporal distance algorithm.

The global relative distance method works as well as many published algorithms, and has the advantage of being quite simple and obvious. An accelerated version is fast enough for actual use, but and has a slightly lower success rate.

The temporal distance method is very successful, giving a solo success rate of 98.6%. It is fast and easy to compute. It can be performed directly on point data without the need to create an image, and the exemplars can be saved as feature vectors, speeding up the recognition process.

A simple direct comparison between signatures is possible, and has a high success rate. It has been combined with other methods by using simple voting to improve the success rate to over 99%.

Acknowledgments

This work was partly supported by the Natural Science and Engineering Research Council of Canada (NSERC). I would like to thank Mark Baumback and Lani Bateman, former students who contributed significantly to the work. Thanks also to Dr. Marina Gavrilova and Dr. Svetlana Yanushkevich.

Bibliography

1. Ammar, M. Yoshida, Y. and Fukumura, T. (1988). Off-line preprocessing and verification of signatures, *Int. J. Pattern Recognition and Artificial Intelligence*, **2**, 4, pp. 589–602.
2. Bajaj, R. and Chaudhury, S. (1997). Signature verification using multiple neural classifiers, *Pattern Recognition*, **30**, 1, pp. 1–7.
3. Baraghimian, G. A. and Klinger, A. (1990). Preference voting for sensor fusion, *SPIE Sensor Fusion III*, Orlando, Fl. April pp. 19–20.
4. Black, D. (1958). *The Theory of Committees and Elections*, Cambridge University Press, 1958.
5. Jean-Charles de Borda, (1781). *Memoire sur les Elections au Scrutin* Histoire de l'Academie Royale des Sciences, Paris.
6. Marquis de Condorcet, (1785). *Essai sur l'application de l'analyse a la probabilite des decisions rendues a la pluralite des voix (Essay on the Application of Analysis to the Probability of Majority Decisions)* Paris.
7. Fishburn, P. (1996). Preference structures and their numerical representations, *Proc. ORDAL'96*, Ottawa, Aug 5–9.
8. Huang, K. and Yan, H. (1995). On-line signature verification based on dynamic segmentation and global and local matching, *Optical Engineering*, **34**, 12, pp. 3480–3487.
9. Kimura, F. and Shridhar, M. (1991). Handwritten numeral recognition based on multiple algorithms, *Pattern Recognition*, **24**, 10, pp 969–983.
10. Kittler, J., Hatef, M. and Duin, R. P. (1996). Combining classifiers, *Proc. ICPR*, pp. 897–901.
11. LeClerc, F. and Plamondon, R. (1994). Automatic signature verification: The state of the art, *Int. J. Pattern Recognition*, **8**, 3, pp. 643–660.
12. Munich, M. E. and Perona, P. (2002). Visual input for pen based computers, *IEEE Trans. Pattern Analysis and Machine Intelligence*, **24,3**, pp. 313–328.
13. Munich, M. E. and Perona, P. (2000). Apparatus and method for tracking handwriting from visual input, *US Patent* 6,044,165, filed 6/15/1995, granted 3/28/2000.
14. Parker, J. R. (1988). A faster method for erosion and dilation of reservoir pore complex images, *Canadian J. Earth Sciences*, July.
15. Parker, J. R. (1996). Voting methods for multiple autonomous agents, *Proc. ANZIIS'96*, Perth, Australia.
16. Parker, J. R. (1998). Histogram methods for character recognition, *SPIE Vision Geometry III*, San Diego.
17. Parker, J. R. (2002). Simple distances between handwritten signatures, *Proc. Conf. Vision Interface*, Calgary, Alberta, May 27–29.
18. Parker, J. R., Bateman, L. and Baumback, M. (2002). Scale effects in shadow masks for signature verification, *Proc. Int. Conf. Artificial Intelligence*, Las Vegas, USA, June 24–27.

19. Riba, J., Carnicer, A., Vallmitjana, S. and Juvells, I. (2000). Methods for invariant signature classification, *Proc. ICPR,* Barcelona, Spain. pp. 957–960.

20. Sabourin, R., Cheriet, M. and Genest, G. (1993). An extended shadow code based approach for off-line signature verification, *Proc. ICDAR'93,* Tsukuba, Japan, pp. 1–5.

21. Sabourin, R. and Genest, G. (1994). An extended shadow code based approach for off-line signature verification: Part I - evaluation of the bar mask definition, *Proc. 12th ICPR,* Jerusalem, Israel, Oct 9–13, pp. 450–453.

22. Srihari, S. N. (1993). Recognition of handwritten and machine-printed text for postal address interpretation, *Pattern Recognition Letters,* **14**, pp 291–302.

23. Straffin, P. D., Jr. (1980). *Topics in the Theory of Voting,* Birkhauser, Boston.

24. Tumer, K. and Ghosh, J. (1998). Classifier combining through trimmed means and order statistics, *Proc. Int. Joint Conf. Neural Networks,* pp. 757–762, May, Anchorage, AL.

25. Tumer, K. and Ghosh, J. (1995). Order statistics combiners for neural classifiers, *Proc. World Congress on Neural Networks,* July, Washington, DC, pp. I:31–34.

26. Wilkinson, T. and Goodman, J. (1990). Slope histogram detection of forged signatures, *Proc. SPIE Conf. High Speed Inspection, Barcoding and Character Recognition,* **1384**, pp. 293–304.

27. Yanushkevich, S. N., Stoica, A., Shmerko, V. P. and Popel, D. V. (2005). *Biometric Inverse Problems,* CRC Press/Taylor & Francis Group, Boca Raton, FL.

28. Yanushkevich, S. N., Stoica, A. and Shmerko, V. P. (2006). Fundamentals of Biometric-Based Training System Design, this issue.

Chapter 7

Force Field Feature Extraction for Ear Biometrics

David Hurley

Analytical Engines Ltd.,
61 Meon Crescent, Chandlers Ford,
Eastleigh, Hampshire, SO53 2PA, UK
djh@analyticalengines.co.uk

The overall objective in defining feature space is to reduce the dimensionality of the original pattern space, whilst maintaining discriminatory power for classification. To meet this objective in the context of ear biometrics a new force field transformation is presented which treats the image as an array of mutually attracting particles that act as the source of a Gaussian force field. Underlying the force field there is a scalar potential energy field, which in the case of an ear takes the form of a smooth surface that resembles a small mountain with a number of peaks joined by ridges. The peaks correspond to potential energy wells and to extend the analogy the ridges correspond to potential energy channels. Since the transform also turns out to be invertible, and since the surface is otherwise smooth, information theory suggests that much of the information is transferred to these features, thus confirming their efficacy.

We describe how field line feature extraction, using an algorithm similar to gradient descent, exploits the directional properties of the force field to automatically locate these channels and wells, which then form the basis of the characteristic ear features. We also show how an analysis of this algorithm leads to a separate closed analytical description based on the divergence of force direction.

The technique is validated by performing recognition on a database of ears selected from the XM2VTS face database, and by comparing the results with the more established technique of Principal Components Analysis (PCA). This confirms not only that ears do indeed appear to have potential as a biometric, but also that the new approach is well suited to their description.

Contents

Glossary

CCR — Correct Classification Rated

FFE — Force Field feature Extraction

Log — Laplacian of Gaussian operator

PCA — Principal Components Analysis

7.1. Introduction

The potential of the human ear for personal identification was recognized and advocated as long ago as 1890 by the French criminologist Alphonse Bertillon [1]. In machine vision, ear biometrics has received scant attention compared to the more popular techniques such as automatic face, eye, and fingerprint recognition [2,3]. However, ears have played a significant role in forensic science for many years, especially in the United States, where an ear classification system based on manual measurements has been developed by Iannarelli, and has been in use for more than 40 years [4], although the safety of ear-print evidence has recently been challenged in the Courts. [5]

Ears have certain advantages over the more established biometrics; as Bertillon pointed out:

> *"The ear, thanks to these multiple small valleys and hills which furrow across it, is the most significant factor from the point of view of identification. Immutable in its form since birth, resistant to the influences of environment and education, this organ remains, during the*

entire life, like the intangible legacy of heredity and of the intra-uterine life."

The ear does not suffer from changes in facial expression and is firmly fixed in the middle of the side of the head so that the immediate background is predictable whereas face recognition usually requires the face to be captured against a controlled background.

Collection does not have an associated hygiene issue, as may be the case with direct contact fingerprint scanning, and is not likely to cause anxiety, as may happen with iris and retina measurements. The ear is large compared with the iris, retina, and fingerprint and therefore is more easily captured.

The technique provides a robust and reliable description of the ear without the need for explicit ear extraction. It has two distinct stages:

(a) Image to force field Transformation, and

(b) force field Feature Extraction.

Firstly, the entire image is transformed into a force field by pretending that each pixel exerts a force on all the other pixels, which is proportional to the pixel's intensity and inversely proportional to the square of the distance to each of the other pixels. It turns out that treating the image in this way is equivalent to passing it through an extremely powerful low pass filter which transforms it into a smooth undulating surface, but with the interesting property that the new surface retains all the original information.

Operating in the force field domain allows us to access to a wealth of established vector calculus techniques to extract information about this surface. For example, we will show how the divergence operator may be used to extend this algorithmic approach leading to an even richer mathematical description.

The powerful smoothing also affords valuable resistance to noise and surface matching is also greatly facilitated when the surfaces are smooth. Also, because it is based on a natural force field, there is the prospect of implementing the transform in silicon hardware by mapping the image to an array of electric charges.

The smooth surface corresponds to the potential energy field underlying the vector force field and the directional properties of the force field can be exploited to automatically locate a small number of potential wells and channels which correspond to local energy peaks and ridges respectively which then form the basis of the new features.

Information theory would also suggest that these are good features as follows. A smooth surface is easier to describe than a complicated one and therefore requires less information; in the extreme, a hemisphere can be described just by its radius. At the other extreme, a surface made up of random noise would be very hard to describe and therefore would require much more information. The total image information is conserved in the transformation, so we argue that there must be a redistribution of information density away from the smoothest parts towards the salient channels and wells which break up the monotony of the otherwise smooth surface.

It is worth noting that the force field is derived directly from the entire image without any pre-processing whatsoever; we effectively smooth the entire image with a gigantic $1/r$ kernel, which is more than four times the area of the original image, thus obliterating all fine detail such as edges, yet preserving all the image information. The target application is ear biometrics where we would expect measurement conditions such as diffuse-lighting and viewpoint to be carefully controlled and we shall assume that the subject will be looking directly ahead, sideways to the camera view.

The resulting camera projection will be taken as the subject's ear image for biometric purposes, even if the ears should protrude excessively. We do not seek pinna-plane tilt invariance simply because there is no reason why the tilted version of a given pinna should not correspond to another subject's ear.

A selection of samples taken from each of sixty-three subjects drawn from the XM2VTS face profiles database [11] has been used to test the viability of the technique. A classification rate of 99.2% has been achieved so far, using just simple template matching on the basic channel shapes, demonstrating the merit of the technique at least at this scale.

It can reasonably be expected that the use of more sophisticated channel matching techniques would lead to even better results. We also do a head-to-head comparison with PCA which also helps to demonstrate the implicit extraction inherent in the technique. As such, a new low-level feature extraction approach is demonstrated with success in a new application domain.

7.2. Ear Topology

The ear does not have a completely random structure, it is made up of standard features just like the face. The parts of the ear are less familiar than the eyes, nose, mouth, and other facial features but nevertheless are always present in a normal ear. These include the outer rim (helix), the ridge (antihelix) that runs inside and parallel to the helix, the lobe, and the distinctive u-shaped notch known as the intertragic notch between the ear hole (meatus) and the lobe. Figure 7.1 shows the locations of the anatomical features in detail.

"The ear does not have a completely random structure, it is made up of standard features just like the face."

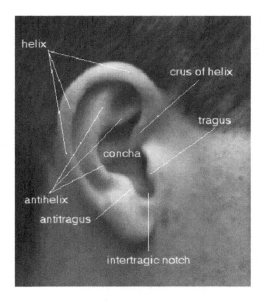

Fig. 7.1. Topology of the human ear. Just like the face, the ear has standard constituent features consisting of the outer and inner helices, the concha — named for its shell-like appearance, the inter-tragic notch and of course the familiar ear-lobe.

7.3. The Force Field Transforms

This section gives a brief description of the mathematical foundations of the new transforms. The basic concepts used can be found in various introductory works on physics [12,13] and electromagnetics. [14,15] We consider how faster computation can be achieved by applying the convolution theorem in the frequency domain. The question of transform invertibility is considered as this establishes that the transforms are information preserving. Further details of invariance, including initialization invariance, scale invariance, and noise tolerance can be found in [6,7,8,9].

7.3.1. *Transformation of the Image to a Force Field*

The image is transformed to a force field by treating the pixels as an array of mutually attracting particles that act as the source of a Gaussian force field, rather like Newton's *Law of Universal Gravitation* where, for example, the moon and earth attract each other as shown in Fig. 7.2.

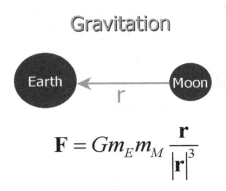

Fig. 7.2.   Newton's Universal law of Gravitation. The earth and the moon are mutually attracted according to the product of their masses m_E and m_M respectively, and inversely proportional to the square of the distance between them. G is the gravitational constant of proportionality.

We use Gaussian force as a generalization of the inverse square laws which govern the gravitational, electrostatic, and magnetostatic force fields, to discourage the notion that any of these forces are in play; the laws governing these forces can all be deduced from Gauss's Law, itself a consequence of the inverse square nature of the forces.

So, purely as an invention, The pixels are considered to attract each other according to the product of their intensities and inversely to the square of the distances between them.

Each pixel is assumed to generate a spherically symmetrical force field so that the total force $\mathbf{F}(\mathbf{r}_j)$ exerted on a pixel of unit intensity at the pixel location with position vector $\mathbf{r}_j$ by a remote pixels with position vector $\mathbf{r}_i$

and pixel intensities $P(\mathbf{r}_i)$ is given by the vector summation,

$$\mathbf{F}(\mathbf{r}_j) = \sum_i \begin{cases} P(\mathbf{r}_i)\frac{\mathbf{r}_i - \mathbf{r}_j}{|\mathbf{r}_i - \mathbf{r}_j|}, & \forall\, i \neq j; \\ 0, & \forall\, i = j. \end{cases} \qquad (7.1)$$

In order to calculate the force field for the entire image, this equation should be applied at every pixel position in the image. Units of pixel intensity, force, and distance are arbitrary, as are the co-ordinates of the origin of the vector field.

Figure 7.3 shows an example of the calculation of the force at one pixel location for a simple 9-pixel image. We see that the total force is the vector sum of 8 forces. It is not just the vector sum of the forces exerted by the neighbouring forces, it is the sum of the forces exerted by all the other pixels. For an n-pixel image there would be $n-1$ forces in the summation.

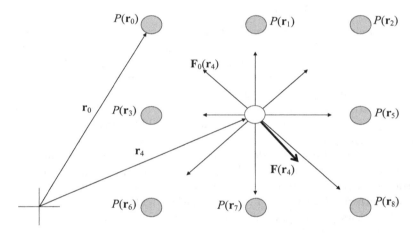

Fig. 7.3. In this simple illustration the force field at the center of a 9-pixel image is calculated by substituting the center pixel with a unit value test pixel and summing the forces exerted on it by the other 8 pixels. In reality, this would not just involve 8 other pixels but hundreds or even thousands of other pixels.

The defining equations could be applied directly, but in practice for greater efficiency, the process can be treated as a convolution of the image with the force field corresponding to a unit value test pixel, and then invoking the *Convolution Theorem* to perform the calculation as a multiplication in the frequency domain, the result of which is then transformed back into the spatial domain. The force field equation for an $M \times N$ pixel image becomes,

$$\texttt{Force field} = \sqrt{MN}\{\Im^{-1}[\Im(\texttt{unit force field}) \times \Im(\texttt{image})]\} \qquad (7.2)$$

where $\Im$ stands for the Fourier transform and $\Im^{-1}$ for its inverse. This applies equally to the energy field which we will presently describe. The usual care must be taken to ensure that dimensions of the unit sample force field are twice those of the image dimensions and that sufficient zero padding is used to avoid aliasing effects.

The code for this calculation in *Mathcad* is shown in Appendix Fig. A.1, and it is hoped that users of other languages will easily be able to convert this to their own requirements.

7.3.2. *The Energy Transform for an Ear Image*

There is a scalar potential energy field associated with the vector force field where the two fields are related by the well-known equation [14,15],

$$\mathbf{F}(\mathbf{r}) = -grad(E(\mathbf{r})) = \nabla E(\mathbf{r}) \qquad (7.3)$$

This equation tells us that the force at a given point is equal to the additive inverse of the gradient of the potential energy field at that point. This simple relationship allows the force field to be easily calculated by differentiating the energy field and allows some conclusions drawn about one field to be extended to the other.

We can restate the force field formulation in energy terms to derive the energy field equations directly as follows. The image is transformed by treating the pixels as an array of particles that act as the source of a Gaussian potential energy field. It is assumed that there is a spherically symmetrical potential energy field generated by each pixel, so that $E(\mathbf{r}_j)$ is the total potential energy imparted to a pixel of unit intensity at the pixel location with position vector $\mathbf{r}_j$ by the energy fields of remote pixels with position vectors $\mathbf{r}_i$ and pixel intensities $P(\mathbf{r}_i)$, and is given by the scalar summation,

$$E(\mathbf{r}_j) = \sum_i \begin{cases} \frac{P(\mathbf{r}_i)}{|\mathbf{r}_i - \mathbf{r}_j|}, & \forall\, i \neq j; \\ 0, & \forall\, i = j. \end{cases} \qquad (7.4)$$

where the units of pixel intensity, energy, and distance are arbitrary, as are the co-ordinates of the origin of the field. Figure 7.4 show the scalar potential energy field of an isolated test pixel.

To calculate the energy field for the entire image Eq. (7.4) should be applied at every pixel position. The result of this process for the energy transform for an ear image is shown in Fig. 7.5, where the same surface has been depicted from a variety of different perspectives below the lobe.

The potential surface undulates, forming local peaks or maxima, with ridges leading into them. These peaks we call *potential energy wells* since,

Fig. 7.4. Potential function for an isolated test pixel. The energy field for the entire image is obtained by locating one of these potential functions at each pixel location and scaling it by the value of the pixel, and then finding the sum of all the resulting functions. In practice, this is done by exploiting the Fourier Transform and the Convolution Theorem.

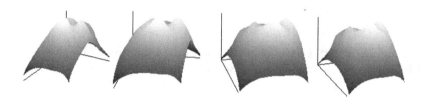

Fig. 7.5. Energy surface for an ear viewed from below the lobe. Notice the peaks corresponding to potential energy wells, and the ridges leading into them corresponding to potential energy channels. At least three wells are clearly visible.

by way of analogy, if the surface were to be inverted and water poured over it, the peaks would correspond to small pockets where water would collect.

Notice that the highest of the three obvious peaks in Fig. 7.5 has a ridge that slopes gently towards it from the smaller peak to its left. This corresponds to a potential energy channel, because to extend the analogy, water that happened to find its way into its inverted form would gradually flow along the channel towards the peak.

The reason for the dome shape of the energy surface can be easily understood by considering the case where the image has just one gray level throughout. In this situation, the energy field at the center would have the greatest share of energy because test pixels at that position would have the shortest average distance between themselves and all the other pixels in the image, whereas test pixels at the corners would have the greatest average distance to all the other pixels, and therefore the least total energy imparted to them.

7.3.3. *An Invertible Linear Transform*

The transformation is linear since the energy field is derived purely by summation which is itself a linear operation. What is less obvious is that the transform is also invertible. For an N-pixel image, the application of Eq. (7.4) at each of the N pixel positions leads to a system of N equations in N unknowns. Now if the N equations are linearly independent, then it follows that the system of equations can be solved for the pixel values, given the energy values. In other words, the transform would be invertible, and the original image could be completely recovered from the energy surface, thus establishing that the transform preserves information. This system of N equations can be expressed as a matrix multiplication of an $N \times 1$ vector of pixel intensities by an $N \times N$ square matrix of coefficients d_{ji} corresponding to the inverse distance scalars given by,

$$d_{ji} = \frac{1}{|\,\mathbf{r}_j - \mathbf{r}_i\,|} \tag{7.5}$$

producing an $N \times 1$ vector of pixel energies. Equation (7.6) shows this multiplication for a simple 2×2 pixel image.

$$\begin{bmatrix} 0 & d_{01} & d_{02} & d_{03} \\ d_{10} & 0 & d_{12} & d_{13} \\ d_{20} & d_{21} & 0 & d_{23} \\ d_{30} & d_{31} & d_{32} & 0 \end{bmatrix} \begin{bmatrix} P(\mathbf{r}_0) \\ P(\mathbf{r}_1) \\ P(\mathbf{r}_2) \\ P(\mathbf{r}_3) \end{bmatrix} = \begin{bmatrix} E(\mathbf{r}_0) \\ E(\mathbf{r}_1) \\ E(\mathbf{r}_2) \\ E(\mathbf{r}_3) \end{bmatrix} \tag{7.6}$$

All the determinants of matrices corresponding to the sequence of square images ranging from 2×2 pixels to 33×33 pixels have been computed and have been found to be non-zero. It has also been verified that all non-square image formats up to 7×8 pixels have associated non-singular matrices.[9] Notwithstanding questions of machine accuracy, these results suggest that the energy transform is indeed invertible for most image sizes and aspect ratios.

7.4. Force Field Feature Extraction

In this section field line feature extraction is first presented followed by the analytic form of convergence feature extraction. The striking resemblance of convergence to the Marr-Hildreth operator [18] is illustrated and the differences highlighted, especially the nonlinearity of the convergence operator. We also investigate how the features are affected by the combination of the unusual dome shape and changes in image brightness. The close correspondence between the field line and convergence techniques is demonstrated by superimposing their results for an ear.

7.4.1. *Field Line Feature Extraction*

The concept of a unit value exploratory test pixel is exploited to assist with the description field lines. This idea is borrowed from physics, where it is customary to refer to unit value test particles when describing force fields associated with gravitational masses and electrostatic charges. When such notional test pixels are placed in a force field and allowed to follow the field direction their trajectories are said to form field lines. When this process is carried out with many different starting points a set of field lines will be generated that capture the general flow of the force field.

Figure 7.6 demonstrates the field line approach to feature extraction for an ear image, by means of a "film-strip" consisting of 12 images depicting the evolution of field lines, where each image represents 10 iterations of evolution. The evolution proceeds from top left to bottom right. We see that in the top left image a set of 40 test pixels is arranged in an ellipse shaped array around the ear and allowed to follow the field direction so that their trajectories form field lines describing the flow of the force field.

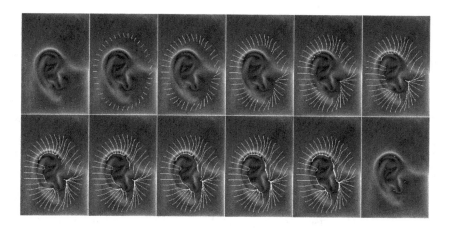

Fig. 7.6. Field line, channel, and well formation for an ear. Field line evolution is depicted as a "film-strip" of 12 images, each depicting 10 iterations of evolution. The top left image shows where 40 test-pixels have been initialized, and the bottom right image shows where 4 wells have been extracted.

The test pixel positions are advanced in increments of one pixel width, and test pixel locations are maintained as real numbers, producing a smoother trajectory than if they were constrained to occupy exact pixel grid locations.

Notice how ten field lines cross the upper ear rim and how each line joins a common channel that follows the curvature of the ear rim rightwards finally terminating in a potential well. The well locations have been extracted by observing clustering of test pixel co-ordinates so that the bottom right image is simply obtained by plotting the terminal positions of all the co-ordinates.

7.4.2. *Dome Shape and Brightness Sensitivity*

As stated in the introduction, we do not seek viewpoint invariance, or illumination invariance, either in intensity or direction, because we have assumed carefully controlled measurement conditions. However, it is still interesting to investigate how the position of features will be affected by the combination of the unusual dome shape and changes in image brightness. The effect will first be analysed and then confirmed by experiment.

Should the individual pixel intensity be scaled by a factor a and also have and an additive intensity component b, we would have,

$$E(\mathbf{r}_j) = \sum_i \frac{aP(\mathbf{r}_i) + b}{|\mathbf{r}_i - \mathbf{r}_j|} = a \sum_i \frac{P(\mathbf{r}_i)}{|\mathbf{r}_i - \mathbf{r}_j|} + \sum_i \frac{b}{|\mathbf{r}_i - \mathbf{r}_j|} \qquad (7.7)$$

We see that:

(a) Scaling the pixel intensity by the factor a merely scales the energy intensity by the same factor a.

(b) Adding an offset b is more troublesome, effectively adding a pure dome component corresponding to an image with constant pixel intensity b. This could be corrected by subtracting the dome component, if b can be estimated.

The effect of the offset and scaling is shown in Fig. 7.7 with the channels superimposed. We see that scaling by a factor of 10 in (e) has no effect as expected.

The original image in (a) has a mean value of 77 and a standard deviation of 47. Images (b) to (d) show the effect of progressively adding offsets of one standard deviation. At one standard deviation the effect is hardly noticeable and even at 3 standard deviations the change is by no means catastrophic as the channel structure alters little.

We therefore conclude that operational lighting variation in a controlled biometrics environment will have little effect. These conclusions are borne out by the results of the corresponding recognition experiments in Table 7.1.

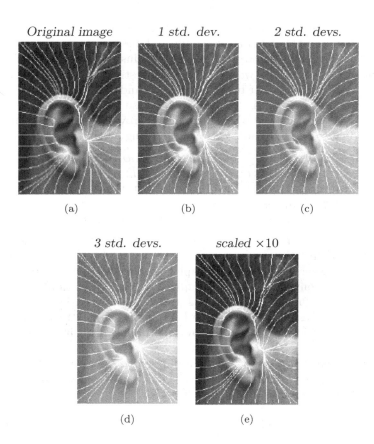

Fig. 7.7. Effect of additive and multiplicative brightness changes.The original image in (a) has a mean value of 77 and a standard deviation of 47. Images (b) to (d) show the effect of progressively adding offsets of one standard deviation. We see that the channel structure hardly alters and we therefore conclude that operational lighting variation in a controlled biometrics environment will have little effect.

7.4.3. *Convergence Feature Extraction*

Here we introduce the analytical method of feature extraction as opposed to the field line method. This method came about as a result of analyzing in detail the mechanism of field line feature extraction. As shown in Fig. 7.9(c), when the arrows usually used to depict a force field are replaced with unit magnitude arrows, thus modeling the directional behavior of exploratory test pixels, it becomes apparent that channels and wells arise as a result of patterns of arrows converging towards each other, at the interfaces between regions of almost uniform force direction.

As the divergence operator of vector calculus measures precisely the opposite of this effect, it was natural to investigate the nature of any relationship that might exist between channels and wells and this operator. This resulted not only in the discovery of a close correspondence between the two, but also showed that divergence provided extra information corresponding to the interfaces between diverging arrows.

> *"Convergence provides a more general description of channels and wells in the form of a mathematical function in which wells and channels are revealed to be peaks and ridges respectively in the function value."*

The concept of the divergence of a vector field will first be explained, and then used to define the new function. Convergence is compared with the *Marr-Hildreth* operator which is a Laplacian operator and the important difference that convergence is not Laplacian, due to its nonlinearity, is illustrated. The function's properties are then analyzed in some detail, and the close correspondence between field line feature extraction and the convergence technique is illustrated by superimposing their results for an ear image.

The divergence of a vector field is a differential operator that produces a scalar field representing the net outward flux density at each point in the field. For the vector force field it is defined as,

$$\text{div}\mathbf{F}\left(\mathbf{r}\right) = \lim_{\Delta V \to 0} \frac{\oint \mathbf{F}(\mathbf{r}) \cdot d\mathbf{S}}{\Delta V} \tag{7.8}$$

where $d\mathbf{S}$ is the outward normal to a closed surface S enclosing an incremental volume ΔV. In two-dimensional Cartesian coordinates it may be expressed as follows [14,15].

$$\text{div}\mathbf{F}\left(\mathbf{r}\right) = \nabla \cdot \mathbf{F}(\mathbf{r}) = \left(\frac{\partial F_x}{\partial x} + \frac{\partial F_y}{\partial y} \right) \tag{7.9}$$

where F_x and F_y are the Cartesian components of $\mathbf{F}$. Figure 7.8 illustrates the concept of divergence graphically. In Fig. 7.8(a) we see an example of positive divergence where the arrows flow outwards from the center, and in Fig. 7.8(b) we see negative divergence, where the arrows flow inwards, whereas in Fig. 7.8(c) there is no divergence because all the arrows are parallel.

Having described divergence we may now use it to define convergence feature extraction. Convergence provides a more general description of channels and wells in the form of a mathematical function in which wells

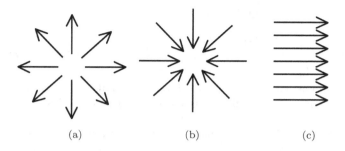

(a)　　　　　　　　(b)　　　　　　　　(c)

Fig. 7.8.   Divergence of a vector field: (a) shows positive divergence where the arrows point outwards, (b) shows negative divergence where the arrows point inwards, and (c) shows zero divergence where the arrows are parallel. (c) could actually exhibit divergence if the strength of the field varied even though the direction is constant.

and channels are revealed to be peaks and ridges respectively in the function value. The new function maps the force field to a scalar field, taking the force as input and returning the additive inverse of the divergence of the force direction. The function will be referred to as the force direction convergence field $C(\mathbf{r})$ or just convergence for brevity. A more formal definition is given by

$$
\begin{aligned}
C(\mathbf{r}) &= -\mathrm{div}\ \mathbf{f}(\mathbf{r}) \\
&= -\lim_{\Delta A \to 0} \frac{\oint \mathbf{f}(\mathbf{r}) \cdot d\mathbf{I}}{\Delta A} \\
&= -\nabla \cdot \mathbf{f}(\mathbf{r}) \\
&= -\left(\frac{\partial f_x}{\partial x} + \frac{\partial f_y}{\partial y} \right)
\end{aligned}
\tag{7.10}
$$

where $\mathbf{f}(\mathbf{r}) = \frac{\mathbf{F}(\mathbf{r})}{|\mathbf{F}(\mathbf{r})|}$, ΔA is incremental area, and $d\mathbf{I}$ is its boundary outward normal.

This function is real valued and takes negative values as well as positive ones where negative values correspond to force direction divergence. It is interesting to compare this function with the *Marr-Hildreth* operator given by,

$$
\begin{aligned}
MH(\mathbf{r}) &= \mathrm{div}\ \mathbf{g}(\mathbf{r}) \\
&= \nabla \cdot \mathbf{g}(\mathbf{r}) \\
&= \left(\frac{\partial g_x}{\partial x} + \frac{\partial g_y}{\partial y} \right)
\end{aligned}
\tag{7.11}
$$

where

$$\mathbf{g(r)} = \mathrm{grad}(G(\mathbf{r}) * I(\mathbf{r})),$$

$I(\mathbf{r})$ is the image and $G(\mathbf{r}) = e^{\frac{|\mathbf{r}|^2}{2\sigma^2}}$ is a Gaussian kernel.

The *Marr-Hildreth* operator, also known as *Laplacian of Gaussian* (LoG), uses a Gaussian kernel which is optimal in Gaussian noise, whereas the $1/r$ kernel is an artifact of its force field nature, and the intrinsic smoothing it affords is merely a fortunate consequence of its formulation.

We must also stress that whilst the *Marr-Hidlreth* operator is linear, convergence is non-linear because it is based on force direction rather than force. This nonlinearity means that we are obliged to perform the operations in the order shown; we cannot take the divergence of the force and then divide by the force magnitude: div (grad/ | grad |) $\neq$ (div grad)/ | grad |. This is easily illustrated by a simple example using the scalar field e^x:

$$\left\{ \begin{array}{l} \mathrm{div}(\mathrm{grad}/|\mathrm{grad}|) \\ \nabla \cdot \left(\frac{\nabla e^x}{|\nabla e^x|} \right) = \nabla \cdot \frac{e^x \mathbf{i}}{e^x} = \nabla \cdot \mathbf{i} = 0 \end{array} \right\} \neq \left\{ \begin{array}{l} (\mathrm{divgrad})/|\mathrm{grad}| \\ \frac{\nabla \cdot \nabla e^x}{|\nabla e^x|} = \frac{e^x}{e^x} = 1 \end{array} \right\} \quad (7.12)$$

where $\mathbf{i}$ is a unit vector in the x direction. The convergence is zero because we have a field of parallel unit magnitude vectors, whereas in the second case the vectors are parallel but the magnitude changes, resulting in a net outflow of flux at any point. This illustrates that even though convergence looks very much like a Laplacian operator, it definitely is not.

Figure 7.9(b) shows the convergence field for an ear image, while Fig. 7.9(a) shows the corresponding field lines. A magnified version of a small section of the force direction field, depicted by a small rectangular insert in Fig. 7.9(a), is shown in Fig. 7.9(c). In Fig. 7.9(b) the convergence values have been adjusted to fall within the range 0 to 255, so that negative convergence values corresponding to *antichannels* appear as dark bands, and positive values corresponding to channels appear as white bands. Notice that the antichannels are dominated by the channels, and that the antichannels tend to lie within the confines of the channels. Also, notice how wells appear as bright white spots.

> "*The convergence map provides more information than the field lines, in the form of negative versions of wells and channels*"

The correspondence between the convergence function and the field line features can be seen by observing the patterns in the force direction field shown in Fig. 7.9(c). Notice the correspondence between the converging arrows and white ridges, and between the diverging arrows and black ridges.

The convergence map provides more information than the field lines, in the form of negative versions of wells and channels or *antiwells* and

Field line features   Convergence map   Magnified insert force field

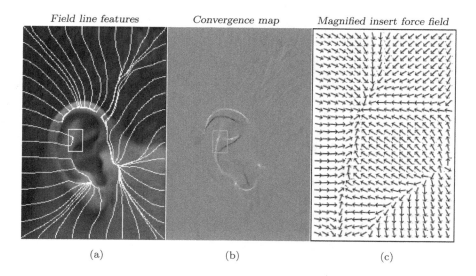

(a)   (b)   (c)

Fig. 7.9. Convergence field for an ear. (b) shows the convergence field whilst (a) shows the corresponding field line feature extraction. (c) depicts a magnified portion of the force field corresponding to the little rectangular insert show in (a).

antichannels. Although it should be possible to modify the field line technique to extract this extra information by seeding test pixels on a regular grid and reversing the direction of test pixel propagation.

Figure 7.10 shows the convergence field of an ear image with the corresponding field lines superimposed. Figure 7.10(a) is the field line map, and Fig. 7.10(b) is the convergence map, while Fig. 7.10(c) is the superposition of one on the other. We can see clearly how channels coincide with white ridges in the convergence map and that potential wells coincide with the convergence peaks. Notice the extra information in the center of the convergence map that is not in the field line map, illustrating an advantage of convergence over field lines.

7.5. Ear Recognition

The technique was validated on a set of 252 ear images taken from 63 subjects selected from the XM2VTS face database [11] achieving almost total correct recognition.

Multiplicative template matching of ternary thresholded convergence maps was used where levels less than minus one standard deviation are mapped to -1, whilst those greater than one standard deviation map to

Field lines   Convergence map   Superposition

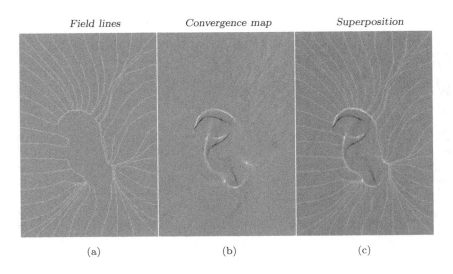

(a)   (b)   (c)

Fig. 7.10.   Correspondence between channels and convergence. (a) shows the field line map whilst (b) shows the convergence map and (c) shows the superposition of these two maps where we can see clearly that the channels coinccide with the ridges in the convergence map.

+1, and those remaining map to 0. A threshold level of one standard deviation was chosen experimentally resulting in the template channel thickness shown in Fig. 7.11. This figure also shows a rectangular exclusion zone centered on the convergence magnitude centroid; the centroid of the convergence tends to be stable with respect to the ear features and this approach has the added advantage of removing unwanted outliers such as bright spots caused by spectacles. The size of the rectangle was chosen as 71 × 51 pixels by adjusting its proportions to give a good fit for the majority of the convergence maps. Notice how for image 000-2 which is slightly lower than the other three, that the centroid-centered rectangle has correctly tracked the template downwards.

The inherent automatic extraction advantage was demonstrated by deliberately not accurately extracting or registering the ears in the sense that the database consists of 141 × 101 pixel images where the ears have only an average size of 111 × 73 and are only roughly located by eye in the center of these images. This can be seen clearly in the top row of Fig. 7.11 where we see a marked variation both in vertical and horizontal ear-location, and also that there is a generous margin surrounding the ears.

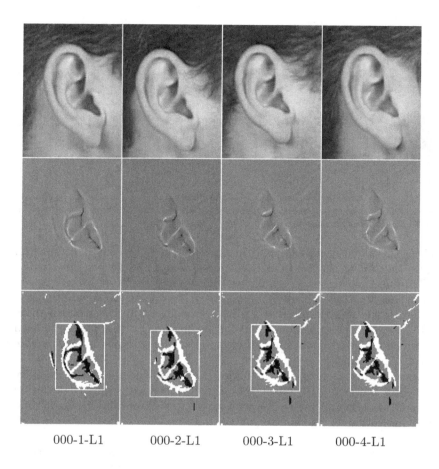

000-1-L1   000-2-L1   000-3-L1   000-4-L1

Fig. 7.11.   Feature extraction for subject 000 at 141 × 101 pixels. The top row shows the four sample ear images for the subject, whilst the middle row shows the corresponding convergence maps. The bottom row shows the maps after ternary thresholding and also depicts a rectangle centered on the convergence centroid. Notice that the ear image for 000-2 is lower than the others but that the rectangle has correctly tracked it downwards.

7.5.1.   *Experiment Results and Analysis*

The force field technique gives a correct classification rate of 99.2% on this set. Running PCA[19] on the same set gives a result of only 62.4%, but when the ears are accurately extracted by cropping to the average ear size of 111 × 73, running PCA then gives a result of 98.4%, thus demonstrating the inherent extraction advantage. The first image of the four samples from

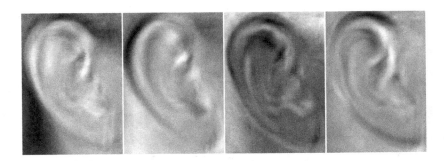

Fig. 7.12. This figure shows the first 4 of the 63 eigenvectors for the 111×73-pixel images. The first image of the four samples from each of the 63 subjects was used in forming the PCA covariance matrix.

each of the 63 subjects was used in forming the PCA covariance matrix. Figure 7.12 shows the first 4 eigenvectors for the 111×73-pixel images.

The effect of brightness change by addition was also tested where we see that in the worst case where every odd image is subjected to an addition of 3 standard deviations the force field results only change by 2%. whereas those for PCA under the same conditions fall by 36%, or by 16% for normalized intensity PCA, thus confirming that the technique is robust under variable lighting conditions.

These results are presented in Table 7.1. We also include the decidability index after Daugman [17] which combines the mean and standard deviation of the intra-class and inter-class measurement distributions giving a good single indication of the nature of the results. This index d' measures how well separated the distributions are, since recognition errors are caused by their overlap. The measure aims to give the highest scores to distributions with the widest separation between means, and smallest standard deviations. If the two means are μ_1 and μ_2 and the two standard deviations are σ_1 and σ_2 then d' is defined as

$$d' = \frac{\mid \mu_1 - \mu_2 \mid}{\sqrt{(\sigma_1^2 + \sigma_2^2)/2}} \qquad (7.13)$$

Notice that the best case index for PCA is slightly higher than the value of 3.43 obtained for the 141×101 images but this could be attributed to the reduction in data set size from 252 to 189 also to the fact that the images have been more fully extracted for PCA.

We have also included noise performance figures where noise has been modeled as additive noise with a zero mean Gaussian distribution. The signal to noise ratios of 6dB, 12dB, and18dB used are calculated as $20log_{10}(S/N)$. We see that the technique enjoys excellent noise tolerance where even for extreme noise of 6dB the performance only falls by about 3.6%. Interestingly at a ratio of 18dB the recognition rate actually improves over the noiseless recognition rate, but this must be put down to the combination of small changes and the random nature of the noise process. For reference we have also included the corresponding noise results for PCA under the same conditions, where we see that PCA also performs well under noisy conditions but not quite as well as FFE at 6dB where the fall is about 4.8%.

Table 7.1. Comparison of force field and PCA recognition results.

METHOD	PASSES	NOISE	CCR	BRIGHT	DECIDABILITY
141 x 101 image with deliberately poor extraction and registration					
FFE	250/252	Nil	99.2%	0	3.432
FFE	251/252	18 dB	99.6%	0	3.488
FFE	249/252	12 dB	98.8%	0	3.089
FFE	241/252	6 dB	95.6%	0	1.886
FFE	250/252	Nil	99.2%	1	3.384
FFE	247/252	Nil	98.0%	2	3.137
FFE	245/252	Nil	97.2%	3	2.846
PCA	118/189	Nil	62.4%	0	1.945
111 x 73 image with accurate extraction and extraction and registration					
PCA	186/189	Nil	98.4%	0	3.774
PCA	186/189	18 dB	98.4%	0	3.743
PCA	186/189	12 dB	98.4%	0	3.685
PCA	177/189	6 dB	93.6%	0	3.606
PCA	130/189	Nil	68.8%	1	1.694
PCA	120/189	Nil	63.6%	2	0.878
PCA	118/189	Nil	62.4%	3	0.476
PCA	181/189	Nil	95.6%	1*	3.171
PCA	172/189	Nil	91.0%	2*	1.910
PCA	166/189	Nil	82.5%	3*	1.140

Note: The table is divided between images which have deliberately poor registration in the upper half and those which have accurate registration in the lower half. The first column indicates whether force field feature extraction (FFE) or PCA extraction has been used. The second column gives the number of passes as a fraction of the total number of trials and this has been converted to a correct classification rated (CCR) in column four. An asterisk indicates that the PCA has been normalized with respect to image intensity.

7.6. Conclusions

In conclusion we may say that in the context of ear biometrics we have developed a new linear transform that transforms an ear image, with very powerful smoothing and without loss of information, into a smooth dome shaped surface whose special shape facilitates a new form of feature extraction that extracts the essential ear signature without the need for explicit ear extraction; and in the process we have also verified the recognition potential of the human ear for biometrics. We have also described the convergence operator and shown that it is a valuable alternative form of the field line feature extraction. We have validated the technique by experiment and in the process we have contributed to the mounting evidence that ears are indeed viable as a biometric. In our future work we will continue to promote the case for ears as a biometric and to extend our work in new directions. For example, this chapter has focused on ear biometrics from an analysis point of view, but we believe that ear synthesis could play an important role in future and some work has already begun in this direction.

Acknowledgments

The author would like to pay tribute to Dr. Vlad P. Shmerko for his generous help in preparing the LaTeX version of this manuscript and for his valuable advice and suggestions regarding its content.

Bibliography

1. Bertillon, A. (1890). La photographie judiciaire, avec un appendice sur la classification et l'identification anthropometriques, *Gauthier-Villars,* Paris.
2. Burge, M. and Burger, W. (2000). Ear Biometrics in Computer Vision, *Proc. ICPR*, pp. 822–826.
3. Chang, K., Bowyer, K. W., Sarkar, S. and Victor, B. (2003). Comparison and Combination of Ear and Face Images in Appearance-Based Biometrics, *IEEE Trans. PAMI*, **25**, 9, pp.1160–1165.
4. Iannarelli,A., (1989). *Ear Identification,* Paramount Publishing Company, Freemont, California.
5. STATE v. David Wayne KUNZE, (1999). Court of Appeals of Washington, Division 2. 97 Wash. App. 832, 988 P.2d 977.
6. Hurley, D. J., Nixon, M. S. and Carter, J. N. (2005). Force Field Feature Extraction for Ear Biometrics, *Computer Vision and Image Understanding* CVIU(98), No. 3, pp. 491–512.
7. Hurley, D. J., Nixon, M. S. and Carter, J. N., (1999). Force Field Energy

Functionals for Image Feature Extraction. *Proc. 10th British Machine Vision Conference,* pp. 604–613.

8. Hurley, D. J., Nixon, M. S. and Carter, J. N., (2002). Force Field Energy Functionals for Image Feature Extraction, *Image and Vision Computing,* 20, pp. 311–317.

9. Hurley, D. J., (2001). Force Field Feature Extraction for Ear Biometrics, PhD Thesis, *University of Southampton,* UK.

10. Hurley, D. J., Nixon, M. S. and Carter, J. N., (2000). A New Force Field Transform for Ear and Face Recognition, *Proc. IEEE Int. Conf. Image Processing,* pp. 25–28.

11. Messer, K., Matas, J., Kittler, J., Luettin, J., and Maitre, G., (1999). XM2VTSDB: The Extended M2VTS Database, *Proc. AVBPA'99* Washington, D.C.

12. Halliday, D., and Resnick, R., (1977). *Physics,* Part I, Wiley, New York, Third Edition.

13. Halliday, D., and Resnick, R., (1962). *Physics,* Part II, Wiley, New York.

14. Grant, I.S. and Phillips, W. R., (1990). *Electromagnetism,* Wiley, New York.

15. Sadiku, M. N. O., (1989). *Elements of Electromagnetics,* Saunders College Publishing, Second Ed.

16. Strang, G., (1988). *Linear Algebra and its Applications,* 3rd Ed. Saunders HBJ.

17. Daugman, J., (1999). Biometric decision landscapes, Technical Report TR482, *University of Cambridge Computer Laboratory.*

18. Marr, D., Hildreth, E., (1980). Theory of edge detection. *Proc. R. Sol.* London, B 207, pp. 187–217.

19. Turk, M., Pentland, A. (1991). Eigenfaces for Recognition, *J. Cognitive Neuroscience,* pp.71–86.

Appendix: Force and energy field computation in *Mathcad*

$$
\begin{aligned}
\text{ff(pic)} :=\ &sr \leftarrow 2\cdot(\text{rows(pic)}-1), sc \leftarrow 2\cdot(\text{cols(pic)}-1) \\
&r \leftarrow \text{rows(pic)}-1, c \leftarrow \text{cols(pic)}-1 \\
&\text{for } rr \in 0\,..\,sr \\
&\quad \text{for } cc \in 0\,..\,sc \\
&\qquad usr_{rr,cc} \leftarrow \frac{(r+c\cdot 1j)-(rr+cc\cdot 1j)+0j}{(|r+c\cdot 1j-(rr+cc\cdot 1j)|)^3} \\
&usr_{3\cdot\text{rows(pic)}-3,\,3\cdot\text{cols(pic)}-3} \leftarrow 0 \\
&pic_{3\cdot\text{rows(pic)}-3,\,3\cdot\text{cols(pic)}-3} \leftarrow 0 \\
&oup \leftarrow \sqrt{\text{rows(pic)}\cdot\text{cols(pic)}}\cdot\text{icfft}\left(\overrightarrow{(\text{cfft(usr)}\cdot\text{cfft(pic)})}\right) \\
&ff \leftarrow \text{submatrix}(oup, r, 2\cdot r, c, 2\cdot c)
\end{aligned}
$$

Fig. A.1. Mathcad code for the force field by convolution in which "j" denotes the complex operator and "cfft" and "icfft" denote the Fourier and inverse Fourier transforms, respectively, and "usr" denotes the unit sample response for a force field. You may need to type "1j" to tell Mathcad that "j" is not a Mathcad variable.

$$
\begin{aligned}
\text{ef(pic)} :=\ &sr \leftarrow 2\cdot(\text{rows(pic)}-1), sc \leftarrow 2\cdot(\text{cols(pic)}-1) \\
&r \leftarrow \text{rows(pic)}-1, c \leftarrow \text{cols(pic)}-1 \\
&\text{for } rr \in 0\,..\,sr \\
&\quad \text{for } cc \in 0\,..\,sc \\
&\qquad usr_{rr,cc} \leftarrow \frac{|(r+c\cdot 1j)-(rr+cc\cdot 1j)|+0j}{(|r+c\cdot 1j-(rr+cc\cdot 1j)|)^2} \\
&usr_{3\cdot\text{rows(pic)}-3,\,3\cdot\text{cols(pic)}-3} \leftarrow 0 \\
&pic_{3\cdot\text{rows(pic)}-3,\,3\cdot\text{cols(pic)}-3} \leftarrow 0 \\
&oup \leftarrow \sqrt{\text{rows(pic)}\cdot\text{cols(pic)}}\cdot\text{icfft}\left(\overrightarrow{(\text{cfft(usr)}\cdot\text{cfft(pic)})}\right) \\
&ef \leftarrow \text{submatrix}(oup, r, 2\cdot r, c, 2\cdot c)
\end{aligned}
$$

Fig. A.2. Mathcad code for the energy field by convolution. Note that the code fragment "usr" or unit sample response is the only difference from the code for the force field shown in Fig. A.1. The "usr" in this case corresponds to the potential energy function shown in Fig. 7.4.

Chapter 8

Nontensor-Product-Wavelet-Based Facial Feature Representation

Xinge You[*,†,§], Qiuhui Chen[*], Patrick Wang[¶], and Dan Zhang[*]

Faculty of Mathematics and Computer Science, Hubei University,
WuHan, HuBei, 430062, China

†*Department of Computer Science,*
Hong Kong Baptist University, Hong Kong

‡*Image Processing Group, College of Computer and Information Science,*
Northeastern University, Boston, MA 02115, USA

§*xyou@hubu.edu.cn*

¶*pwang@ccs.neu.edu*

Biometrics are automated methods of recognizing a person based on a physiological or behavioural characteristic. Among the features measured are: face, fingerprints, hand geometry, handwriting, iris, retinal, vein, and voice. Facial Recognition technologies are becoming one of the foundation of an extensive array of highly secure identification and personal verification solutions. Meanwhile, a smart environment is one that is able to identify people, interpret their actions, and react appropriately. one of the most important building blocks of smart environments is a person identification system. Face recognition devices are ideal for such systems, since they have recently become fast, cheap, unobtrusive, and, when combined with voice identification, are very robust against changes in the environment.

As with all biometrics, facial recognition follows the same four steps below: sample capture, feature extraction and representation, template comparison, and matching. Feature representation of facial image play a key role in facial recognition system. In other words, the performance of facial recognition technology is very closely tied to the quality of the representation of facial image. In this chapter, we develop a new method for facial feature representation by using a nontensor product bivariate wavelet transform. A new nontensor product bivariate wavelet filter banks with linear phase are constructed from the centrally symmetric matrices. Our investigations demonstrate that these filter banks have a matrix factorization and they are capable of describing the features of face image. The implementations of our algorithm are made of three parts: first, by perform 2-level wavelettransform with a new nontensor product wavelet filter, a face images are represented by the lowest resolution subbands after decomposition. Second, the Principal Component Analysis (PCA) feature selection scheme is adopted to reduce the computational complexity of feature representation. Finally, to test the robustness of the proposed facial feature representation, the Support Vector Machines (SVM) is applied for classification. The experimental results show that our method is superior to other methods in terms of recognition accuracy and efficiency.

Contents

Glossary

DNWT — Discrete Nontensor product Wavelet Transform

DSWT — Discrete tensor product Wavelet Transform

KFDA — Kernel Fisher's discriminant Analysis

KPCA — Kernel Principle Component Analysis

LDA — Linear discriminant Analysis

MRA — Multiresolution analysis

PCA — Principal Component Analysis

NFL — Nearest Feature Line algorithm

NN — Nearest Neighbor algorithm

SVM — Support Vector Machines

8.1. Introduction

Face recognition is a active research area, and they can be used in wide range of applications such as surveillance and security, telecommunication and digital libraries, human-computer intelligent interaction, and smart environments. Compared to classical pattern recognition problems such as fingerprint recognition, face recognition is much more difficult because there are usually many individuals(classes), only a few images (samples) per person, so a face recognition system must recognize faces by extrapolating from the training samples. Various changes in face images also present a great challenge, and a face recognition system must be robust with respect to the many variabilities of face images such as viewpoint, illumination, and facial expression conditions.

Many novel attempts have been made to face recognition research since the late 1970s [25,26]. There are two major approaches for vision research: geometrical local feature-based (e.g. relative positions of eyes, nose, and mouth.) schemes and holistic template-based systems and their variations [27]. The geometrical-based approach performs successfully in accurate facial feature detection scheme. However, it remains limited applications because of its difficult implementation and its unreliability in some cases. Compared to this approach, template-based approach is more promising due to its ease of implementation and robustness. In holistic template-matching systems, attempts are made to capture the most appropriate representation of face images as a whole and exploit the statistical regularities of pixel intensity variations. Principal Component Analysis (PCA) [22] and Linear discriminant Analysis (LDA) [13,22] are the two most classical and popular methods. The PCA is a typical method, which faces are represented by a linear combination of weighted eigenvectors, known as eigenfaces [1]. The LDA obtain features through eigenvector analysis of scatter matrices with the objective of maximizing between-class variations and minimizing within-class variations. These two methods both provides a small set of features that carry the most relevant information for classification purposes. However, the PCA usually give high similarities indiscriminately for two images from a single person or from two different persons and the LDA is also complex as there is a lot of within-class variation due to differing facial expressions, head orientations, lighting conditions, etc. Although many improving approaches have been proposed based on the two methods such as kernel PCA (KPCA) [16] and kernel Fisher's discriminant analysis (KFDA) [8,12,24] which used kernel skills, the essence problem has not been solved.

As we all known, the main challenge in feature representation is to represent the input data in a reduced low-dimensional feature space, in which, the most facial features are revealed or kept. Wavelet transform has much more advantages on this point. Compared to the PCA and LDA projections, wavelet subband coefficients can efficiently capture substantial facial features while keeping computational complexity low. It is well known to all that wavelet transform has a robust multi-resolution capability which accords well with human visual system. Moreover, it provides a spatial and a frequency decomposition of a image simultaneously. Using wavelet transforms as multiscale orthogonal representations of face images, different components of wavelet decomposition capture different visual aspects of a gray-scale image. This multi-resolution analysis is provided in the form of coefficient matrix.

Each face is described by a subset of band filtered images containing waveletcoefficients. At each level of decomposition there are four orthogonal subimages corresponding to LL, LH, HL, and HH. By spatial frequency analysis, the image is represented as a weighted combination of basis functions, in which high frequencies (LH, HL, HH) carry finely detail information and low frequency (LL) carry coarse, shape-based information. Only a change in face will affect all frequency components. Earlier published research [27] demonstrated that: the effect of different facial expressions can be attenuated by removing the high-frequency components and the low-frequency components only are sufficient for recognition. Subsequently, an appropriate wavelet transform can result in robust representations with regard to lighting changes and be capable of capturing substantial facial features while keeping computational complexity low. It is increasingly popular used in face representation in recent years and good results are obtained for race and gender classification [21,27].

Almost all the literatures of wavelet based face recognition use two-dimensional tensor product ones, which is the tensor product of one-dimensional wavelets. However, the property of anisotropic makes tensor product wavelet not attractive most for the purpose of facial representation [9,4]. Nontensor product wavelet, which is, the corresponding scaling function and associated wavelet function can't be written in the form of products of one-dimensional ones, can reveal more features than that of the common used tensor product wavelet transform [4,5,10,11,20]. Therefore, we suggest to represent facial features by discrete nontensor product wavelet transform (DNWT) in this paper.

Many efforts have been spent on constructing nontensor product wavelets. However, up to now, there is no systematic method to construct two-dimensional nontensor product wavelets [9,4,20]. In this chapter, we present a novel method for constructing nontensor product wavelet filters. A new nontensor product bivariate waveletfilter banks with linear phase are constructed from the centrally symmetric matrices. Our investigations demonstrate that these filter banks have a matrix factorization and they are capable of describing the features of face image. The new nontensor product filters derived from our method are applied for the feature representation of face image. To test the effect of representation based on the nontensor product wavelet transform, we design a set of scheme for we use Support Vector Machines (SVM) for classification. The SVM is a newly powerful machine learning approach, owing to its remarkable characteristics such as good generalization performance, the absence of local minimal and sparse representation of solution. It has become a popular research method in anomaly detection, and good application are reported [8,19][a]. Experiments are tested by using popular used ORL face database. The efficiency of our approach produced a significant improvement which includes a substantial advance in correctness and in time of processing comparing with those obtained by the discrete tensor product wavelet transform (DSWT) and the well-known conventional PCA and LDA methods.

This chapter is organized as follows: The construction of new nontensor product bivariate wavelet filter banks are briefly described in Section 8.2. The feature vector selection algorithm and experiment result are demonstrated in Section 8.3. Finally, conclusions are drawn in Section 8.4.

8.2. Construction of Nontensor Product Wavelet Filters Banks

Multiresolution analysis (MRA) theory provides a natural framework for understanding wavelets and filter banks. According to MRA, refinable functions (scaling functions) and wavelets are completely determined by a low-pass filter and high-pass filters, respectively. In subband code schemes, a low-pass filter and high-pass filters are respectively used as analysis filter and synthesis filters which form perfect reconstruction filter banks. Daubechies [7] designed univariate two-channel perfect reconstruction filter banks having finite impulse response (FIR) corresponding to a univariate orthonormal wavelet having a compact support and vanishing moments. It is well known that there does not exist an orthonormal symmetric wavelet with a compact support in the univariate dyadic dilation case, that is, two-channel perfect reconstruction FIR banks having a linear phase are not available in the univariate case.

Our interest here is in multivariate filter banks [23].

> *A commonly used method builds multivariate filter banks by the tensor products of univariate filters. This construction of filter banks focuses excessively on the coordinate direction.*

Therefore, nontensor product approaches for construction of multivariate filter banks or wavelets are desirable. Much interest has been given

---

[a]The SVM was developed by *V. N. Vapnic* (V. N. Vapnic, "The Nature of Statistical Learning Theory", Springer, 1995), *Remarks of Editors.*

to the study of nontensor product wavelets in $L^2(R^d)$ [5,20], and also [4] for constructions of multivariate wavelets on invariant sets) as well as to multiwavelets and corresponding vector-valued filter banks [6,17,18].

However, it is not easy to design multivariate filter banks.

> At present, no general method is available for designing multivariate filter banks and vector-valued filter banks. There are two fundamental difficulties that one encounters in the design of a low-pass filter and high-pass filters which are used for the construction of refinable functions and wavelets, respectively.

Most of the current study in multivariate wavelets is given to a dilation matrix with determinant two [11,23], since in this case, only one high-pass filter is needed to be construct and the matrix extension is the same as the univariate two-channel case [5].

Often, one seeks filter banks leading to smooth wavelets. However, in the application of filter banks to texture analysis, experiments show that "smooth" filter banks are not suitable because texture images is not smooth. Here we describe a general construction of bivariate nontensor product wavelet filter banks with linear phase by using centrally symmetric matrices in [3,4]. The family of filter banks given in this paper are suitable in this context although it is difficult to achieve smoothness. These filter banks have a matrix factorization and can be applied to facial representation.

8.2.1. *Characteristics of Centrally Symmetric Orthogonal Matrix*

We consider the following $n \times n$ centrally symmetric matrix

$$B = (b_{j,l})_{j,l=1}^n,$$

where

$$b_{j,l} = b_{n+1-j,n+1-l}, \quad j, l = 1, 2, \ldots, n.$$

Similarly, centrally anti-symmetric matrix $\tilde{B} = \left(\tilde{b}_{j,l}\right)_{j,l=1}^n$ is defined if

$$\tilde{b}_{j,l} = -\tilde{b}_{n+1-j,n+1-l}, \quad j, l = 1, 2, \ldots, n.$$

The centrally symmetric and central anti-symmetric matrices of order n are closed related to the special matrix H_n, which is defined by

$$H_n := \begin{pmatrix} 0 & 0 & \cdots & 1 \\ 0 & \cdots & 1 & 0 \\ \vdots & \vdots & \vdots & \vdots \\ 1 & 0 & \cdots & 0 \end{pmatrix}. \tag{8.1}$$

B is a centrally symmetric matrix of order n if and only if it satisfies the matrix equation

$$H_n B H_n = B. \tag{8.2}$$

Similarly, $\tilde{B}$ is a centrally anti-symmetric matrix if and only if the following matrix equation holds

$$\tilde{B} = -H_n \tilde{B} H_n. \tag{8.3}$$

Note that, for centrally anti-symmetric matrix $\tilde{B} = \left(\tilde{b}_{j,l}\right)_{j,l=1}^{n}$ with odd number n, the $\left(\left[\frac{n+1}{2}\right], \left[\frac{n+1}{2}\right]\right)$ elements $\tilde{b}_{\left[\frac{n+1}{2}\right],\left[\frac{n+1}{2}\right]} = 0$. Throughout this note, we use the notation $[x]$ to denote the integer no more than the real number x.

To construct two channel filter banks suitable for image processing, we need to consider the concrete construction of centrally symmetric orthogonal matrix of order 4, which corresponds the case $n = 4$. In this case, any centrally symmetric orthogonal matrix B has the general form

$$B = \frac{1}{2} \begin{pmatrix} I_2 & -H_2 \\ H_2 & I_2 \end{pmatrix} \begin{pmatrix} Z_1 & 0 \\ 0 & Z_2 \end{pmatrix} \begin{pmatrix} I_2 & H_2 \\ -H_2 & I_2 \end{pmatrix} \tag{8.4}$$

with Z_1 and Z_2 being orthonormal matrices of order 4. Let

$$S = \begin{pmatrix} 1 & 0 & 0 & -1 \\ 0 & 1 & -1 & 0 \\ 0 & 1 & 1 & 0 \\ 1 & 0 & 0 & 1 \end{pmatrix},$$

more precisely, we have the following equivalent form

$$B = \frac{1}{2} S \begin{pmatrix} a & b & 0 & 0 \\ c & d & 0 & 0 \\ 0 & 0 & e & f \\ 0 & 0 & g & h \end{pmatrix} S^T \tag{8.5}$$

with real numbers a, b, c, d, e, f, g, h satisfying

$$a^2 + b^2 = c^2 + d^2 = 1, \quad ac + bd = 0,$$

$$e^2 + f^2 = g^2 + h^2 = 1, \quad eg + fh = 0.$$

The parametrization solutions of above equations for these real numbers are $a = d = \cos\alpha$, $b = -\sin\alpha$, $c = \sin\alpha$, $e = h = \cos\beta$, $f = -\sin\beta$ and

$g = \sin\beta$ for any real numbers α and β. Therefore, any centrally symmetric orthogonal matrix B has the more simple parametrization representation

$$B = \frac{1}{2}\begin{pmatrix} 1 & 0 & 0 & -1 \\ 0 & 1 & -1 & 0 \\ 0 & 1 & 1 & 0 \\ 1 & 0 & 0 & 1 \end{pmatrix}\begin{pmatrix} \cos\alpha & -\sin\alpha & 0 & 0 \\ \sin\alpha & \cos\alpha & 0 & 0 \\ 0 & 0 & \cos\beta & -\sin\beta \\ 0 & 0 & \sin\beta & \cos\beta \end{pmatrix}\begin{pmatrix} 1 & 0 & 0 & 1 \\ 0 & 1 & 1 & 0 \\ 0 & -1 & 1 & 0 \\ -1 & 0 & 0 & 1 \end{pmatrix}$$

$$(8.6)$$

Further, we defined

$$Q_{(\alpha,\beta)} := B$$

$$= \frac{1}{2}\begin{pmatrix} \cos\alpha + \cos\beta & -\sin\alpha + \sin\beta & -\sin\alpha - \sin\beta & \cos\alpha - \cos\beta \\ \sin\alpha - \sin\beta & \cos\alpha + \cos\beta & \cos\alpha - \cos\beta & \sin\alpha + \sin\beta \\ \sin\alpha + \sin\beta & \cos\alpha - \cos\beta & \cos\alpha + \cos\beta & \sin\alpha - \sin\beta \\ \cos\alpha - \cos\beta & -\sin\alpha - \sin\beta & -\sin\alpha + \sin\beta & \cos\alpha + \cos\beta \end{pmatrix}.$$

$$(8.7)$$

Thus we get an constructive characterization of centrally symmetric orthogonal matrix of order 4.

In the following,we will offer some examples of construction of centrally symmetric orthogonal matrix of order 4,which play an crucial role in the design of nontensor product bivariate filter banks with two channels. The first case is to let $\alpha = \beta$. Letting $\alpha = \beta = \frac{\pi}{2}$ and $\alpha = \beta = \frac{\pi}{4}$ respectively, we have the following 4 order centrally symmetric orthogonal matrix

$$Q_{(\frac{\pi}{2},\frac{\pi}{2})} = \begin{pmatrix} 0 & 0 & -1 & 0 \\ 0 & 0 & 0 & 1 \\ 1 & 0 & 0 & 0 \\ 0 & -1 & 0 & 0 \end{pmatrix}, \qquad Q_{(\frac{\pi}{4},\frac{\pi}{4})} = \frac{\sqrt{2}}{2}\begin{pmatrix} 1 & 0 & -1 & 0 \\ 0 & 1 & 0 & 1 \\ 1 & 0 & 1 & 0 \\ 0 & -1 & 0 & 1 \end{pmatrix}.$$

We consider the case with $\alpha = 0$ and $\beta = \frac{\pi}{2}$, we get that

$$Q_{(0,\frac{\pi}{2})} = \frac{1}{2}\begin{pmatrix} 1 & 1 & -1 & 1 \\ -1 & 1 & 1 & 1 \\ 1 & 1 & 1 & -1 \\ 1 & -1 & 1 & 1 \end{pmatrix}$$

8.2.2. *Nontensor Product Wavelet Filter*

Next by using the above, we will develop a general method for constructing nontensor product bivariate waveletilter banks.

Given a bivariate trigonometric polynomials

$$m_0(\xi,\eta) = \sum_{j\in\mathbb{Z}}\sum_{k\in\mathbb{Z}} c_{j,k} e^{-i(j\xi+k\eta)}, \quad (\xi,\eta) \in \mathbb{R}^2,$$

its polyphase factors are the bivariate trigonometric polynomials $m_{0,l}$ defined for $l = 0, 1, 2, 3$ as

$$m_{0,0}(\xi, \eta) = \sum_{j \in \mathbb{Z}} \sum_{k \in \mathbb{Z}} c_{2j,2k} e^{-i(j\xi + k\eta)},$$

$$m_{0,1}(\xi, \eta) = \sum_{j \in \mathbb{Z}} \sum_{k \in \mathbb{Z}} c_{2j+1,2k} e^{-i(j\xi + k\eta)},$$

$$m_{0,2}(\xi, \eta) = \sum_{j \in \mathbb{Z}} \sum_{k \in \mathbb{Z}} c_{2j,2k+1} e^{-i(j\xi + k\eta)},$$

$$m_{0,3}(\xi, \eta) = \sum_{j \in \mathbb{Z}} \sum_{k \in \mathbb{Z}} c_{2j+1,2k+1} e^{-i(j\xi + k\eta)}.$$

Reversing the process, we can construct the bivariate trigonometric polynomials m_0 from its polyphase factors $m_{0,l}$, $j = 0, 1, 2, 3$ by the formula

$$m_0(\xi, \eta) = m_{0,0}(2\xi, 2\eta) + e^{-i\xi} m_{0,1}(2\xi, 2\eta) + e^{-i\eta} m_{0,2}(2\xi, 2\eta)$$

$$+ e^{-i(\xi+\eta)} m_{0,3}(2\xi, 2\eta),$$

where $(\xi, \eta) \in R^2$.

The construction of multivariate compactly supported orthonormal multiwavelets using MRA is equivalent to the design of orthogonal FIR and QMF filter banks, which leads to the following two questions.

(i) Find low-pass filter $m_0(\xi, \eta)$ satisfying the orthogonal condition

$$|m_0(\xi, \eta)|^2 + |m_0(\xi + \pi, \eta)|^2 + |m_0(\xi, \eta + \eta)|^2$$
$$+ |m_0(\xi + \pi, \eta + \eta)|^2 = 1, \quad (\xi, \eta) \in R^2;$$

(ii) Find 3 high-pass filter m_1, m_2, m_3 such that the matrix $M = (\alpha_0, \alpha_1, \beta_0, \beta_1)$ is unitary, where α_i and β_i are the row vector

$$(m_0(\xi + i\pi, \eta), m_1(\xi + i\pi, \eta), m_2(\xi + i\pi, \eta), m_3(\xi + i\pi, \eta))^T,$$

and $(m_0(\xi+i\pi, \eta+\pi), m_1(\xi+i\pi, \eta+\pi), m_2(\xi+i\pi, \eta+\pi), m_3(\xi+i\pi, \eta+\pi))^T$, $(i = 0, 1)$.

Given the orthogonal filter banks $m_l, l = 0, 1, 2, 3$ at hand, one can use the Pyramid algorithm to decompose and reconstruct the image. It will benefit us from the point view of polyphase to understand the conditions (i) and (ii).

Both of the problems (i) and (ii) (equivalently a and b) are nonlinear problem in mathematics, which is essentially quadratic algebraic equations with multiple variables. There is no general solution for this problem presently. Now we will offer a class of solutions of (i) and (ii) starting from centrally symmetric matrix.

Let

$$V_0 = (1,1,1,1)^T, V_1 = (1,-1,1,-1)$$
$$V_2 = (1,1,-1,-1)^T, V_3 = (1,-1,-1,1) \tag{8.8}$$

and denote by $D(\xi,\eta)$ the matrix of trigonometric polynomial

$$D(\xi,\eta) = \begin{pmatrix} 1 & 0 & 0 & 0 \\ 0 & e^{-i\xi} & 0 & 0 \\ 0 & 0 & e^{-i\eta} & 0 \\ 0 & 0 & 0 & e^{-i(\xi+\eta)} \end{pmatrix}, \quad (\xi,\eta) \in \mathbb{R}^2.$$

For any fixed positive integer N, arbitrarily chosen real number pairs (α_k, β_k), $k = 1, 2, \ldots, N$ (for $k \neq j$, (α_k, β_k) may equal to (α_j, β_j)), The low-pass filter $m_0(\xi, \eta)$ is defined as follows:

$$\frac{1}{4} \left(1, e^{-i\xi}, e^{-i\eta}, e^{-i(\xi+\eta)}\right) \left(\prod_{k=1}^{N} Q_{(\alpha_k,\beta_k)} D(2\xi, 2\eta) U_{(\alpha_k,\beta_k)}^T\right) V_0, \tag{8.9}$$

where $Q_{(\alpha_k,\beta_k)}$ is centrally symmetric orthogonal matrix defined previously, and $(\xi, \eta) \in \mathbb{R}^2$. It is easy to see that $m(0,0) = 1$, which means that m_0 is a low-pass filter.

Correspondingly, three high-pass filters $m_j, j = 1, 2, 3$ with respect to the above low-pass filter $m_0(\xi, \eta)$ are defined as follows:

$$m_1(\xi, \eta) = \frac{1}{4} \left(1, e^{-i\xi}, e^{-i\eta}, e^{-i(\xi+\eta)}\right) \left(\prod_{k=1}^{N} Q_{(\alpha_k,\beta_k)} D(2\xi, 2\eta) U_{(\alpha_k,\beta_k)}^T\right) V_1,$$

$$m_2(\xi, \eta) = \frac{1}{4} \left(1, e^{-i\xi}, e^{-i\eta}, e^{-i(\xi+\eta)}\right) \left(\prod_{k=1}^{N} Q_{(\alpha_k,\beta_k)} D(2\xi, 2\eta) U_{(\alpha_k,\beta_k)}^T\right) V_2,$$

$$m_3(\xi, \eta) = \frac{1}{4} \left(1, e^{-i\xi}, e^{-i\eta}, e^{-i(\xi+\eta)}\right) \left(\prod_{k=1}^{N} Q_{(\alpha_k,\beta_k)} D(2\xi, 2\eta) U_{(\alpha_k,\beta_k)}^T\right) V_3,$$

with V_j defined in (8.8), where $j = 1, 2, 3$, $(\xi, \eta) \in \mathbb{R}^2$. It is easy to check that $m_j(0,0) = 0$, $j = 1, 2, 3$. That is to say, $m_j, j = 1, 2, 3$ are high-pass filters.

8.2.3.   *Examples*

Now we will provide two concrete important examples of filter banks obtained by our approach.

Example 1. Using the previous matrix

$$Q_{(0,\frac{\pi}{2})} = \frac{1}{2} \begin{pmatrix} 1 & 1 & -1 & 1 \\ -1 & 1 & 1 & 1 \\ 1 & 1 & 1 & -1 \\ 1 & -1 & 1 & 1 \end{pmatrix}$$

and setting $N = 1$, we get the following filter banks

$$\begin{cases} m_0(x,y) = 1/8(-x + y + 1 + x^2y^2 + x^2 - y^2 + xy - x^3y + xy^3 + x^3y^3 \\ \quad +xy^2 + x^3y^2 + yx^2 - x^2y^3 + x^3 + y^3), \\ m_1(x,y) = 1/8(-x + y + 1 - x^2y^2 + x^2 + y^2 + xy - x^3y - xy^3 - x^3y^3 \\ \quad -xy^2 - x^3y^2 + yx^2 + x^2y^3 + x^3 - y^3), \\ m_2(x,y) = 1/8(x - y + x^2y^2 + x^2 + y^2 - xy - 1 - x^3y - xy^3 + x^3y^3 \\ \quad -xy^2 + x^3y^2 + yx^2 - x^2y^3 + x^3 - y^3), \\ m_3(x,y) = 1/8(-x + y + 1 + x^2y^2 - x^2 + y^2 + xy + x^3y - xy^3 + x^3y^3 \\ \quad -xy^2 + x^3y^2 - yx^2 - x^2y^3 - x^3 - y^3), \end{cases}$$

with $x = e^{-i\xi}, y = e^{-i\eta}$.

Example 2. Letting $\alpha = -\beta$, and $\alpha = \frac{\pi}{2}$ leads to

$$Q_{(\frac{\pi}{2}, -\frac{\pi}{2})} = \frac{\sqrt{2}}{2} \begin{pmatrix} 1 & -1 & 0 & 0 \\ 1 & 1 & 0 & 0 \\ 0 & 0 & 1 & 1 \\ 0 & 0 & -1 & 1 \end{pmatrix}.$$

Further, setting $N = 2$, the matrices $Q_{(0,\frac{\pi}{2})}$ and $Q_{(\frac{\pi}{2},-\frac{\pi}{2})}$ lead to the following filter banks

$$\begin{cases} m_0(x,y) = 1/4x^3y^3 + 1/4x^2y^2 + 1/8x^2 + 1/8yx^2 - 1/8x^3y + 1/8y^5x^2 \\ \quad +1/8y^4x^3 + 1/8x^3y^5 + 1/8x^3 - 1/8y^4x^2, \\ m_1(x,y) = -1/4x^3y^3 + 1/4x^2y^2 + 1/8x^2 + 1/8yx^2 + 1/8x^3y + 1/8y^5x^2 \\ \quad -1/8y^4x^3 - 1/8x^3y^5 - 1/8x^3 - 1/8y^4x^2, \\ m_2(x,y) = -1/4x^2y^3 + 1/8x^2 + 1/8yx^2 + 1/4x^3y^2 - 1/8x^3y - 1/8y^5x^2 \\ \quad -1/8y^4x^3 - 1/8x^3y^5 + 1/8x^3 + 1/8y^4x^2, \\ m_3(x,y) = -1/4x^2y^3 + 1/8x^2 + 1/8yx^2 - 1/4x^3y^2 + 1/8x^3y - 1/8y^5x^2 \\ \quad +1/8y^4x^3 + 1/8x^3y^5 - 1/8x^3 + 1/8y^4x^2. \end{cases}$$

8.3. Experimental Results

By using the proposed facial feature representation, we develop a new nontensor-product-wavelet-based face recognition scheme by combining the techniques of PCA and SVM. To test the robustness of the proposed nontensor-product-wavelet-based facial feature representation, the following experiments are conducted by ORL face database, which contains a set of faces taken between April 1992 and April 1994 at the Olivetti Research Laboratory in Cambridge, UK. There are 10 different images of 40 distinct subjects. For some of the subjects, the images were taken at different times, varying lighting slightly, facial expressions (open/closed eyes, smiling/non-smiling) and facial details (glasses/no-glasses). All the images are taken against a dark homogeneous background and the subjects are in up-right, frontal position (with tolerance for some side movement). Some sample images randomly chosen from this database are shown in Fig. 8.3.

Fig. 8.1.   Some sample images of ORL face database.

The proposed algorithm based-on nontensor product bivariate wavelet is conducted as follows.

Algorithm for Nontensor-Product-Wavelet-Based face recognition

Step 1 Input training face images **I**, perform two level DNWT with the new nontensor product wavelet transform constructed in Section 8.2 on each face;

Step 2 Select low-frequency approximate components as the facial feature representation and the PCA is applied for reduction of computation complexity;

Step 3 Lengthen the feature matrix from columns and save them as the input training data of the SVM;

Step 4 Input test face images, the same processing as Step 1 and Step 2 are applied. Finally, input these face data into the SVM systems and compute the correctness.

Here, we test the proposed method in both the performance against the conventional PCA method [15], the PCA+LDA+NN [28] method and the recent face recognition scheme such as the DSWT+SVM and

DSWT+PCA+SVM [21]. To draw comparisons with others methods, the same training and test sizes are used as in [21], that is, five images chosen randomly of each subject for training and the remaining for testing.

Table 8.1. Comparison of the computational time and recognition rate among other algorithms.

Method	Computation time (seconds)	Recognition rate (%)
PCA+NN	11.5	89.88
PCA+LDA+NN	12.8	92.55
PCA+SVM	13.43	93.9
DSWT+SVM	7.28	96.5
DSWT+PCA+SVM	5.14	97.0
DNWT+PCA+SVM	4.40	98.6

As shown in Table 8.1, our algorithm produced a significant improvement which includes a substantial advance in recognition rate and in time of processing comparing the other illustrated algorithms. The result shows that our approach: with a 98.6 recognition rate compares favorably against the recent face recognition schemes that apply the same ORL database for performance evaluation,such as the ULLELDA+NFL method with a recognition rate 95.94 [28], the Gabor face+LDA+NFL method with recognition rate 95.8 [29], and the KPCA method with recognition rate 97.75 [16].

Detailed evaluations are illustrated in Fig. 8.2. When the number of features is 20, the proposed method gives the accuracy of 92.6, while the DSWT+PCA+SVM, PCA+SVM and the PCA+NN give just 90.1, 92.5 and 88, respectively. When 85 features are used, the recognition rate of the proposed method attains 96.5, and it is higher than that of DSWT+PCA+SVM, PCA+SVM and PCA+NN. The good performance of the proposed method demonstrate that DNWT+PCA+SVM is more effective than other four methods in extracting and representing facial features of real face images for face recognition.

In Fig. 8.3, the DSWT+PCA can represent the most features of real face images. It is still inferior to the proposed the DNWT+PCA, perhaps because it can reveal more facial features than tensor product ones do. Figure 8.3 shows the effects of training sample size. The DSWT+PCA obtains recognition rate 96.5 while the DNWT obtains the same recognition rate at just 4 training faces. This is 2.1 lower than the accuracy of the DNWT at 5 train samples. Especially, the DNWT can reach 100 when the training sample is more than 7 while the DSWT never reaches.

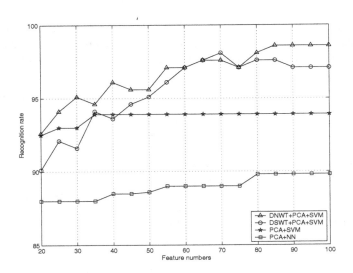

Fig. 8.2. Recognition rates versus the feature space dimension.

In a sense, it implies our method can obtains a perfect recognition rate so long as the training sample is sufficient.

Many factors effect the correctness, such as feature representation, training sample number. Besides these factors, the selection of classifier also plays an important role. Conventional face recognition scheme commonly adopted Nearest Feature Line algorithm (NFL) and Nearest Neighbor algorithm (NN) as classifier. They have given a good evaluation under the ORL testing framework [14,15,28,29]. However, in lately research work, the SVM classifier has demonstrated to be better than the algorithms mentioned above. In this paper, the SVM is applied to final classification. The polynomial kernel function is often used and it was found that its performance was better than Gauss kernel and other kernels in [21]. Here we focus on the effects of SVM (Fig. 8.4) applied in the DSWT+PCA+SVM method and our proposed method (DNWT+PCA+SVM). The influence of the degree d upon polynomial kernel function is shown in Fig. 8.4 with degree $d = 1, 2$ and 3. It is illustrated that:

(a) The performance of the polynomial with $d = 1$ obviously outperform $d = 3$ and a slightly better than $d = 2$.
(b) Our scheme focused on the degree 1 which got the best recognition rate.

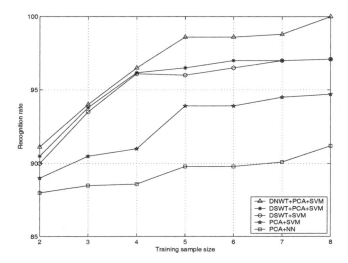

Fig. 8.3.   Recognition rates versus the train sample size.

Actually, we tried to take the high frequency sub-bands into account. The detail information they carried may be those very discriminated. We tried to plus one high frequency part to low frequency part after threshold, and the entire compressed subband is represented as the facial feature, but no better result has been obtained in such data set. It proves that nontensor product wavelet transform reveals more facial features in face image processing. The low-frequency components only are sufficient for recognition.

8.4.   Conclusions

Like other biometric systems, the primary stages employed by facial recognition system to identify and verify subjects include facial image capture, feature analysis, verification and identification. Feature analysis is perhaps the most widely utilized facial recognition technology. Facial feature extraction and representation are frequently used as the basis of feature analysis. In this chapter, a new approach for facial feature extraction and representation are developed based on the a constructed nontensor product bivariate wavelet filter banks with linear phase, which are constructed from the centrally symmetric matrices. By performing nontensor product wavelet transform, the frequence components of the wavelet subband are used for facial representation. It is demonstrated that

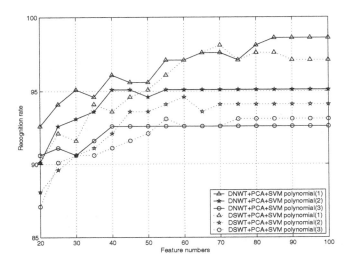

Fig. 8.4.   Recognition rates versus different kernels.

the proposed approach based the nontensor product wavelet transform is more capable of representing facial image for recognition purpose than the tensor product one. We also proposed a nontensor-product-wavelet-based face recognition scheme by combining the technique of PCA and SVM. Some compared experiments with main traditional methods are conducted. The efficiency of our scheme has been demonstrated on the ORL database.

Acknowledgments

This work was partially supported by a grant (60403011) from the National Natural Science Foundation of China and a grant (2003ABA012) and (20045006071-17) from the Science & Technology Department, Hubei province and Wuhan respectively, China.

Bibliography

1. Atalay, I. (1996). Face recognition using eigenfaces, M.sc. thesis, Institute of Science and Technology, Istanbul Technical University, Istanbul, Jan.
2. Cavaretta, A. S., Dahmen, W. and Micchelli C. A. (1991). Stationary subdivision, *Mem. Amer. Math. Soc.*, **93**, 453, pp. 1–186.
3. Qiuhui Chen, Charles A. Micchelli, Silong Peng, and Yuesheng Xu (2003).

Multivariate filter banks having a matrix factorization. *SIAM Journal on Matrix Analysis and its Application*, **25**,2, pp. 517–531.

4. Qiuhui Chen, Charles A. Micchelli, and Yuesheng Xu.(2005). Biorthogonal wavelet filter banks from block centrally symmetric matrix. *Linear Algebra and Its Application*, **402**, pp. 111–125.

5. A. Cohen and I. Daubechies.(1993). Non-separable bidimensional wavelet bases. *Rev. Mat. Iberoamericana*, **9**,pp. 51–137.

6. W. Dahmen and C. A. Micchelli.(1997). Biorthogonal wavelet expansions. *Constr. Approx.*, **13**,pp. 293–328.

7. I. Daubechies. *Ten Lectures on Wavelets*. (1992). Society for Industrial and Applied Mathemathics, Philadelphia.

8. Hai-Hua Gao, Hui-Hua Yang, and Xing-Yu Wang.(2005). "Kernel PCA Based Network Inrusion Feature Extraction and Dectection Using SVM". In *Proc. IEEE NPL Conf.ICNC -2005- LNCS,3611*, pp. 89–94.

9. W. He and M. J. Lai.(2000). "Examples fo bivariate nonseparable compactly supported continuous wavelets". *IEEE Trans. Image Process*, **9**,8, pp. 949–953.

10. R. Q. Jia and C. A. Micchelli.(1992). Using the refinement equation for the construction of prewavelets v: extensibility of trigonometric polynomials. *Computing*, **48**, pp. 61–72.

11. J. Kovacevic and M. Vetterli.(1992). Nonseparable multidimensional perfect reconstruction filter banks and wavelet bases for r^n. *IEEE Tran. on Information Theory*, **38**, pp. 533–555.

12. Zhizheng Liang and Pengfei Shi.(2005). "Kernel direct discriminant analysis and its theoretical foundation". *Elsevier Pattern Recognition Society 38*, pp. 445–447.

13. Hsien-Jen Lin.(2002). *On Improving Linear Discriminant Analysis for Face Recognition with Small Sample Size Problem*. Master thesis, Department of Computer Science and Information Engineering, National ChengKung University.

14. L.J.Cao and W.K.Chong.(2002). "Feature Extraction in Support Vector Machine: A Comparison of PCA,KPCA and ICA". *Pros. 9th Int. Conf. on Neural Information Processing (ICONIP'02)*.

15. M.A.Turk and A.P.Pentland.(1991). "Face recognition using eigenfaces". *J. Cognitive Neuroscience*, **3**,1 pp. 71–86.

16. M.H.Yang, N.Ahuja, and D.Kriegman.(2000). "Face recognition using kernel eigenfaces". *Proc. IEEE Int. Conf. Image Processing.* **1**pp. 37-40

17. C. A. Micchelli.(2000).(1992). Using the refinement equation for the construction of pre-wavelets vi: Shift invariant subspaces. *Approximation Theory, Spline Functions and Applications,(Maratea, 1991), NATO Adv. Sci. Inst. Ser. C Math. Phys. Sci., 356, Kluwer Acad. Publ., Dordrecht.*

18. C. A. Micchelli and T. Sauer.(1997) Regularity of multiwavelets. *dv. Comput. Math..*

19. Vo Dinh Minh Nhat and Sungyoung Lee.(1997) "Line-Based PCA and LDA Approaches for Face Recognition". *Proc. IEEE NPL Conf. ICNC -2005- LNCS,3611*, pp. 101–104.

20. S. D. Riemenschneider R. Q. Jia and D. X. Zhou.(1999). Smoothness of multiple refinable functions and multiple wavelets. *SIAM J. Matrix Anal. Appl.*, **21**, pp. 1–28.

21. Majid Safari, Meshrtash T.Harandi, and Babak N.Araabi.(2004). A svm-based method for face recognition using a wavelet-pca representation of faces. *Pros. IEEE Int. Conf. on Image Processing*, pp. 24-27.

22. Wendy S. Yambor. (2000). Analysis of PCA-based and fisher discriminant-based image recognition. Technical Report CS-00-103, Computer Science Department,Colorado State University.

23. M. Vetterli.(1986). Filter banks allowing perfect reconstruction. *Signal Proc.*, **10** (3):219–244.

24. Jian Yang, Alejandro F.Frangi, Jing yu Yang, David Zhang, and Zhong Jin.(2005). "KPCA Plus LDA: A Complete Kernel Fisher Discriminant Framework for Feature Extraction and Recognition". *IEEE Trans. Pattern Anal. Mach. Intell.*, **27** (2).

25. Yanushkevich, S. N., Stoica, A., Shmerko, V. P. and Popel, D. V. (2005). *Biometric Inverse Problems.* CRC Press/Taylor & Francis Group, Boca Raton, FL.

26. Yanushkevich, S. N., Shmerko, V. P. and Stoica, A. (2006). Fundamentals of Biometric-Based Training System Design, this issue.

27. BaiLing Zhang, HaiHong Zhang, and Shuzhi Sam Ge.(2004). "Face Recognition by Applying Wavelet Subband Representation and Kernel Associative Memory". *IEEE Trans. on Neural Networks.*, **15** (1).pp. 166–177

28. Junping Zhang, HuanXing Shen, and Zhi-Hua Zhou.(2004) "Unified Locally Linear Embedding and Linear Discriminant Analysis Algorithm (ULLELDA) for Face Recognition". *Advances in Biometric Personal Authentication.*, LNCS 3338: pp.209–307.

29. Jianke Zhu, Mang I Vai, and Peng Un Mak.(2004) "Gabor Wavelets Transform and Extended Nearest Feature Space Classifier for Face Recognition". In *Proc.Third International Conference on Image and Graphics, ICIG 2004.*, Hong Kong.

Chapter 9

Palmprint Identification by Fused Wavelet Characteristics

Guangming Lu[*,‡,§], David Zhang[†,¶], Wai-Kin Kong[†,‖], Qingmin Liao[*,**]

Visual Information Processing Laboratory,
Graduate School at Shenzhen, Tsinghua University, Shen Zhen, China
†*Biometric Research Center,*
Department of Computing, The Hong Kong Polytechnic University,
Kowloon, Hong Kong
‡*School of Computer Science and Engineering,*
Harbin Institute of Technology, Harbin, China
§*lugmsz.tsinghua.edu.cn*
¶*csdzhangcomp.polyu.edu.hk*
‖*cswkkongcomp.polyu.edu.hk*
**liaoqmtsinghua.edu.cn*

The wavelet theory has become a hot research topic in the last few years for its important relative characteristics, such as sub-band coding, multi-resolution analysis and filter banks. In this Chapter, we propose a novel method of feature extraction for palmprint identification based on the waveletransform, which is very efficient to handle the textural characteristics of palmprint images at low resolution. In order to achieve a high accuracy, four sets of statistical features (mean, energy, variance, and kurtosis) are extracted based on the wavelet transform. Five classifier combination strategies are then represented. The experimental comparison of various feature sets and different fusion schemes shows that the individual feature set of energy has the best classification ability and the fusion schemes of Median rule as well as the Majority rule have the best performance.

Contents

9.1.   Introduction

Biometrics based identification/verification technology has been greatly emphasized and developed during the last few years. Lots of interesting and meaningful research results have been achieved at the same time. Although fingerprint identification is widely popular and extensively accepted by the public, but some people do not have clear fingerprint because of their physical work or problematic skin. Iris and retina recognition can provide a very high accuracy, but they either suffer from the high cost input devices or pose intrusion to users [1], [2], [3]. Recently, many researchers focus on face and voice identification, nevertheless, their accuracy are still far from being satisfactory. The accuracy and uniqueness of 3-D hand geometry is still an open question [4]. Therefore, the investigation and development in palmprint identification become particularly valuable.

First of all, the surface area of palmprint is very large when comparing with fingerprint image. As the palm is the inner surface of the hand between the wrist and the fingers, the stability of palmprint is high and the chance of getting damaged is lower than that of the fingerprint [5], [6]. The data of palmprint would be helpful, as more information is available to verify or identify the identity of a person. Many features of a palmprint can be used to uniquely identify a person (as shown in Figure 9.1). In general, six major types of features can be observed on a palm as follows.

Geometry Features. According to the palm's shape, we can easily get the corresponding geometry features, such as width, length and area.

Principal Line Features.   Both location and form of principal lines in a palmprint are very important physiological characteristics for identifying individuals because they vary little over time.

Wrinkle Features. In a palmprint, there are many wrinkles which are different from the principal lines in that they are thinner and more irregular. These are classified as coarse wrinkles and fine wrinkles so that more features in detail can be acquired.

Datum Points. Two end points called datum points are obtained by using the principal lines (see Figure 9.1). These intersect on both sides of a palm and provide a stable way to register palmprints. The size of a palm can be estimated by using the Euclidean distance between these end points.

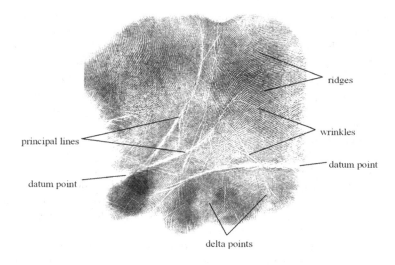

Fig. 9.1. Different features in the palm.

Delta Point Features. The delta point is defined as the center of a delta-
like region in the palmprint. Usually, there are delta points located in
the finger-root region. These provide stable and unique measurements
for palmprint authentication.

Minutiae Features. A palmprint is basically composed of the ridges,
allowing the minutiae features to be used as another significant
measurement.

Since

> A palmprint contains different features including principal lines,
> wrinkes, minutiae points, datum points and texture, which constitute
> a set of multiscale features.

It motivates us to apply multi-resolution based approach to extract the
multiscale features. Different features require different image resolution
and have different applications. For minutiae points, and datum points,
extracting them requires a resolution of 400 dpi to 600 dpi [7]; for
principal lines and wrinkes, they can be extracted from images of
100 dpi [8], [9].   In fact, we have tried to extract texture feature
from lower resolution images [10,11].   The high-resolution images are

suitable for criminal related applications since ridges, datum points and minutiae points can be extracted and matched to the features in latent prints. For civil and commercial applications that require real-time identification/verification, low-resolution palmprints images are more suitable than high-resolution images since low-resolution palmprints images product small file size, which results in short computation time for preprocessing and feature extraction and therefore, they are useful for real-time applications.

The wavelet theory has become hot in the last few years. It has important relative characters, such as sub-band coding, multi-resolution analysis and filter banks. It has been widely used in signal and image processing, and has better local performance in the time domain and the frequency domain. We also find that it has good quality for textural feature extraction. Texture description using the wavelet transform provides sub-bands (LL, HL, LH, HH), which have spatial and orientational selectivity [12]. The features of a palmprint image extracted by the wavelet transform represent an approximation of the image energy distribution over different scales and orientations.

This chapter is organized as follows: the palmprint image collection and preprocessing are described in Section 9.2. Section 9.3 presents a brief introduction to wavelet-based palmprint feature extraction. Palmprint matching and decision combination are given in Section 9.4. Experimental results and some conclusions are given in Section 9.5 and Section 9.6, respectively.

9.2. Palmprint Images Collection and Preprocessing

Palmprint images collection, especially online collection, is the fundamental work for palmprint algorithm research and palmprint identification system development. During the passed several years, we have put great efforts into designing and implementing an online palmprint acquisition device. Such a device must be able to online acquire a good quality palmprint image and have good interface to users.

The architecture of the newest version of palmprint acquisition device is shown in Fig. 9.2. There is a user interface for the input of a palm. A set of optical components work together to obtain the data from the palm. The analog signal is converted into digital signal by the video frame grabber using an A/D converter, and the digitized signals are stored in the system main memory [5]. Figure 9.3 shows some typical palmprint images captured by our proposed device.

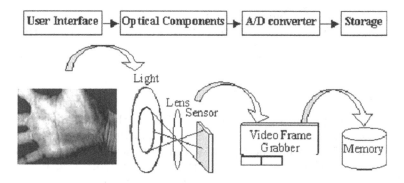

Fig. 9.2. The architecture of our palmprint capture device.

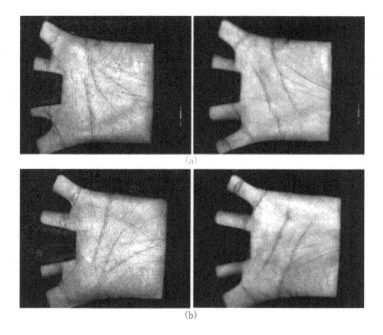

Fig. 9.3. Several typical palmprint images captured by our proposed device: (a) is the images from the Right hand of two different persons, and (b) is the corresponding Left hand palmprints.

When capturing a palmprint, the position, direction and stretching degree may vary from time to time. As a result, even the palmprints from the same palm could have a little rotation and translation.

Also the sizes of palms are different from one another. It is necessary to align all palmprints and normalize their sizes for further feature extraction and matching [5]. In our palmprint acquisition device, a panel, which can locate the palms by several pegs, is designed to constrain both the rotation and translation of the palms. Then the preprocessing algorithm can locate the coordination system of the palmprints quickly. Sub-images with fixed size (128×128) are extracted from the captured palmprint images (384×284) based on the coordination system. So that different palmprints are converted into the same image size for further processing. The size and location of the sub-image are determined through many observations. The principle used to decide the size and location of the sub-image is simply based on making sure that most palmprint features are retained within this area, and that all palmprints contain that sub-image. The basic steps of palmprint alignment are shown in Fig. 9.4, and described in detail as below.

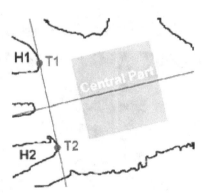

Fig. 9.4. The major steps on palmprint preprocessing: $H1$ and $H2$ are boundary of the holes between the two fingers, where $T1$ and $T2$ are the tangent point of $H1$ and $H2$, respectively. The central box is the sub-image of a palm.

Step 1. Use a threshold to convert the original grayscale image into a binary image, then using a low-pass filter to smooth the binary image.

Step 2. Trace the boundary of the holes $H1$ and $H2$ between those fingers.

Step 3. Compute the tangent of the holes $H1$ and $H2$. $T1$ and $T2$ are the tangent points of $H1$ and $H2$, respectively.

Step 4. Align $T1$ and $T2$ to determine the Y-axis of the palmprint coordination system and making a line passing through the midpoint of the two points ($T1$ and $T2$), which is perpendicular to this Y-axis to determine the origin of the system.

Step 5. Extract a sub-image of a fixed size based on the coordinate system (see Fig. 9.4). The sub-image is located at a certain area of the palmprint image for feature extraction.

Some typical original palmprints and their corresponding sub palmprint images are shown in Fig. 9.5.

9.3. Wavelet Based Feature Extraction

Feature extraction is to describe a palmprint in a concise feature set other than the original image. How to define the feature is the key point of palmprint identification. A good feature must well distinguish the palmprint from different persons and at the same time, be similar to each other from the same persons' palmprints. The sub-images derived from the palmprint preprocessing have a great deal of different features, such as principal lines, wrinkes, ridges, minutiae points, singular points and texture, which constitute a set of multi-scale features. Then, it is reasonable for us to take account of the multi-resoulution analysis based method to represent the multi-scale features [13]. The wavelet transform is implemented as filter banks, and is an effective tool for multi-resolution analysis. In this Chapter, we extend the use of the wavelet transform for palmprint feature extraction and propose a 2-D wavelettransform scheme for palmprint feature representation.

Generally, the one-dimensional (1-D) wavelet function and 1-D wavelettransform can be defined as follows:

If function $\psi(x) \in L^2(R)$, and $\psi(\omega)$ stands for the Fourier transform of $\psi(x)$, the square integrable functions $\psi(\omega)$ satisfies the admissibility

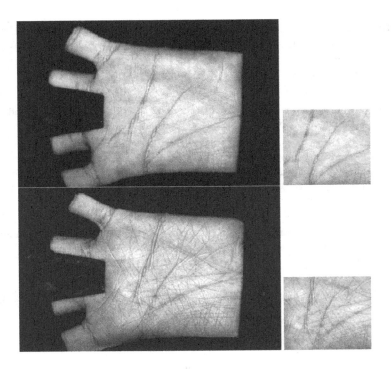

Fig. 9.5.   Some typical original palmprints and their corresponding sub palmprint images.

condition

$$C_{\omega} = \int_R \frac{|\psi(\omega)|}{|\omega|} d\omega \prec \infty, \tag{9.1}$$

we call $\psi(\omega)$ the mother wavelet. In the case of continuous wavelet transform, the wavelet function can be expanded from the mother wavelet by various translations and dilations. In other words, the transform makes use of the functions

$$\psi_{a,b}(x) = \frac{1}{\sqrt{|a|}} \psi\left(\frac{x-a}{b}\right), a, b \in R, a \neq 0. \tag{9.2}$$

These functions are scaled so that their $L^2(R)$ norms are independent of a. The continuous wavelet transform of a function $f \in L^2(R)$ is now defined by

$$W(a,b) = \langle f, \psi_{a,b} \rangle = \frac{1}{\sqrt{|a|}} \int_R f(x) \psi\left(\frac{x-a}{b}\right) dx \tag{9.3}$$

Initially, wavelet transform just focused on the 1-D situation. However, multidimensional wavelets are also available, especially the two dimensional (2-D) wavelets. In the same way, the 2-D wavelet transform can be treated as two 1-D wavelet transforms: one 1-D wavelet transforms along the row direction and the other 1-D wavelet transforms along the column direction. Thus, the 2-D wavelet transform can be computed in cascade by filtering the rows and columns of images with 1-D filters [9]. Figure 9.6 shows a one-octave decomposition of an image into four components: low-pass rows, low-pass columns (LL); low-pass rows, high-pass columns (LH); high-pass rows, low-pass columns (HL); high-pass rows, high-pass columns (HH). The coefficients obtained by applying the 2-D wavelet transform on an image are called the sub images of the wavelet transform [14]–[18]. Figure 9.7 shows the original palmprint and its corresponding wavelet sub images after 4 octaves wavelet transform.

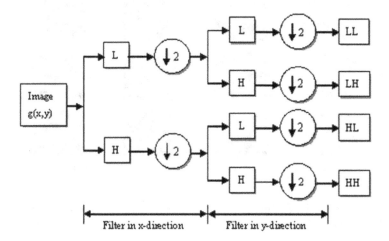

Fig. 9.6.   Decomposition algorithm of 2-D wavelet transform.

As mentioned above, the 2-D wavelet decomposition of I octaves of a sub palmprint image $g(m, n)$ represents the image by $3I + 1$ sub images

$$\left[g_I, \{g_i^1, g_i^2, g_i^3\}_{i=1...I}\right],\tag{9.4}$$

where

g_I is a low resolution of the original image;

g_i^d are the wavelet sub images containing the image details at different scales (2^j) and orientations (d) at high resolution;

The wavelet decomposition coefficient g_i^1 corresponds to the vertical high frequencies, g_i^2 corresponds to the horizontal high frequencies, and g_i^3 corresponds to the high frequencies in both directions [19].

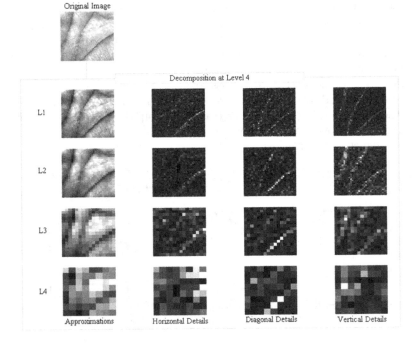

Fig. 9.7.   The sub-palmprint image, and the feature images obtained after 4 octaves wavelet transform.

As features, four statistical features are extracted from each sub-band at each octave by the wavelet transform. The four statistical features are mean, energy, variance, and kurtosis. Their definitions are shown below:

Mean magnitude:

$$M_{i,d} = \frac{1}{m_i * n_i} \sum_{j=1}^{m} \sum_{k=1}^{n} \left| g_i^d(j,k) \right|, \tag{9.5}$$

Energy:

$$E_{i,d} = \frac{1}{m_i * n_i} \sum_{j=1}^{m} \sum_{k=1}^{n} \left(g_i^d(j,k) \right)^2, \tag{9.6}$$

Variance:

$$V_{i,d} = \frac{1}{m_i * n_i} \sum_{j=1}^{m} \sum_{k=1}^{n} (g_i^d(j,k) - M_{i,d})^2, \tag{9.7}$$

Kurtosis:

$$K_{i,d} = \frac{1}{m_i * n_i} \sum_{j=1}^{m} \sum_{k=1}^{n} \frac{(g_i^d(j,k) - M_{i,d})^4}{V_{i,d}^4} \tag{9.8}$$

where

$d = 0, 1, 2, 3$ refers to four sub-bands;

$f_i = [f_{i,1}, f_{i,2}, f_{i,3}, f_{i,4}]$ (f refers to M, E, V, or K, respectively) and is the feature vector at the ith octave;

m_i and n_i denote the width and height of sub-bands at the ith octave in pixels.

So we can get a set of features with length $4 * i$, for example, after the 4 octaves wavelet transform a feature vector of length 16 is created. In order to get a longer feature set, a feasible way is to divide the original sub-palmprint image into smaller blocks. We try to divide the sub-palmprint into 16 blocks, and for each block four octaves wavelet transform is applied, then a feature set of length 256 is obtained.

9.4. Palmprint Feature Matching and Decision Fusion

Palmprint feature matching is based on measuring the distance between the corresponding feature vectors. In our work, the Euclidean distance is used to cluster those features. Since four sets of texture feature vectors of each sample have been extracted, four distance vectors can be acquired by the Euclideandistance classifier. In order to get better recognition accuracy, a considerable scheme is to combine the multiple classifiers. The multiple classifier combination makes the decision of which class from a set of candidate is to be the final result in the most reliable sense. It can be considered as a special classification problem. In this chapter, five kinds of combination methods are proposed, including the minimum distance algorithm, the mean distance algorithm and the majority voting. Before the decision fusion, it is essential to normalize all the distances for each classifier. The well-known formula is used here:

$$d_{\mathrm{norm}} = \frac{d - d_{\min}}{d_{\max} - d_{\min}} \tag{9.9}$$

where d is the distance vector from the candidate to the templates in the database, the d_{min} and d_{max} are the minimum and maximum value of the vector d, respectively.

Then, each output of all classifiers $C_i(i = 1, \ldots, m)$ is denoted as a distance vector $r_i(i = 1, \ldots, m)$ associated with each class, where m denotes the number of the classifiers. The discriminant function is defined as:

$$S = f(R) \quad R = \{r_i, (i = 1, \ldots, m)\} \tag{9.10}$$

Five forms of the discriminant function $f(R)$ are proposed:

$$Sumrule : f(R) = sum(r_1, r_2, \ldots, r_m);$$

$$Maxrule : f(R) = max(r_1, r_2, \ldots, r_m);$$

$$Minrule : f(R) = min(r_1, r_2, \ldots, r_m);$$

$$Medianrule : f(R) = med(r_1, r_2, \ldots, r_m);$$

$$Majorityvoting : f(R) = maj(r_1, r_2, \ldots, r_m).$$

In the case of Majority voting, each classifier $C_i(i = 1, \ldots, m)$ is associated with a single label. The majority voting method selects the class that gets more votes than any other class. Each candidate has the same vote and weight [20]–[25].

9.5. Experimental Results

Palmprint images were collected in our laboratory from 191 people using our self-designed capture device. Since the two palmprints (right-hand and left-hand) of each person are different, we captured both and treated them as palmprints from different people. Eight samples were captured for each palm with different rotation and translations. Thus, a palmprint database of 382 classes was created, which includes a total of 3056 ($=191\times2\times8$) images with size 384×284 pixels in 256 gray levels.

Identification test is a one-against-many, N comparison process. In this experiment, N is set to 382, which is the total number of different palms in our database. The palmprint database is divided into two sub databases: registration and testing databases. The registration database

Table 9.1.   The recognition rate of each classifier.

Individual feature set	Identification rate (%)
Mean features (Eq. (5))	95.75
Energy features (Eq. (6))	95.81
Variance features (Eq. (7))	95.35
Kurtosis features (Eq. (8))	95.62

contains 1528 palmprint images. Four images per palm are used to calculate the mean feature set for registration. The testing database includes the remaining 1528 palmprint images. Each palmprint image in the testing database is matched to all of the palmprint images in the registration database. Therefore, each testing image generates one correct and 381 incorrect matchings. The minimum distances of correct matching and incorrect matchings are regarded as the identification distances of genuine and impostor, respectively.

As to the wavelet bases, the Symmlet wavelet is used with 9 vanishing moments. Based on these schemes, the imposter and genuine distributions of different statistical features are presented by means of Receiver Operating Characteristic (ROC) curves (as Fig. 9.8), which are plotted by the genuine acceptance rates against the false acceptance rates, and measure the overall performance of the method. Generally, a biometric system operates at a low false acceptance rate and therefore, we only plot the range of false acceptance rate between 0 and 5%. In this range, the feature of mean and the feature of kurtosis obtain similar performance. The feature of energy is the best and the feature of variance is the worst for all operating points. According to the ROC curves, the features of energy can operate at a point with 94% genuine acceptance rate and 0.05% false acceptance rate and its equal error rate is about 4.2%.

In the decision fusion experiment, five different combination rules are applied and their results are compared. Table 9.1 shows the results of the classification of individual classifiers while the results of different combination rules are shown in Table 9.2. From the outcome of our experiments, we find that a higher recognition accuracy can be obtained when the proper combination strategy is used. Among the five different combination rules, the Median rule has the best results. The majority voting rule is very close to the Median rule in the performance.

Compared with the approach in [26], which used a set of feature points along the prominent palm lines and the associated line orientation of palmprint images to identify the individuals, and a matching rate about 95% was achieved. But only 30 palmprint samples from three persons were

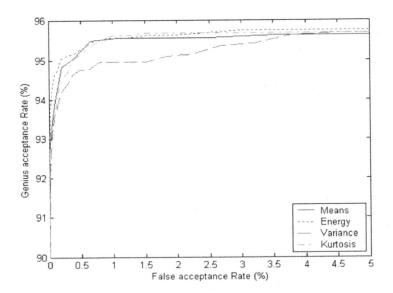

Fig. 9.8. ROC curves for the four statistical features.

Table 9.2. The recognition rate of different combination rules.

Fusion rules	Identification rate (%)
Sum Rule (Eq. 11)	96.99
Max Rule (Eq. 12)	97.32
Min Rule (Eq. 13)	96.99
Medin Rule (Eq. 14)	97.64
Majority Voting (Eq. 15)	97.51

collected for testing. It seems that the testing set is too small to cover the distribution of all palmprints. An average recognition rate 91% was achieved by the technology proposed in [27], which involved a hierarchical palmprint recognition fashion. The global texture energy features were used to guide the dynamic selection for a small set of similar candidates from the database at coarse level for further processing. An interesting point based image matching was performed on the selected similar patterns at fine levels for the final confirmation. Nevertheless, the recognition rate of our method is more efficient, as illustrated in Table 9.3.

Table 9.3. Comparison of different palmprint recognition methods.

Method	Feature Points	Hierarchical Identification	Our Method
Database	30 Samples	200 Samples	3056 Samples
Features	Feature points	Global texture feature	Statistical features
Accuracy	95%	91%	97.64%

9.6. Conclusions

The chapter proposed a wavelet-based palmprint identification method. Using the wavelet transform to decompose the normalized sub images and extracting statistical features from the sub-bands, four features (mean, energy, variance and kurtosis) are tested. The feature of energy provides the best performance at all potential operating points. It can operate at high genuine acceptance rate of 94% and low false acceptance rate of 0.05%. Its equal error rate is 4.2%. In order to get a higher recognition accuracy, fives kinds of decision fusion strategies are used to combine the outcomes of each classifier, and the testing results show that the Median rule and the Majority voting rule can greatly improve the identification performance. A satisfied recognition rate (97.64%) can be achieved by using the Median rule.

Acknowledgments

The work is partially supported with NSFC Project (90209020) funds from the China Government, the UGC/CRC & CERG funds from the Hong Kong SAR Government and the central fund from The Hong Kong Polytechnic University.

Bibliography

1. Jain, R. Bolle and S. Pankanti, *Biometrics: Personal Identification in Networked Society* (Kluwer Academic Publishers, 1999).
2. D. Zhang,*Automated Biometrics: Technologies and Systems* (Kluwer Academic Publishers, 2000)
3. A.K. Jain, A. Ross and S. Prabhakar, An introduction to biometric recognition, *IEEE Transactions on Circuits and Systems for Video Technology, Special Issue on Image- and Video- Based Biometrics*, vol. 14, no. 1, January (2004)
4. R. Sanchez-Reillo, C. Sanchez-Avilla and A. Gonzalez-Marcos, Biometric identification through hand geometry measurements, *IEEE Transactions on*

Pattern Analysis and Machine Intelligence, vol. 22, no. 18, pp.1168-1171 (2000)

5. D. Zhang, *Palmprint Authentication*, (Kluwer Academic Publishers, 2004)
6. Jain, H. Lin, and R. Bolle, On-Line Fingerprint Verification, *IEEE Transactions on Pattern Analysis and Machine Intelligence*, vol. 19, no. 4 (1997)
7. W. Shi, G. Rong, Z. Bain and D. Zhang, Automatic Palmprint Verification, *International Journal of Image and Graphics*, vol. 1, no. 1, pp. 135-152 (2001)
8. D. Zhang and W. Shu, Two Novel Characteristics in Palmprint Verification: Datum Point Invariance and Line Feature Matching, *Pattern Recognition*, vol. 32, no. 4, pp. 691-702 (1999)
9. W. Shu and D. Zhang, Automated Personal Identification by Palmprint, *Optical Engineering*, vol. 37, no. 8, 2659-2362, (1998)
10. W. Li, D. Zhang, Palmprint Identification by Fourier Transform, *International Journal of Pattern Recognition and Artificial Intelligence*, vol. 16, no. 4, pp. 417-432 (2002)
11. D. Zhang, W.K. Kong, J. You and M. Wong, On-line palmprint identification, *IEEE Transactions on Pattern Analysis and Machine Intelligence*, vol. 25, no. 9, pp. 1041-1050 (2003)
12. Yu Tao, Ernest C.M. Lam, Yuan Y. Tang, Feature Extraction Using Wavelet and Fractal, *Pattern Recognition*, 22, pp. 271-287 (2001)
13. His-Chin Hsin, Texture Segmentation Using Modulated Wavelet Transform, *Transactions on Image Processing*, 9, pp. 1299-1302 (2000)
14. A.K. Jain, S. Prabhakar, L. Hong and S. Pankanti, Filterbank-based fingerprint matching, *IEEE Transactions on Image Processing*, vol. 9, no. 5, pp. 846-859 (2000)
15. Vasily Strela, Peter Niels Heller, The Application of Multiwavelet Filterbanks to Image Processing, *IEEE Transactions on Image Processing*, 8, 548-563 (1999)
16. Q. Xiao, H. Raafat, Fingerprint Image Postprocessing: A Combined Statistical and Structural Approach, *Pattern Recognition*, vol. 24, no. 10, pages 985-992, (1991)
17. L. Zhang and D. Zhang, 2004, Characterization of Palmprints by Wavelet Signatures via Directional Context Modeling, *IEEE Transactions on System, Man, and Cybernetics Part-B*, Volume: 34, Issue: 3, 1335-1347 (2004)
18. Thierry Blu and Michael Unser, Wavelets, Fractals, and Radial Basis Functions, *IEEE Transactions on Signal Processing, VOL.* 50, NO. 3, 543-553 (2002)
19. M. Tico, P. Kuosmanen, J. Saarinen, Wavelet Domain Features for Fingerprint Recognition, *Electronics Letters*, 37, 21-23 (2001)
20. Yvette Mallet, Danny Coomans, Jerry Kautsky, and Olivier De Vel, "Classification Using Adaptive Wavelets for Feature Extraction, *IEEE Transactions on Pattern Analysis And Machine Intelligence*, VOL. 19, NO. 10 (1997)
21. Josef Kittler, Mohamad Hatef, Robert P.W. Duin, and Jiri Matas, On Combining Classifiers, *IEEE Transactions on Pattern Analysis and Machine*

Intelligence, vol. 20, pp. 226-239 (1998)

22. John Sheffeld, The Future of Fusion, *Nuclear Instruments and Methods in Physics Research*, A 464, 33-37 (2001)

23. Ravi P. Ramachandrana, Kevin R. Farrellb, Roopashri Ramachandrana, Richard J. Mammonec, Speaker recognition-general classifier approaches and data fusion methods, *Pattern Recognition*, 35, 2801-2821 (2002)

24. Andrew Senior, A Combination Fingerprint Classifier, *IEEE Transactions on Pattern Analysis and Machine Intelligence*, vol. 23, pp. 1165-1174 (2001)

25. Lei Xu, Adam Krzyzak, and Ching Y. Suen, Methods of Combining Multiple Classifiers and Their Applications to Handwriting Recognition, *IEEE Transaction on System, Man, and Cybernetics*, vol. 22, no. 3, pp. 418-435 (1992)

26. N. Duta, Anil K. Jain, K.V. Mardia, Matching of palmprint, *Pattern Recognition Letters*, pp. 477-485 (2002)

27. J. You, W. Li and D. Zhang, Hierarchical Palmprint Identification via Multiple Feature Extraction, *Pattern Recognition*, vol. 35, no. 4, pp. 847-859 (2002)

Chapter 10

Behavioral Biometrics for Online Computer User Monitoring

Ahmed Awad E. Ahmed*, Issa Traoré[†]

*Information Security and Object Technology Laboratory,
Department of Electrical and Computer Engineering,
University of Victoria, Canada*
* aahmed@ece.uvic.ca
[†] itraore@ece.uvic.ca

Traditional biometrics technologies such as fingerprints or iris recognition systems require special hardware devices for biometrics data collection. This makes them unsuitable for online computer user monitoring, which to be effective should be non-intrusive, and carried out passively. Behavioural biometrics based on human computer interaction devices such as mouse and keyboards do not carry such limitation, and as such are good candidates for online computer user monitoring. We present in this chapter artificial intelligence based techniques that can be used to analyze and process keystroke and mouse dynamics to achieve passive user monitoring.

Contents

Glossary

FAR — False Acceptance Rate
FRR — False Rejection Rate
FTE — Failure to Enroll
FTC — Failure to Capture
MLP — Multi-Layer Perceptrons

10.1. Introduction

Biometrics can be defined as a set of distinctive, permanent and universal features recognized from human physiological or behavioural characteristics [17], [18]. As such, biometrics systems are commonly classified into two categories: physiological biometrics and behavioural biometrics. Physiological biometrics, which include finger-scan, iris-scan, retina-scan, hand-scan, and facial-scan use measurements from the human body. Behavioural biometrics such as signature or keystroke dynamics use measurements based on human actions [3], [4], [13], [14]. Due to their strong variability over time, so far, behavioural biometric systems have been less successful compared to physiological ones [3]. Despite such limitation, behavioural biometrics such as mouse dynamics and keystroke dynamics, carry the greatest promise in the particular field of online computer user monitoring [7]. For such application, passive or non-intrusive monitoring is essential. Unfortunately most biometrics systems require special hardware device for biometrics data collection, restricting their use to only networks segments where such devices are available. Behavioural biometrics such as mouse dynamics and keystroke dynamics are appropriate for such context because they only require traditional human-computer interaction devices.

In this chapter, we present some techniques for extracting and analyzing mouse and keystroke dynamics data for online computer user monitoring. While the use of mouse dynamics for online monitoring is straightforward, the use of keystroke dynamics for such purpose faces the important challenges underlying the need for free-text detection.

10.2. Biometrics Modes and Metrics

Biometric systems operate in two modes, the enrollment mode and the verification/identification mode. In the first mode, biometric data is acquired using a user interface or a capturing device, such as a fingerprints scanner. Raw biometric data is then processed to extract the biometric features representing the characteristics, which can be used to distinguish between different users. This conversion process produces a processed biometric identification sample, which is stored in a database for future identification/verification needs. Enrolled data should be free of noise and any other defects that can affect its comparison with other samples. In the second mode, biometric data is captured, processed and compared against the stored enrolled sample. According to the type of application, a verification or identification process will be conducted on the processed sample as follows.

> *Verification process: conducts one-to-one matching by comparing the processed sample against the enrolled sample of the same user. For example, user authentication at login: the user declares his identity by entering his login name. He then confirms his identity by providing a password and biometric information, such as his signature, voice password, or fingerprint. To verify the identity, the system will compare the user's biometric data against his record in the database, resulting with a match or non-match.*
>
> *Identification process: matches the processed sample against a large number of enrolled samples by conducting a 1 to N matching to identify the user; resulting in an identified user or a non-match.*

In order to evaluate the accuracy of a biometric system, the following metrics must be computed:

False Acceptance Rate (FAR), the ratio between the number of occurrences of accepting a non-authorized user compared to the number of access trials.

False Rejection Rate (FRR), the ratio between the number of false alarms caused by rejecting an authorized user compared to the number of access trials.

Failure to Enroll (FTE), the ratio characterizing the number of times the system is not able to enroll a user's biometric features. This failure is caused by poor quality samples during enrollment mode.

Failure to Capture (FTC), the ratio characterizing the number of times the system is not able to process the captured raw biometric data and extract features from it. This occurs when the captured data does not contain sufficient information to be processed.

FAR and FRR values can vary significantly depending on the sensitivity of the biometric data comparison algorithm used in the verification/identification mode; FTE and FTC represent the sensitivity of the raw data processing module.

In order to tune the accuracy of the system to its optimum value, it is important to study the effect of each factor on the other. If the system is designed to minimize FAR to make the system more secure, FRR will increase; on the other hand, if the system is designed to decrease FRR by increasing the tolerance to input variations and noise, FAR increases. By considering the current utilization of biometrics in the market, we note that they are widely used for secure identification and verification purposes. Other security systems, such as intrusion detection systems, do not use this technology because the application requires continuous monitoring and recording of biometric information, which can sometimes be unavailable during the detection process. Mouse dynamics biometrics, which we introduce in this paper, happens to be a good fit for such an application [1].

> *Mouse dynamics biometric system does not need any training, operating totally transparent from the user, as its enrollment mode does not require the user to present any particular biometric information by conducting specific actions.*

Figure 10.1 shows a generic architecture, which covers the components involved in the implementation of a behavioural biometric based detector. The flow of control and data is shown for the three possible scenarios: enrolling a user, verifying his identity, and identifying an unknown user.

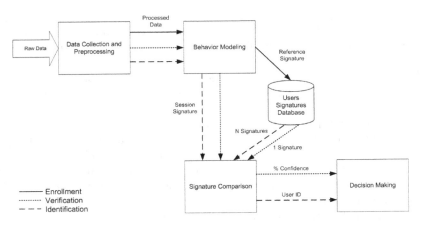

Fig. 10.1. Generic architecture of a behavioural biometric system.

10.3. Mouse dynamics

10.3.1. *Overview*

In contrast with keystroke dynamics, which has been widely studied in computer security, previous works on mouse dynamics have, so far, been limited to user interface design improvement [16], [15], [12]. In particular, mouse movement analysis has been the purpose of extensive research works. Studies have been conducted to establish the applicability of Fitts' law in predicting the duration of a movement to a target based on the size of the target and the distance from the starting point to the target [16]. According to Fitts' law, the mean movement time for a movement with distance A to a target with width W is as follows: $MT = a + b(\log_2(2A/W))$ where a and b are empirically determined parameters [16].

Researches conducted on mouse dynamics have focused mainly on the formalization of the measured data after fixing the environment variables. In our research, we target the biometric identification problem by focusing on extracting the behavioural features related to the user and using these features in computer security. Mouse dynamics is a new behavioural biometric introduced at the Information Security and Object Technology (ISOT) research lab at the University of Victoria [1], [2]. Mouse dynamics can be described as the characteristics of the actions received from the mouse input device for a specific user, while interacting with a specific graphical user interface. The raw data collected for each mouse movement consist of the *distance, time,* and *angle*. We broadly refer to the angle as the direction of movement. *Silence* periods between mouse movements also carry valuable information that can be used for biometric recognition.

The user characteristics can be described by a set of factors generated as a result of analyzing the recorded mouse actions. Those factors represent the Mouse Dynamics Signature, which can be used in verifying the identity of the user.

The architecture described in Fig. 10.1 can be applied for the implementation of any detector based on this biometric. Figure 10.2 shows sample data provided to the behavior-modeling component for processing. Each point in the figure represents a movement of the mouse with a specific distance, which was completed in a specific time. By examining the output of the data collection and processing component and comparing it to its output for different sessions for the same user, one can find a pattern characterizing the data. In some cases one can even differentiate between readings for two different users.

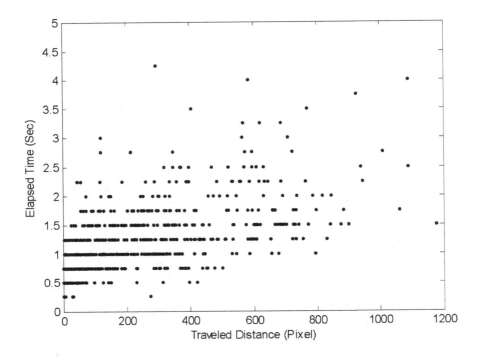

Fig. 10.2.   Graph showing a relationship between speed and distance based on sample intercepted data.

10.3.2. *Detection Process*

In order to automate the detection process, however, it is important to formalize the data in such a way that it can be used for comparison. Various statistical analysis packages can be used to achieve this goal, according to the characteristic of each factor [6], [17]. We choose to use Neural Networks to approximate the collected data to a curve that can be used to identify the user behavior. Function approximation is considered one of the most common uses of neural networks. It was shown by Hecht-Nielsen that for any continuous mapping of f with n inputs and m outputs, there must exist a three layer neural network with an input layer of n nodes, a hidden layer with $2n + 1$ nodes, and an output layer with m nodes that implements f exactly. According to those results, it is expected that neural networks can approximate any function in the real world. Hecht-Nielsen established that back propagation neural network is able to implement any function to any desired degree of accuracy. For our neural network, we use a feed-forward multi-layer perceptrons (MLP) network. MLP is one of the most popular network architectures; it is widely used in various applications. Our network involves a number of nodes organized in a layered feed-forward topology consisting of an input layer, an output layer and one hidden layer. All connections between nodes are feeding forward from inputs toward outputs. The MLP network uses a linear Post Synaptic Potential (PSP) function; the PSP function used is the weighted sum function. The transfer function used in this network is the log-sigmoid function. A linear transfer function is used for the input and output layers to allow the expected input and output range. For faster training, the network is initialized with the weights and biases of a similar network trained for a straight line. The output of our neural network can be described by the following equation:

$$y = \sum_{j=1}^{N} \left(\frac{w_{2j}}{1 + e^{(\sum_{i=1}^{N} w_{1i}x) - b_{1j}}} \right) - b_{21}, \tag{10.1}$$

where w_{ij} and b_{ij} represent the weights and biases of the hidden and output layers respectively, x is the input to the network, and N represents the number of nodes in the hidden layer (which is set to $N = 5$ in our design). We use the back propagation algorithm to train the network. The error criterion of the network can be defined as follows:

$$E = \frac{1}{2} \sum_{i=1}^{p} (t_i - y_i(x_i, w))^2, \tag{10.2}$$

where w represents the network weights matrix and p is the number of input/output training pairs set (x_i, y_i). During the behavior modeling stage, the neural network is trained with processed raw data. Input vectors and their corresponding target vectors are used. The back propagation-training algorithm is used to train a network until it can approximate a function describing the collected data. Curve over fitting is one of the common problems related to this approach; this problem occurs when the resulted curve has high curvatures fitting the data points. The main cause of this problem is when the network is over trained with a high amount of data; such a problem increases the noise effect and reduces the generalization ability of the network.

In order to avoid the over fitting problem, first the right complexity of the network should be selected. In our design, we tested different network configurations and concluded that a network with a single hidden layer containing five perceptrons produces the best results. Second, the training of the network must be validated against an independent training set.

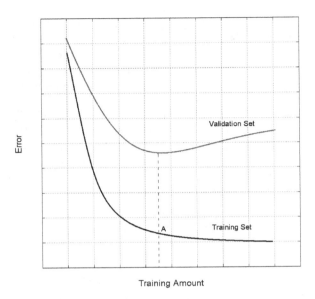

Training Amount

Fig. 10.3.   Determining the training stop point for curve approximation neural network.

At the beginning of the training, the training error and the validation error will decrease until we reach a point where the validation error will start to increase. We call this point the *stop point* (corresponds to point A in Fig. 10.3), where the training should stop to obtain the desired generalization. After the network-training curve reaches the stop point, the network will be fed with a test stream presenting the spectrum of the input data; the result is a curve approximation of the training data. This curve is considered as a biometric factor in the Mouse Dynamic Signature.

Figure 10.4 shows the mouse dynamics signature for three different users, the figure shows the relation between the traveled distance and the elapsed time for any performed actions. Figure 10.5 shows the same curves computed over five different sessions for the same user. Each of those curves was produced as a result of training the neural network with session raw data similar to what is shown in Fig. 10.2. The neural network approximates the collected data to a curve, which can be used to check the user identity by comparing it to a reference signature. Deviations between the curves can be used to detect sessions belonging to different users, or to recognize sessions belonging to the same user.

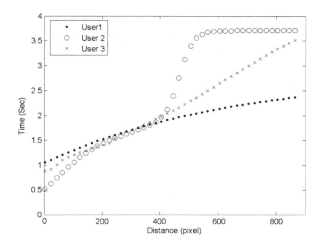

Fig. 10.4.   Mouse signature for three different users.

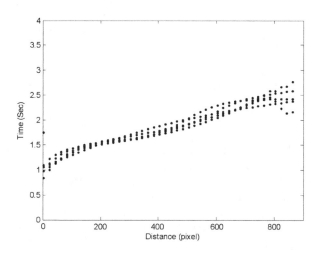

Fig. 10.5.   Mouse signature computed over five different sessions for the same user.

10.3.3. *Silence Analysis*

As mentioned above, *silence* periods can also be considered as valid and useful mouse biometrics data. Analysis of the silence periods can lead to the detection of a number of distinctive characteristics for each user. In this analysis we only consider the short silence periods (less than 20 sec), which happens between movements. Longer silence periods may occur as a response to a particular action like reading a document, and usually contain noise.

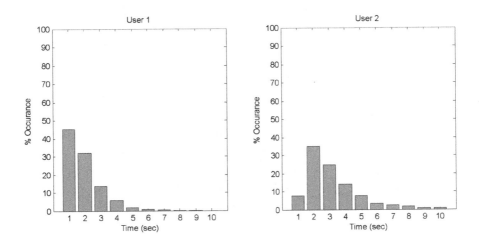

Fig. 10.6.   Silence Analysis: histogram of the silence time for two different users.

Figure 10.6 illustrates one of the signature factors that can be derived from silence analysis. This factor is based on the histogram of the short silence periods. Each bar in the figure represents the number of silence periods reported in a user session where the silence period is within a 2 sec interval covering a spectrum of 20 sec. The figure shows the histogram for two different users, we can easily notice the difference in the behavior. From the first bar $(0 < t < 2)$ we can notice that for the first user 45% of his silence periods are in this category, while for the second user only 8% of his silence periods last less than 2 sec.

Figure 10.7 illustrate silence histogram computed over three different sessions for the same user. We can notice the similarity of the behavior across these different sessions.

Many other factors can be computed from the raw mouse data, and used for biometrics recognition. We refer the reader to [1], [2] for more about these parameters.

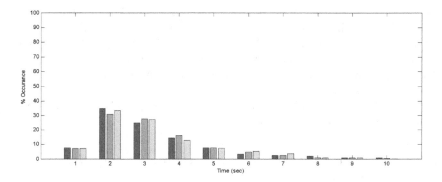

Fig. 10.7. Silence Analysis: histogram of the silence time based on three different sessions for the same user.

10.4. Keystroke Dynamics

10.4.1. *Overview*

Keystroke dynamics recognition systems measure the dwell time and flight time for keyboard actions, and use such data to construct a set of digraphs, tri-graphs or n-graphs producing a pattern identifying the user. User authentication is the most suitable application for such technology.

Since the 1980s, many researches have been done in this area. Most of the researches focus on using this technology for user authentication or access control [3], [4], [5], [8], [10], [11]. The various works reported in the keystroke literature involves using a wide range of statistical methods to analyze keystroke dynamics. For instance, Brown and Rogers used neural networks to solve the problem of identifying specific users through the typing characteristics exhibited when typing their own names [5]. In [11], Monrose and Rubin developed a technique to harden passwords based on keystrokes dynamics. More recently, Bergadano and colleagues presented a new technique based on calculating the degree of disorder of an array to quantify the similarity of two different samples [3]. Most of these works focus on fixed text detection. Because of the time limitation of the identification process, the user is asked to type a pre defined word or set of words in order to get reasonable amount of data for the identification. During the enrollment process, the user is also required to enter the same fixed text. For passive monitoring, we need to be able to detect the user without requiring him to enter a predefined message or text. So free text detection is essential for our purpose. However, free text detection presents huge challenges, which explain the limited number of related work published in the literature. So far, one of the most significant works in this area was authored recently by Guneti and Picardi who adapt for free text detection the technique based on the degree of disorder of an array introduced in [3] for fixed text detection. Still the importance of this issue warrants investigating alternative techniques. In the rest of this section, we discuss the different factors underlying this issue and propose three different techniques, which can be used to tackle them.

10.4.2. *Free Text Detection Using Approximation Matrix Technique*

The first approach we propose is based on digraph analysis. The approach utilizes a neural network to simulate the user behavior based on the detected digraphs.

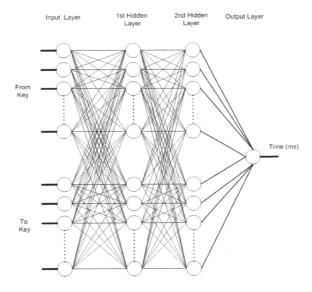

Fig. 10.8.   Neural network model used for free text detection based on approximation matrix.

The neural network used for this approach depicted by Fig. 10.8 is a feedforward multilayer perceptrons network. The training algorithm is back propagation. The network consists of four layers: an input layer, two hidden layers, and a single node output layer. The input layer consists of

N number of nodes where $N = 2 \times n$, with n corresponding to the *Number of Monitored Keyboard keys*. Input to the nodes is binary 0 or 1, as each node in the input layer represents a key. The first n nodes represent the key where the action is started at, and the second n nodes represent the key where the action ends. Each batch of nodes should have only one input set to 1 while the other inputs are set to 0; the node set to 1 represents the selected key.

During the enrollment phase, a batch of M actions will be collected and fed to the behavior modeling neural network as training data. A simulation will run after the neural network has been trained with this batch, this simulation will consist of the set of non-redundant actions collected from the enrollment data. The result of this simulation will be stored for each user as well as the training data, which will be used also in the verification stage. During the verification mode a small batch of actions will be used in this stage to verify the user identity. This batch will be added to the training batch of the user's neural network, resulting a network with different weights. The effect of the small batch on the network weights represents a deviation from the enrollment network. In order to measure this deviation, another simulation will run on this network with the same batch prepared for the enrollment process for the specific user. By comparing the result of this simulation to the enrollment stage result, the deviation can be specified. An approach that can be used here is to calculate the sum of the absolute difference of the two results, if this deviation is low then the collected sample is for the same user, if not then this sample is for another user.

Figure 10.9 shows the detector architecture and the flow of data in enrollment and detection modes. In enrollment mode extracted monographs and digraphs are encoded with a mapping algorithm. This process is needed in order to convert key codes into another representation, which is relevant and meaningful as an input to the neural network.

Since this detector is based on free input text it is very important to be able to evaluate if the collected data is enough during enrollment mode. The aim of this research is to develop a technique to help in minimizing the amount of data needed for the enrollment process, by extracting the needed information from the information detected so far.

In order to approximate unavailable digraphs, we use a matrix-based approximation techniques. Specifically we use a pair of matrix named *coverage matrix* and *approximation matrix*. *Coverage matrix* is a two dimensional matrix, which is used to store the number of occurrences of the observed graphs in the enrollment mode. Keeping track of such information helps in different areas such as in evaluating the overall coverage of the enrollment process and the development of a customized enrollment

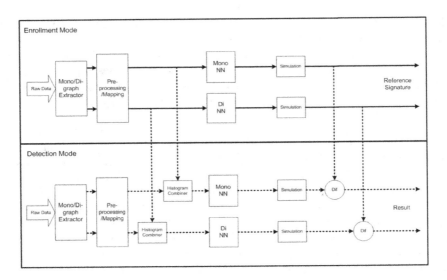

Fig. 10.9.   Keystroke dynamics detector architecture.

scenario, which can be used in case of low coverage. *Approximation matrix*, which is a two dimensional matrix represents the relations between the keys and how close or far they are from each other; the matrix will be initialized with numbers representing the relative distances between the keys on the keyboard.

Figure 10.10 illustrates how the approximation process is performed. Let's assume that an approximation for the EB digraph is needed. We can detect that directly from its value corresponding to 0 in the coverage matrix, depicted by Fig. 10.10(b). The approximation matrix, depicted by Fig. 10.10(a) will be used to locate alternative entries (for each key), which have the lowest distance in the matrix; in this case these correspond to (D, F) and (G, H) respectively. From this step we can enumerate the tentative approximations, which correspond in this case to DG, DH, FG, and FH. In the next step the distance of each combination will be calculated from the approximation matrix (underlined numbers in Fig. 10.10(a)), where they will be sorted according to their closeness to the original distance of the approximated digraph ($AppMatrix(EB) = 3$). The sorted result is (FH, DG, DH, FG). The Coverage matrix will be used to make the final decision out of the sorted result. The matrix in Fig. 10.10(b) shows only the weights of the tentative combinations. Notice that digraph FH has a coverage of 30, which means that it is a good candidate (the best fit in this case, since it is also the closest fit in the approximation matrix). The

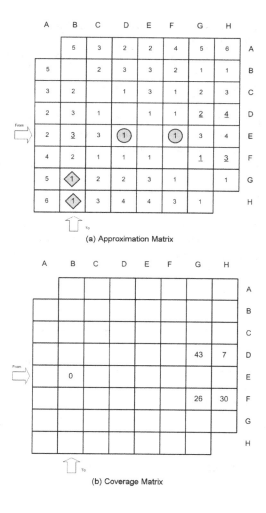

(a) Approximation Matrix

(b) Coverage Matrix

Fig. 10.10.   Example of how to approximate unavailable digraphs.

second alternative *DG* also has good coverage, while *DH* has a relatively low coverage.

10.4.3.   *Free Text Detection Based on Keyboard Layout Mapping*

One of the important factors to be considered in the enrollment phase is the amount of data needed to enroll the user and create a signature modeling his behavior. The aim is to minimize the enrollment time as much as possible without affecting the accuracy of the system in detection mode. Since the key codes do not reflect the relation between the keys like their absolute or relative positions, the pre processing stage (Fig. 10.9) should include a mapping mechanism in order to convert those sets of keystrokes into numbers, which are suitable to train the network with. Such numbers should reflect a specific characteristic for each key and its relation to other keys. The mapping method used in the previous technique is a binary mapping since each key is represented by its own network input, and the effect of other keys was removed by restricting the input to only one high input at a time. This technique is considered to involve high computation power requirement, since the number of nodes in the neural network is large.

The second approach we propose is based on a keyboard layout mapping technique. The function of this technique is to replace each key code by its location on a previously identified keyboard layout. In our implementation we use the QWERTY keyboard layout depicted by Fig. 10.11

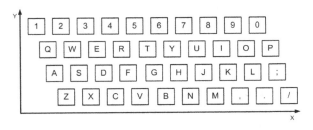

Fig. 10.11.   QWERTY Keyboard layout can be used to encode keys as input to the neural network.

Each key will be presented by a pair of numbers representing its x and y location. Neural networks can still be used to model the relation between the keys and the time used to move between them.

Figures 10.12(a) and 10.12(b) show the neural networks used for the digraph and monograph analysis respectively. The networks in this case are lighter than the one used in the previous technique. The numbers of input nodes are 2 and 4 for each network respectively as only 2 network inputs are needed to represent a key. The hidden layers consist of 5 nodes for the monograph network and 12 nodes for the digraph network.

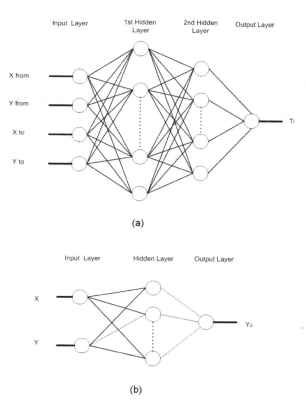

Fig. 10.12. Neural Networks used for free text detection based on keyboard layout mapping: (a) Digraph Analysis and (b) Dwell Time Analysis.

10.4.4. *Free Text detection Based on Sorted Time Mapping*

Another mapping technique, which can be used to prepare the data for the neural network is to sort the key codes based on associated average time, and accordingly to map them with corresponding order. For the monographs set, the average time is the dwell time. So, in this case the key codes will be sorted according to the average dwell time. Key codes will be mapped to their order in the sorted list before being fed to the neural network. In this case the mono network will have only one input node. For the digraph set the sorting order is different. Digraphs will be sorted two times: the first time they will be sorted according to the average of the times of all digraphs with the same code in the from-key; the second time they will be sorted according to the average of the to-key. A key code will be mapped to its sort order according to whether it is *to* or *from*. The input layer for the neural network in this case consists of two nodes.

The keyboard layout mapping technique doesn't require a full coverage of all possible set of keys as it is based on a very definite reference scheme, which is the keyboard layout. So missing a number of keys during enrollment phase does not prevent this technique from working. However, for the sorting technique it is mandatory that the user provide input at enrollment phase covering the whole set of monitored keys. This criterion makes the approximation matrix technique and the keyboard layout technique more suitable if the system is designed to hide the enrollment process from the user. The sorting technique will be very effective if a controlled enrollment procedure is used.

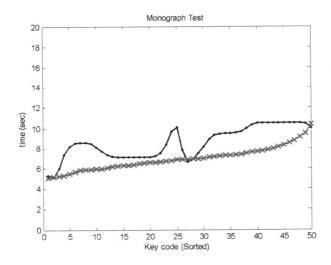

Fig. 10.13.   Comparing signatures (Monograph Analysis): User 2's session compared to user 1's reference signature.

Figure 10.13 shows monographs signatures for two different users. The figure shows the relation between the key code, and the dwell time involved. The sequence of the keys represented by those codes is different for each user as they were sorted for each user individually. The Y-axis represents the output of the neural network after it has been trained with the sorted key order. Taking a look again at Fig. 10.9, the output of the behavior-modeling component in this case is equivalent to the curve shown in Fig. 10.13 for monograph analysis and to a set of curves or a 3-D matrix for digraph analysis.

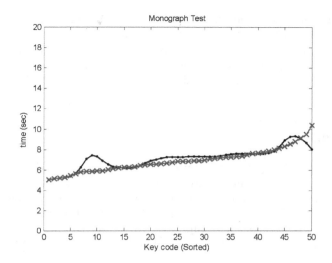

Fig. 10.14. Comparing signatures (Monograph Analysis): User 1's session compared to user 1's reference signature.

Figure 10.14 shows Monograph signature calculated from one of user 1's sessions compared to his reference signature. Figure 10.13 shows the signature calculated for one of user 2's sessions during detection mode as a result of supplying his session data to User 1's neural network. We can note from the figure that the difference between this curve and user 1's reference signature is high compared to the difference between user 1's curves.

10.5. Conclusion

Biometric systems are widely deployed to ensure security in a large spectrum of today's industries. Some implementations rely solely on biometrics; others utilize it to increase their level of security. The choice to implement a specific biometric technology in an environment is ruled by a number of factors. Examples of those factors are its accuracy, cost, user acceptance, and stability on the long term.

In this chapter we present methods based on statistical and artificial intelligence techniques targeting the use of behavioural biometrics for online computer user monitoring. The techniques presented in this chapter work on a large set of features from a newly introduced biometric (mouse dynamics) and the keystroke dynamics biometric without putting any restriction on the user at any stage of operation making it suitable for online identity verification and detection applications. As those biometrics gain high acceptability they also gain a relatively high accuracy. In experiments conducted in our lab, we achieve an overall accuracy of $FAR = 0.00651$, and $FRR = 0.01312$. This is based on a number of experiments conducted locally and remotely through our lab involving a set of 22 participants.

Acknowledgments

This work was partly supported by the Natural Science and Engineering Research Council of Canada (NSERC) through an Idea-to-Innovation grant. We would like to thank all the participants in the experimental evaluation of our framework, as well the Innovation Development Corporation of the University of Victoria for their support. We would like also to thank Dr Svetlana Yanushkevich for her review and valuable comments on this work.

Bibliography

1. Ahmed, A. A. E. and Traoré, I. (2005). Detecting computer intrusions using behavioural biometrics, *Proc. 3rd Annual Conf. Privacy, Security and Trust*, New Brunswick, Canada, pp. 91–98.
2. Ahmed, A. A. E. and Traoré, I. (2003). System and method for determining a computer user profile from a motion-based input device. *PCT patent application with the World Intellectual Property Organization*, filed by the University of Victoria (PCT/CA2004/000669) filing date: 3 May 2004 priority date: 2 May 2003.
3. Bergadano, F., Guneti, D. and Picardi, C. (2002). User authentication through keystroke dynamics, *ACM Trans. Information and System Security*, **5**, 4 pp. 367–397.
4. Bleha, S., Slivinsky, C. and Hussein, B. (1990). Computer-access security systems using keystroke dynamics, *IEEE Trans. Patt. Anal. Mach. Int.* **12**, pp. 1217–1222.
5. Brown, M. and Rogers, S. J. (1993). User Identification Via Keystroke Characteristics of Typed Names Using Neural Networks. *Int. J. Man-Machine Studies*, **39**, pp. 999–1014.
6. Coello Coello, C. A. (2006). Evolutionary algorithms: basic concepts and applications in biometrics, this issue.
7. Denning, D. (1987). An intrusion detection model, *IEEE Trans. Software Engineering*, **13**, 2, pp. 222–232.
8. Gaines, R., Lisowski, W., Press, S. and Shapiro, N. (1980). *Authentication by keystroke timing: some preliminary results*, Rand. Report R-256-NSF. Rand Corporation.

9. Guneti D. and Picardi, C. (2005). Keystroke analysis of free text, *ACM Trans. Information and System Security*, **8**, 3, pp. 312–347.

10. Legget, J. and Williams, G. (1988). Dynamic identity verification via keystroke characteristics, *Int. J. Man-Mach. Stud.* **35**, pp. 859–870.

11. Monrose, F. and Rubin, A. (1997). Authentication via keystroke dynamics, *Proc. 4th ACM Conf. Computer and Communications Security*, pp. 48–56.

12. Oel, P., Schmidt, P. and Shmitt, A. (2001). Time prediction of mouse-based cursor movements, *Proc. Joint AFIHM-BCS Conf. Human-Computer Interaction*, **2**, pp. 37–40.

13. Parker, J. R. (2006). Composite systems for handwritten signature recognition, this issue.

14. Popel, D. V. (2006). Signature analysis, verification and synthesis in pervasive environments, this issue.

15. Whisenand, T. G. and Emurian, H.H. (1999). Analysis of cursor movements with a mouse. *Computers in Human Behavior*, **15**, pp. 85–103.

16. Whisenand, T. G. and Emurian, H. (1996). Effects of angle of approach on cursor movement with a mouse: consideration of fitts' law, *Computer in Human Behavior*, **12**, 3, pp. 481–495.

17. Yanushkevich, S. N., Stoica, A. and Shmerko, V. P. (2006). Fundamentals of biometric-based training system design, this issue.

18. Yanushkevich, S. N., Stoica, A., Shmerko, V. P. and Popel, D. V. (2005). *Biometric Inverse Problems*, Taylor & Francis/CRC Press.

PART 3
BIOMETRIC SYSTEMS AND APPLICATIONS

LARGE-SCALE BIOMETRICS

- *Positive identification and verification*
- *Negative identification and verification*
- *Errors*
- *Performance evaluation*
- *Reliability and selectivity*

EVOLUTIONARY ALGORITHMS

- *Evolutionary strategies*
- *Applications to biometrics*

MEASUREMENTS OF BIOMETRICS

- *Segmentation*
- *Disambiguation*
- *Semantics*

HUMAN BIOMETRIC SENSOR INTERFACE

- *Biometric system performance*
- *Biometric sample quality*
- *Ergonomics in biometric system design*

BIOMETRIC-BASED TRAINING SYSTEM DESIGN

- *Physical access security systems*
- *Techniques of training estimations*
- *Prototyping*
- *Knowledge-based decision making support*

Chapter 11

Large-Scale Biometric Identification:
Challenges and Solutions

Nalini K. Ratha*, Ruud M. Bolle†, Sharath Pankanti‡

*IBM Thomas J. Watson Research Center,
Exploratory Computer Vision Group,
19 Skyline Drive, Hawthorne, NY 10532, USA
*fratha@us.ibm.com
†bolle@us.ibm.com
‡sharatg@us.ibm.com*

In addition to law enforcement applications, many civil applications will require biometrics-based identification systems and a large percentage is predicted to rely on fingerprints as an identifer. Even though fingerprint as a biometric has been used in many identification applications, mostly these applications have been semi-automatic. The results of such systems often require to be validated by human experts. With the increased use of biometric identification systems in many real-time applications, the challenges for large-scale biometric identification are significant both in terms of improving accuracy and response time. In this paper, we briey review the tools, terminology and methods used in large-scale biometrics identification applications. The performance of the identification algorithms need to be significantly improved to successfully handle millions of persons in the biometrics database matching thousands of transactions per day.

Contents

Glossary

FTA — Failure to Acquire

FTE — Failure to Enroll

FAR — False Accept Rate

FRR — False Reject Rate

11.1. Introduction

Biometrics-based identification systems use machine representations (templates) of the biometrics of the multiple, potentially many, enrolled users in the identification system. When a user presents an input biometric to the identification system, a template is extracted from the input signal and it is determined which of the enrolled biometric templates, if any, matches with the query. Such positive identification systems in effect try to answer the question,

Who is this?

Identification systems base the answers only on biometric data. No other means of identification, like an identification number or a user-ID, is supplied to the system.

There are two identification scenarios: *positive* and *negative* identification, distinguished principally by whether the enrolled subjects are cooperative users who wish to be identified to receive some benefit or access, or users who are to be denied access or benefit and consequently may not want to be recognized (e.g. the watch-list scenario). As noted above, a positive biometric system answers the question,

Who am I?

A negative biometric system, on the other hand, confirms the truth of the statement

I am not who I say I am not.

That is, the subject is a legitimate user who is not on the watch list. The important difference between these scenarios is in the enrolled subjects, i.e. voluntary versus involuntary enrollment.

Table 11.1. There is no generally accepted definition of scale of identification systems.

Scale	# of Individuals	# of Stored Samples
Small	330	1,000
Medium	3,300	10,000
Large	33,000	100,000
Very large	330,000	1,000,000
Extremely large	3,300,000	10,000,000

Identification systems by necessity contain a centralized database since upon each query every enrolled template needs to be accessible for matching with the input template. There is no general agreement on what database size would make an identification system "large scale." Phillips [1] gives some approximate definitions for the scale of biometric identification systems, which are shown in Table 11.1 Of course, there is a theoretical (if non- constant) upper bound on the scale of the ultimate large-scale system, the world scale a system which could identify every human being, perhaps by any finger. The largest human identification system would have around 60 billion enrolled biometrics.

11.2. Biometrics Identification versus Verification

Figure 11.1 shows the basic building blocks of a biometric authentication system:

(1) A database where one template, in the case of a verification system, or *multiple* templates, in the case of an identification system are stored.
(2) An input signal acquisition devise like a fingerprint reader.
(3) A signal processing module that extracts a template representation from the input signal.
(4) A biometric matcher that compares the two templates and reports the degree of match in terms of a (normalized) match score or similarity $s \in [0; 1]$.

Biometric identification is based solely on biometric characteristics of a subject, i.e. it is based only on the biometric signal. Such a system has access to a biometric database containing biometric samples or

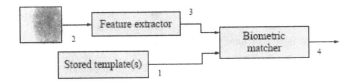

Fig. 11.1. A simple block diagram of the central components of an authentication system.

representations of the biometrics of the enrollees. A biometric identification system further has the capability of searching the biometric database to determine if there are any database entries that resemble the input subject sample. In the simplest form, the database templates are matched to the input sample one by one. The output of the matching phase of an identification system is some candidate list of subject identifiers from the database that resemble the input biometric.

Biometric verification, on the other hand, differs from identification in that the presented biometric is only compared with a single enrolled biometric entity. There may still be a large enrolled population, but the user presents a token which indicates only the user's biometric template from the database for comparison. The statement that is verified by a verification system is:

"I am who I say I am."

The user lays a claim on a certain identity and the biometric systems checks the statement's identity.

Hence biometric authentication is achieved in one of two ways, both are illustrated in Fig. 11.2. To the right we have an identification system where the user presents a biometric and the system establishes the identity of the user. The left side shows a verification system, which, like an identification system, has access to a biometric database. This database contains biometric templates, associated with subjects. However, unlike in a biometric identification system, a distinct identifier (ID number) is associated with each biometric template. Hence, from the database, the biometric template associated with some subject is easily retrieved using the unique ID. The input to the verification system is a biometric input sample of the subject in addition to some identifier ID associated with the identity that the subject claims to be. The output of the matching phase is an "Accept/Reject" decision — the biometric system either accepts or rejects the subject's claim.

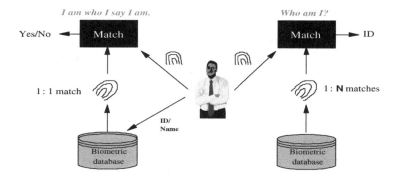

Fig. 11.2. The difference of identification and verification.

As opposed to identification applications where a central database is a requirement, for verification applications, there are two possible database configurations:

Centralized database. A central database stores the biometric information of the enrolled subjects. The user presents some identity token (swipes a card, types a user ID) stating an identity, which allows the retrieval of the corresponding biometric template. This is compared with the newly presented biometric sample. This includes applications where one subject or a small group of subjects is authorized to use some application, such as, laptop, PDA, and cell phones.

Distributed database. The biometric information is stored in a distributed fashion (e.g. smartcards). There is no need to maintain a central copy of the biometric database. A subject presents some biometric device containing a single biometric template directly to the system, for instance by swiping a mag-stripe card or presenting a smartcard. The biometric system can compare this to the newly presented biometric sample from the user and confirm that the two match.

Typically some additional information, such as an ID or name, is also presented for indexing the transaction, and the presentation must be made through a secure protocol.

In practice, many systems may use both kinds of database — a distributed database for of-line day-to-day verification and a central database for card-free, on-line verification or simply to allow lost cards to be reissued without reacquiring the biometric.

11.3. Terminology

As noted above, a biometric identification system can be operated in two different modes. These modes, in turn, define different error terminologies for positive and negative identification applications.

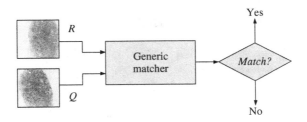

Fig. 11.3. A generic biometric matcher determines if two biometrics look alike.

The generic biometric matcher of Fig. 11.3 takes a query biometric Q and a reference biometric R as input and gives as output "MATCH/NO-MATCH" or "Yes/No." Consequently, such a matcher can make two types of errors:

(1) *A False Match.* Decide that the biometrics Q and R are the same, while in reality they are not.

For the two applications a false match results in a false accept for positive identification and a false reject for negative identification:

(a) The biometric Q has impersonated biometric R in the member database and is falsely accepted to enter the system.

(b) Biometric Q mistakenly "looks like" biometric R in the surveillance database ("watch list") and is falsely rejected from entering the system.

(2) *A False Non-Match.* Decide that the biometrics Q and R are not the same, while in reality they are.

A false non-match results is a false reject for positive identification and a false accept for positive identification:

(a) A legitimate biometric Q is falsely rejected from entering the system because it fails to look like the biometric R in the member database.

(b) Biometric Q has disguised himself not to look like biometric R in the watch-list and is falsely accepted to enter the system.

11.4. Positive Identification and Verification

Positive identification refers to determining that a given individual is enrolled in a (member) database. The individual is accepted or rejected to enter some premises or make use of services. Therefore, in this application, the possible errors

"False accept" and "False reject"

are used:

- False Match → *False Accept:* A subject is falsely accepted, causing an intruder to enter the system.
- False Non-Match → *False Reject:* A legitimate subject is denied service, is falsely rejected from entering the system.

These are the same errors that can occur in a biometric verification system. In fact, positive identification is functionally similar to verification.

11.5. Negative Identification and Screening

Negative identification amounts to determining that a subject is not in some negative databases. This database could for instance be some "most wanted" database. This is also called "screening," because the input subject is in effect screened against the biometric database. This is a very different biometric system where the possible errors are often referred to as

"False negative" and "False positive":

- False Match → *False Positive:* A genuine subject is erroneously flagged as being on the watch list and is denied service or access.
- False Non-Match → *False Negative:* A subject on the watch list is mistakenly not identified as such and is given access.

Biometric identification systems may return multiple candidate matches. It is required for positive identification that the list of candidate matches is of size 1, or at least that the candidate list can be quickly narrowed down to size 1 by some other match mechanism. For negative identification, it is desired that the candidate list is small so it can be examined by human operators. For each identification there should be a small number of false positives.

11.6.   Biometric Specific Errors

In addition to the above misclassification errors, there are error rates associated with error conditions that are more specific to biometrics.

(1) The *Failure to Acquire* (FTA) rate is the percentage of the target population that does not possess a particular biometric, i.e. does not deliver a usable biometric sample.

   Here what the exact definition of a FTA is subtle. It can be that a subject does not possess the biometric that is needed for enrollment, i.e. the subject is missing an eye; or it can be that a subject's biometric cannot be measured, say, the fingerprint of a brick layer (the ridges have been worn away). Technology may well be improved so that this latter subject can be enrolled at some future point.

(2) The *Failure to Enroll* (FTE) rate.   Another variable is the FTE rate, which is the proportion of the population that somehow cannot be enrolled because of limitations of the technology or procedural problems.

These two *failure errors* are specific to biometrics and therefore are very basic error rates that are encountered in any biometric scenario, and are errors that are encountered both in verification and identification systems. Both FTA and FTE are partially due to intrinsic biometric properties and limitations in state of the art of the biometrics.

11.7.   Additional Terminology

There is some more terminology that is specifically related to biometric identification:

Filtering. Narrowing down the search by using non-biometric parameters like sex, age group, nationality, etc.

Binning. Cutting down the search by using an additional biometric, e.g. fingerprint ridge flow pattern (whorl, arch, etc.).

Penetration rate. This is the expected proportion of the database to be used for the search operation when binning or filtering is employed. For example, the use of gender to filter the database will result in a 50 percent penetration rate.

   A related definition is the identification rank when a identification system returns a candidate list from the database.

Identification rank. It is the position of the correct entry in the probable candidate list. In an ideal identification system, the identification rank should be always 1. However, in real-world systems, the correct identity occurs in the top K, with some $K << M$.

11.8.   Identification Methods

We describe how a practical identification system operates to see the types of answers it returns. Typically an identification system will run in one of three modes of operation, depending on the application for which it is being used. In each mode, some subject presents a biometric to the system and that biometric is compared to the biometric samples enrolled in the database **M**. In some cases every enrolled sample in the database will be compared, but in others only some subset of **M** is compared.

For simplicity we consider the former case, with a database **M** of M enrolled subjects, i.e. the $1 : M$ search problem in its most general case. The three modes of operation are related to the three primary criteria for choosing the subset of **M**:

Threshold-based. This approach is effectively the same as repeating the operation of 1:1 verification for each person in the database. The query biometric template B is compared with each of the enrolled biometrics to obtain a match/non-match decision. This is typically done by computing the scores $s(B; B_m)$, $m = 1, \ldots, M$ for all enrolled templates $B_m \in$ **M** and considering as matches all those candidates with scores exceeding some threshold t_o. The complete list of all matching idenfitities is returned. If no candidate matches (e.g. no score exceeds the threshold), the person is presumed not to be in the database.

Rank-based. The system always returns some vector of fixed size, K, of the enrolled identities that best match the presented biometric. This requires an ability to rank (sort) the items in the database. The ranking might be based on scores. With $K = 1$ the system returns a single candidate corresponding to the person in the database most likely to be the same as the input query biometric Q. Note that it is usually not necessary to rank all the identities in **M**. Producing a ranked short-list of the best K items can be accomplished more efficiently than sorting all M enrollees. However, in the most general (and most computational complex case) the output vector is just a permutation or re-ordering or ranking of database vector **M**. It is a vector whose meaning depends upon the relation between the input biometric and the enrolled biometrics.

Hybrid. Here the K highest scoring candidates are returned, provided that their scores are above a threshold. When many candidates are above a threshold this acts like a rank-based system, but when a single (or fewer) candidates are above the threshold it acts like a threshold-based system. Thus, this approach can be viewed as a combination of the previous two approaches. It is the one most commonly used. We can also put in this category systems that dynamically change the criterion for results being in the candidate list — by varying the list length or threshold. Here the length K of output candidate vector is determined by some operating threshold on a sorted list of candidates and we have back a threshold based trade-off system.

11.9. The Closed-World Assumption

An important issue in identification is the closed world assumption. Systems are often designed as closed-world systems, but this impacts how well they operate under both open and closed-world assumptions. In a closed world it is assumed that the only people who will ever try to use the system are people that have previously been enrolled in the system. The "world" of possible users consists solely of the known database $\mathbf{M}$, and no imposter ever tries to "break in."

By contrast, in an open world, a completely unknown user may attempt to access the system. This has ramifications for the kind of answers identification systems should return:

- In an open-world situation, a rank-based system, which always returns K candidates, will necessarily be making an error when presented with an imposter.
- Conversely, a threshold-based system can return no answer, which would always be an error in a closed-world situation.
- To overcome the former problem, operating in an open-world, a rank-based system needs an additional rejection mechanism — in practice this generally means turning it into a hybrid system.
- One rejection mechanism is a generic imposter model — an extra template that is ranked or scored in the same way as the enrolled templates, i.e. "an anti-user" so to speak. If the impostor model is similar enough to the test template, it will appear on the candidate list and indicate that the input biometric is an impostor (or at least a low confidence in the reliability of the identification).

Binning and filtering are two common meta-methods for reducing the size of the biometrics database. That is, they help minimize the number of required biometric 1:1 matches.

11.10.   Performance Evaluation

The basic problem of identification is to generate an output identifier in response to some input biometric signal. Note that the desired output here is typically a multi-valued ID, not a binary "Yes/No" decision as in a verification system. Note also that no other credentials or claimed identity are submitted to the system   the mapping of biometric input identifier to ID is performed on the basis of the biometric information only. Given some input B, an identification system searches a database of biometric templates B_m each with corresponding identity ID_m, to determine if the input B occurs in the the database. Ideally, an identification system returns a candidate set $C(B)$ of matching identities $C(B) = \{\text{ID}_1, \text{ID}_2, \ldots, \text{ID}_K\}$. This set then ideally consists only the true identity of the user B with rank 1, or no answer when the user is not enrolled in the database.

At this point, we feel that it is warranted that we discuss how these large-scale systems can be characterized in terms of their accuracy. The measures below can be applied to threshold-based, rank-based, or hybrid systems. But first let us determine the False Accept Rate and False Reject Rate of identification system.

11.10.1.   *Simple* **FAR(M)** *and* **FRR(M)**

Previous researchers [2,3,4] have simplified the situation by ignoring the case where multiple (correct or incorrect) candidates are matched. They declare an imposter to be falsely accepted if one or more scores for incorrect candidates exceeds the threshold. Under this assumption the chance of *correctly* rejecting an imposter is

$$\text{Prob(correct reject)} = \prod_{i=1}^{M}(1 - \text{FAR}_i) \qquad (11.1)$$

Here the FAR_i are the separately measurable False Accept Rates for each of the M identities ID_i in database **M**. This equation just computes the probability that the system will not falsely accept the imposter as any of the M identities in the database. Although the FAR_i are non-identically but independently distributed random variables, we substitute the expectation of $\text{FAR}_i = \text{FAR}$ (the overall system performance parameter) to obtain

$$\texttt{Prob(correct reject)} = (1 - \text{FAR}_i)^M. \qquad (11.2)$$

This implies that the probability of a false accept is just the complement:

$$\text{FAR}(M) = 1 - \texttt{Prob(correct reject)} = 1 - (1 - \text{FAR})^M. \quad (11.3)$$

For FAR small, $(1 - \text{FAR})^M \approx 1 - M \times \text{FAR}$. Thus we obtain the major result that $\text{FAR}(M)$ is approximately linear in M:

$$\text{FAR}(M) \approx M \times \text{FAR}. \qquad (11.4)$$

A correct identification is considered to occur when the proper candidate score is matched (e.g. its score exceeds the threshold), regardless of what happens with the other candidates. Thus,

$$\texttt{Prob(correct identification)} = 1 - \text{FRR}. \qquad (11.5)$$

To find the probability of *failed* identification, we take the complement and get

$$\begin{aligned} \text{FAR}(M) &= 1 - \texttt{Prob(correct identification)} \\ &= 1 - (1 - \text{FRR}) \\ &= \text{FRR}. \end{aligned} \qquad (11.6)$$

Note that $\text{FRR}(M)$ is thus independent of M; it is just the same as the FRR of the underlying matcher.

11.10.2. *Reliability and Selectivity*

Reliability and *Selectivity* are accuracy measures used to characterize identification systems, particularly when the database is so large or the biometric so weak that the system cannot reliably pick a unique identity, and it is being used for "winnowing" — determining a candidate list of people who *might* be the subject. This is particularly the case in forensic applications, or when we have a *watch list* of "wanted" people, i.e. for negative authentication applications.

The reliability, *Rel*, is the correct detect rate for tests when the true identity is actually in the database. That is, as defined in detection theory [5], it is the proportion of times that the proper enrolled identity is a member of the output candidate list. Thus, for a threshold-based system:

$$Rel = 1 - \text{FRR}. \qquad (11.7)$$

Selectivity, *Sel*, is the average number of incorrect matches (FA) that the identification system reports per match against the entire database.

This is a somewhat counter-intuitive use of the term *selectivity* because *lower* selectivity (few False Accepts on average) is a desirable system characteristic, while high selectivity is not. The selectivity can also be thought of as corresponding to the amount of work it will take a subsequent expert to select the correct answer from the list presented. Selectivity is easier to define assuming closed-world test conditions but will be slightly higher if estimated with imposter test data. For a rank-based system, returning K candidates, there will be $K - 1$ incorrect matches in the list except when the correct candidate is in the list, so

$$Sel = K - Rel. \tag{11.8}$$

For a threshold-based system, the selectivity Sel can be written as a function of M (the size of the identification database $\mathbf{M}$) and the False Accept Rate of the underlying matcher at the chosen operating point:

$$Sel = (M - 1) \times \text{FAR}. \tag{11.9}$$

By varying the criterion for an enrolled identity being returned in the candidate list, the identification system can be tested at a variety of operating points. In a threshold-based system the threshold is varied, in a rank-based system, K can be varied and in a hybrid system the list length, threshold, or other list cut-off criteria can be changed.

Selectivity and Reliability can be measured at each of these operating points and plotted in a graph. As can be seen in Fig. 11.4, there is a trade-off between Sel and Rel as the operating point is varied. If Rel is raised to give a higher chance of retrieving a matching biometric (e.g. by lowering the decision threshold to or increasing K), a higher selectivity Sel will result, i.e. more False Matches per database query. Reliability increases toward one as selectivity increases toward the database size M. An ideal matcher would have selectivity of zero with reliability of one.

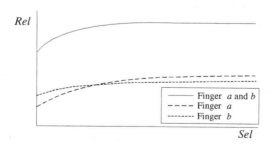

Fig. 11.4. Characterization of a system in terms of the traditional large-scale system parameters selectivity and reliability.

An identification system can only work effectively to identify individuals (as opposed to winnowing down to a short-list) if False Alarms rarely occur for any search, that is, if the selectivity is significantly less than one.

The Selectivity/Reliability curve can help in picking an operating point where this is true. As with the ROC for verification systems, the Reliability-Selectivity curve is a tool to allow us to choose an operating point for an identification system. A system specification might prescribe a desired reliability or selectivity, or the selectivity might be constrained by the amount of (machine or human) labor available to further process the candidate lists. In each case, the graph tells us the performance achievable at the necessary operating point and how the selection criterion must be set to achieve this performance. For more details see [6].

11.10.3. *Cumulative Match Curve (CMC)*

A rank-based identification system, returns the top K most similar enrolled candidates to the presented biometric in some order. Some other secondary decision making process (secondary matcher), for example, a human operator decides on the actual strength of the similarities. There are biometric evaluation scenarios where rank-order statistics are used to evaluate biometric match engines. In particular, they have been used in a number of evaluations of face recognition systems (FERET and FRVT [1,7,8,9]). The CMC(k) shows for any given list size, K, the probability of a test subject's correct identity showing up on the short-list (if the test subject has been enrolled). With the CMC, we can choose the candidate list size, K, so that this probability meets a performance goal, or determine this probability for a determined by other factors. The CMC increases with K and when the whole database) this probability is one. We can see that this is very similar to the Reliability-Selectivity curve — with almost identical axes — and it can be used in the same way. For more details see [10].

11.10.4. *Recall-Precision Curve*

The field of document retrieval is attempting to solve a similar task to biometric identification (finding all matching entries in a database using an inexact matching function). For completeness and comparison we present here the performance measures used in document retrieval. The crucial difference now is that in response to some query, many documents may be considered as correct matches, and a document retrieval system must return as many of these as possible, while returning as few of the unrelated documents as possible. In contrast, a biometric identification system may have several biometric templates enrolled for a single identity, generally it is sufficient for any one of these to be returned, not as many as possible.

Following the definitions of Witten, Moffat, and Bell [11], the precision, *Pre*, of a document retrieval system is the fraction of the returned documents that are relevant to the query

$$Pre = \frac{\texttt{Number retrieved that are relevant}}{\texttt{Total number retrieved}}. \qquad (11.10)$$

This can be calculated for a particular query, or averaged over many queries to estimate expected performance. The recall, *Rec*, is the proportion of relevant documents in the database that are found by the system

$$Rec = \frac{\texttt{Number of relevant that are retrieved}}{\texttt{Total number relevant}}. \qquad (11.11)$$

Recall and precision are often plotted against each other as a *recall-precision* curve. Figure 11.5 shows an example, loosely based on [11].

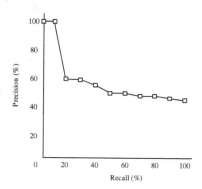

Fig. 11.5.   A recall-precision curve example for document retrieval.

The *recall, precision*, and the associated *recall-precision* trade-off curve, do not apply very well to the $1 : M$ biometric search problem. After all, we do not necessarily need the biometric search engine to return *all* the enrolled biometric identifiers in database **M** that are associated with an input query biometric Q (as we would with documents). The requirement is usually only that *at least one* correct identifier is returned.

11.11. Fingerprint Recognition

Fingerprinting has a long and interesting history as biometric [12]. Finger-prints are part of an individual's phenotype and hence are only weakly determined by genetics in that the prints are mainly formed after conception. It is shown, for instance, that identical twins have fingerprints that are quite different [13].

Fingerprints indeed are distinctive to a person. Within the forensic community it is widely believed that no two people have identical ridge details (see the ridge flow pattern of the fingerprint in Fig. 11.6). By this it is meant that no two fingerprints from different fingers have ever passed the test for identity as applied by an expert. There is evidence that the Chinese were aware of the individuality of fingerprints well over 5,000 years ago [14]. The belief in the uniqueness of fingerprints over the years led to their widespread use in law-enforcement identification applications.

Fig. 11.6.   Fingerprint image and extracted ridges/minutiae.

Early in the 20th century, an ingenious recognition system based on ten prints developed by Sir Edward Henry was brought into operation [14]. This system is now known as the "Henry System" and was adopted and refined by the FBI [15]. It allows for correct identification of offenders by manual indexing into databases of known criminals. It classifies the overall flow pattern of the fingerprint into a number of distinct patters such as "arch," "left whorl," "tented arch," etc. These classes are not uniformly distributed over the population [14,15]. For an individual, all ten fingers are classified in this way to yield a signature vector of the form

$$[\text{Arch},\ \text{Whorl}_{\text{Left}},\ \text{Whorl}_{\text{Right}},\ \text{Arch}_{\text{Tented}},\ \text{Loop}_{\text{Left}}, \ldots],$$

which is marked on a "ten-print card" and forms an index into a smaller set of subjects. While not unique to each person, this sequence can at least be used to rule out some suspects. Much research has been done on this sort of automated fingerprint classification [16,17,18], for Automated Fingerprint Identification System (AFIS) applications. In the early sixties, however, the number of searches that needed to be done on a daily basis simply became too large for manual indexing and AFIS started to be developed [19]. An AFIS still requires much manual labor because a search may result in many False Positives. Human experts use many details, called minutiae features, of the ridge flow pattern to determine if two impressions are from the same finger.

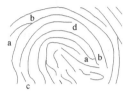

a:	ridge ending
b:	bifurcation
c:	independent ridge
d:	ambiguous ridge ending / bifurcation

Fig. 11.7.   Minutia features from a grey scale image.

Figure 11.7 shows thinned fingerprint structure with a few examples of these features: (a) ridge endings, (b) ridge bifurcations, and (c) an independent ridge (more complete listings include features such as, lake, spur, crossover [14,15]). Automated fingerprint matching algorithms attempt to match fingerprint impressions in a similar fashion. However, the most commonly used fingerprint features are only ridge bifurcations and ridge endings, collectively known as minutiae, which are extracted from the digitized print. A fingerprint image with its corresponding minutiae is shown in Fig. 11.6.

However, many matching algorithms do not even distinguish between bifurcations and endings because during acquisition and fingerprint image processing, depending on the amount of pressure exerted by a subject, a ridge ending may change into a bifurcation and vice versa. For example, for feature (d) in Fig. 11.7, it is unclear if it is an ending or a bifurcation. The process of fingerprint feature extraction typically starts by examining the quality of the input image [20]. Virtually every published method of feature extraction [21,22] then proceeds by computing orientation of the flow of the ridges, which reflects the local ridge direction at each pixel. The local ridge orientation is then used to tune filter parameters for image enhancement and ridge segmentation. From the segmented ridges, a thinned image (Fig. 11.6) is computed to locate the minutia features. Usually, a minutia post-processing stage cleans up several spurious minutiae resulting from fingerprint imperfections (dirt, cuts), enhancement, ridge segmentation or thinning artifacts. The machine representation of a fingerprint is critical to the success of the matching algorithm. A minimal representation of a processed fingerprint is a set (x_i, y_i, θ_i) of minutiae, i.e. a set of points

(x_i, y_i) expressed in some coordinate system with a ridge direction at this point θ_i.

11.12. Conclusions

Biometric identifiers, systems and databases are really put to the test when 1:*many* searches of large biometric databases are part of the enrollment policy or authentication protocol. In such identification problems, not only low error rates are desired, high 1:1 match rates are quite often also required.

Bibliography

1. P. J. Phillips, P. Grother, R. Micheals, D. M. Blackburn, T. Elham, and J. Mike Bone. FRVT 2002: Facial recognition vendor test. Technical report, DoD Counterdrug Technology Development Office, Defence Advance Research Project Agency, National Institute of Justice, Dahlgren, VA; Crane, IN; Arlington, VA, April 2003.
2. J. G. Daugman. High confidence visual recognition of persons by a test of statistical independence, *IEEE Trans. on Pattern Analysis and Machine Intelligence*, 15(11):1148-1161, November 1993.
3. R. Germain. Large scale systems. In A.K. Jain, R.M. Bolle, and S. Pankanti, editors, *Biometrics: Personal Identification in Networked Society*, pages 311-326. Kluwer Academic Press, 1999.
4. L. O'Gorman. Seven issues with human authentication technologies. *Proc. IEEE AutoID 2002*, pages 185-186, March 2002.
5. H. V. Poor. *An Introduction to Signal Detection and Estimation.* Springer-Verlag, 1988.
6. N. K. Ratha and R. M. Bolle, editors, *Automatic Fingerprint Recognition Systems.* Springer, October 2003.
7. P. J. Phillips, H. Moon, P. J. Rauss, and S. A. Rizvi. The FERET September 1996 database and evaluation procedure. In J. Bigun, G. Chollet, and G. Borgefors, editors, *Audio and Video-based Biometric Person Authentication,* volume 1206 of Lecture Notes in Computer Science. Springer-Verlag, Heidelberg, Germany, April 1997.
8. P. J. Phillips, H. Moon, S. A. Rizvi, and P. J. Rauss. The FERET evaluation methodology for face-recognition algorithms. *IEEE Trans. on Pattern Analysis and Machine Intelligence,* 22(10):1090-1104, October 2000.
9. D. M. Blackburn, M. Bone, and P. J. Phillips. FRVT 2000: Facial recognition vendor test. Technical report, DoD Counterdrug Technology Development Office, Defence Advance Research Project Agency, *National Institute of Justice,* Dahlgren, VA; Crane, IN; Arlington, VA, December 2000.

10. R. M. Bolle, J. H. Connell, S. Pankanti, N. K. Ratha and A. W. Senior, *Guide to Biometrics.* Springer, October 2003.

11. I. Witten, A. Moffat, and T. Bell. *Managing Gigabytes: Compressing and Indexing Documents and Images.* Van Nostrand Reinhold, 1994.

12. H. C. Lee and R. E. Gaensslen, editors. *Advances in Fingerprint Technology.* CRC Press, 1994.

13. S. Pankanti, S. Prabhakar, and A.K. Jain. On the individuality of fingerprints. *Proc. of the IEEE Conf. on Computer Vision and Pattern Recognition,* pages I:805-812, December 2001.

14. J. Bery. The history and development of fingerprinting. In H. C. Lee and R.E. Gaensslen, editors, *Advances in Fingerprint Technology,* pages 1-38. CRC Press, 1994.

15. FBI, U.S. Department of Justice, Washington, D.C. 20402. The Science of Fingerprints, Classification and Uses, 1984.

16. R. Cappelli, A. Lumini, D. Maio, and D. Maltoni. Fingerprint classification by directional image partitioning. *IEEE Trans. on Pattern Analysis and Machine Intelligence,* 21(5):402-421,May 1997.

17. A. K. Jain, S. Prabhakar, and L. Hong. A multichannel approach to fingerprint classification. *IEEE Trans. on Pattern Analysis and Machine Intelligence,* 21(4):348-359, April 1999.

18. A. K. Jain, L. Hong, and R. M. Bolle. On-line fingerprint verification. *IEEE Trans. on Pattern Analysis and Machine Intelligence,* 19(04):302- 313, April 1997.

19. R. T. Moore. Automatic fingerprint identification systems. In H. C. Lee and R. E. Gaensslen, editors, *Advances in fingerprint technology,* pages 163-191. CRC Press, 1994.

20. M. Y. -S. Yao, S. Pankanti, N. Haas, N. Ratha, and R. M. Bolle. Quantifying quality: A case study in fingerprints. *Proc. IEEE AutoID 2002,* pages 126-131, March 2002.

21. D. Maio and D. Maltoni. Direct gray-scale minutiae detection in fingerprints. *IEEE Trans. on Pattern Analysis and Machine Intelligence,* 19(1):27-40, January 1997.

22. N. K. Ratha, S. Chen, and A. K. Jain. Adaptive flow orientation based texture extraction in finger print images. *Pattern Recognition,* 28(11):1657-1672, November 1995.

Chapter 12

Evolutionary Algorithms: Basic Concepts and Applications in Biometrics

Carlos A. Coello Coello

CINVESTAV-IPN,
Evolutionary Computation Group,
Dpto. de Ing. Elect./Secc. Computación,
Av. IPN No. 2508, Col. San Pedro Zacatenco,
México, D.F. 07300, Mexico
ccoello@cs.cinvestav.mx

In this chapter, we provide a short introduction to the main concepts related to evolutionary algorithms, including some basic concepts and terminology, a brief description of their main paradigms, and some of their representative applications reported in the specialized literature. In the second part of the chapter, we discuss several case studies on the use of evolutionary algorithms in both physiological and behavioural biometrics. The case studies include fingerprint compression, facial modeling, hand-based feature selection, handwritten character recognition, keystroke dynamics identity verification, and speaker verification. These case studies show the success that different evolutionary algorithms have had in a variety of biometrics-related applications, either as standalone approaches, or combined with other heuristics (e.g. neural networks or Support Vector Machines. In the final part of the chapter, we provide a few guidelines regarding potential research trends for the near future. Such research trends include the use of alternative metaheuristics (e.g. particle swarm optimization, artificial immune systems and the ant system), as well as the use of alternative approaches to model the problems (e.g. through the use of multiobjective optimization or genetic programming).

Contents

Glossary

EA — Evolutionary Algorithm

GA — Genetic Algorithm

AI — Artificial Intelligence

12.1. Introduction

Nature has inspired man since ancient times. Although sometimes this inspiration was simply imitated, many of the real breakthroughs of humankind occurred when nature's principles were understood rather than copied. Many of such examples are well-documented in engineering [41] (e.g. dams, tunnels, and airplanes).

Thus, it is not by any means surprising to find that the evolution of species has served as inspiration to propose search and optimization techniques. After all, nature has evolved (sometimes rather complex) solutions to a wide variety of difficult problems over millions of years [22]. Early suggestions of the connections between evolution and optimization can be traced as long back as the 1930s [14]. However, it was until the 1960s when the three main techniques based on the evolution of species were developed [36]. These approaches (genetic algorithms, evolutionary programming and evolution strategies), which are now collectively denominated "evolutionary algorithms," have been found to be very effective for single-objective optimization and are now widely used in a great variety of disciplines [45,83,35,57,32].

Biometrics is a discipline that measures and statistically analyses biological data. Recently, and in the context of information technology, the term has been adopted to refer to the technologies for measuring and analyzing human body characteristics such as fingerprints, eye retinas and irises, voice patterns, facial patterns and hand measurements, especially for authentication purposes [97]. Biometric applications involve several complex problems. For example, many current biometric applications are closely related to pattern recognition and image analysis [84]. The complexity of these problems (which tend to be approached using statistical techniques) makes attractive the use of heuristics such as evolutionary algorithms, which have been found to be very powerful in a wide variety of optimization and classification tasks [35,45,32,6].

The remainder of this chapter is organized as follows. Section 12.2 provides some basic concepts related to evolutionary algorithms. Section 12.3 attempts to summarize the material from the previous section, by providing a more general framework for studying evolutionary algorithms. This includes a discussion of some of the main advantages offered by evolutionary algorithms.   Section 12.4 discusses a few representative case studies of applications of evolutionary algorithms in biometrics. After that, we provide some possible future research directions and our conclusions in Section 12.5.

12.2.   Basic Notions of Evolutionary Algorithms

The famous naturalist Charles Darwin defined *Natural Selection* or *Survival of the Fittest* as the *preservation of favorable individual differences and variations, and the destruction of those that are injurious* [20].

> *In nature, individuals have to adapt to their environment in order to survive in a process called* evolution, *in which those features that make an individual more suited to compete are preserved when it reproduces, and those features that make it weaker are eliminated. Such features are controlled by units called* genes *which form sets called* chromosomes. *Over subsequent generations not only the fittest individuals survive, but also their fittest genes which are transmitted to their descendants during the sexual recombination process which is called* crossover.

Early analogies between the mechanism of natural selection and a learning (or optimization) process led to the development of the so-called "evolutionary algorithms" (EAs) [5], in which the main goal is to simulate the evolutionary process in a computer. There are three main paradigms within evolutionary algorithms, whose motivations and origins were totally independent from each other: evolution strategies [83], evolutionary programming [39], and genetic algorithms [57]. Additionally, some authors consider genetic programming [66] as another paradigm, although genetic programming can also be seen as a special type of genetic algorithm. Each of these four types of evolutionary algorithm will be discussed next in more detail.

12.2.1. *Evolution Strategies*

 When working towards his PhD degree in engineering at the Technical University of Berlin, Ingo Rechenberg came across some optimization problems in hydrodynamics that could not be solved using traditional mathematical programming techniques [78]. This led him to the development of a very simple optimization algorithm which consisted of applying a set of random changes to a reference solution. The approach was later called "evolution strategy" and it was formally introduced in 1964 [36]. The original evolution strategy was called $(1+1)$-ES, because it consisted of a single parent that was mutated (i.e. subject to a random change) to produce an offspring. Then, the parent was compared to its offspring and the best from them was selected to become parent for the following iteration (or generation).

In the original $(1+1)$-EE, a new individual was produced using $\vec{x}^{t+1} = \vec{x}^t + N(0, \vec{\sigma})$, where t refers to the current *generation* (or iteration) and $N(0, \vec{\sigma})$ is a vector of independent Gaussian numbers with median zero and standard deviation $\vec{\sigma}$. It is important to emphasize that

> An "individual" in an evolution strategy contains the set of decision variables of the problem. No encoding is used in this case. So, if the decision variables are real numbers, such real numbers are directly put together as a single vector for each individual.

Rechenberg [79] stated a rule for adjusting the standard deviation in a deterministic way such that the evolution strategy could converge to the global optimum. This is now known as the "1/5 success rule," and it consists of the following:

$$\sigma(t) = \begin{cases} \sigma(t-n)/c & \text{if } p_s > 1/5 \\ \sigma(t-n) \cdot c & \text{if } p_s < 1/5 \\ \sigma(t-n) & \text{if } p_s = 1/5 \end{cases}$$

where n is the number of decision variables, t is the current generation, p_s is the relative frequency of successful mutations (i.e. those mutations in which the offspring replaced its parent because it had a better fitness) measured over a certain period of time (e.g. at every $10 \times n$ individuals) and $c = 0.817$ (this value was theoretically derived by Hans-Paul Schwefel [83]). $\sigma(t)$ is adjusted at every n mutations.

Over the years, several other variations of the original evolution strategy were proposed, after the concept of population (i.e., a set of solutions) was introduced [5]. The most recent versions of the evolution strategy are the $(\mu + \lambda)$-ES and the (μ, λ)-ES. In both cases, μ parents are mutated to produce λ offspring. However, in the first case (+ selection), the μ best individuals are selected from the union of parents and offspring. In the second case (, selection), the best individuals are selected only from the offspring produced.

In modern evolution strategies, not only the decision variables of the problem are evolved, but also the parameters of the algorithm itself (i.e. the standard deviations). This is called "self-adaptation" [83,5]. Parents are mutated using:

$$\sigma'(i) = \sigma(i) \times \exp(\tau' N(0,1) + \tau N_i(0,1))$$
$$x'(i) = x(i) + N(0, \sigma'(i)),$$

where τ and τ' are proportionality constants that are defined in terms of n. Also, modern evolution strategies allow the use of recombination (either *sexual*, when only 2 parents are involved, or *panmictic*, when more than 2 parents are involved in the generation of the offspring).

Some representative applications of evolution strategies are [83,5,35]: nonlinear control, structural optimization, image processing and pattern recognition, biometrics, classification, network optimization, and airfoil design.

12.2.2. *Evolutionary Programming*

Lawrence J. Fogel introduced in the 1960s an approach called "evolutionary programming," in which intelligence is seen as an adaptive behavior [38,39]. The original motivation of this paradigm was to solve prediction problems using finite state automata. The basic algorithm of evolutionary programming is very similar to that of the evolution strategy. A population of individuals is mutated to generate a set of offspring. However, in this case, there are normally several types of mutation

operators and no recombination (of any type), since evolution is modeled at the species level and different species do not interbreed. Another difference with respect to evolution strategies is that in this case, each parent produces exactly one offspring. Also, the decision of whether or not a parent will participate in the selection process is now determined in a probabilistic way, whereas in the evolution strategy this is a deterministic process. Finally, no encoding is used in this case (similarly to the evolution strategy) and emphasis is placed on the selection of the most appropriate representation of the decision variables.

> *Evolutionary programming emphasizes the behavioral links between parents and offspring, instead of trying to emulate some specific genetic operators (as in the case of the genetic algorithm [45,71]).*

In its original version, evolutionary programming didn't have a mechanism to self-adapt its parameters. However, in the early 1990s, Fogel and some of his co-workers realized of the importance of such mechanism and proposed the so-called "meta-evolutionary programming" [37] for continuous optimization. Over time, other self-adaptation mechanisms were proposed also for discrete optimization [4].

Interestingly, it was until the early 1990s that the evolution strategies community met the evolutionary programming community, despite the fact that both paradigms share very evident similarities [5,39].

Some representative applications of evolutionary programming are [35,39]: forecasting, games, route planning, pattern recognition, and neural networks training.

12.2.3. *Genetic Algorithms*

Genetic algorithms (originally denominated "genetic reproductive plans") were introduced by John H. Holland in the early 1960s [55,56]. The main motivation of this work was the solution of machine learning problems [57]. Before describing the way in which a genetic algorithm (GA) works, we will provide some of the basic terminology adopted by the researchers from this area [51] in Tables 12.1 and 12.2.

> *Genetic algorithms emphasize the importance of sexual recombination (which is the main operator) over the mutation operator (which is used as a secondary operator). They also use probabilistic selection (like evolutionary programming and unlike evolution strategies).*

The basic operation of a Genetic Algorithm is illustrated in the following segment of pseudo-code [13]:

Table 12.1.   Basic terminology of genetic algorithms.

Term	Definition
A chromosome	A data structure that holds a "string" of task parameters, or genes. This string may be stored, for example, as a binary bit-string (binary representation) or as an array of integers (floating point o real-coded representation) that represent a floating point number. This chromosome is analogous to the base-4 chromosomes present in our own DNA. Normally, in the GA community, the haploid model of a cell is assumed (one-chromosome individuals). However, diploids have also been used in the specialized literature [45].
A gene	A subsection of a chromosome that usually encodes the value of a single parameter (i.e. a decision variable).
An allele	The value of a gene. For example, for a binary representation each gene may have an allele of 0 or 1, and for a floating point representation, each gene may have an allele from 0 to 9.
A schema	A pattern of gene values in a chromosome, which may include "do not care" states (represented by a # symbol). Thus, in a binary chromosome, each schema can be specified by a string of the same length as the chromosome, with each character being one of { 0, 1, # }. A particular chromosome is said to "contain" a particular schema if it matches the schema (e.g. chromosome 01101 matches schema #1#0#).
The fitness of an individual	A value that reflects its performance (i.e. how well solves a certain task). A **fitness function** is a mapping of the chromosomes in a population to their corresponding fitness values. A **fitness landscape** is the hypersurface obtained by applying the fitness function to every point in the search space.
A building block	A small, tightly clustered group of genes which have co-evolved in such a way that their introduction into any chromosome will be likely to give increased fitness to that chromosome. The **building block hypothesis** [45] states that GAs generate their solutions by first finding as many building blocks as possible, and then combining them together to give the highest fitness.
Deception	A condition under which the combination of good building blocks leads to reduced fitness, rather than increased fitness. This condition was proposed by Goldberg as a reason for the failure of GAs on certain tasks [45].

Table 12.2. Basic terminology of genetic algorithms (Continuation of Table 12.1).

Term	Definition
Elitism (or an elitist strategy)	A mechanism which ensures that the chromosomes of the highly fit member(s) of the population are passed on to the next generation without being altered by any genetic operator. The use of elitism guarantees that the maximum fitness of the population never decreases from one generation to the next, and it normally produces a faster convergence of the population. More important yet is the fact that it has been (mathematically) proven that elitism is necessary in order to be able to guarantee convergence of a simple genetic algorithm towards the global optimum [82].
Epistasis	The interaction between different genes in a chromosome. It is the extent to which the contribution to fitness of one gene depends on the values of other genes. Geneticists use this term to refer to a "masking" or "switching" effect among genes, and a gene is considered to be "epistatic" if its presence suppresses the effect of a gene at another locus (or position in the chromosome). This concept is closely related to deception, since a problem with high degree of epistasis is deceptive, because building blocks cannot be formed. On the other hand, problems with little or no epistasis are trivial to solve (hill climbing is sufficient).
Exploitation	The process of using information gathered from previously visited points in the search space to determine which places might be profitable to visit next. Hill climbing is an example of exploitation, because it investigates adjacent points in the search space, and moves in the direction giving the greatest increase in fitness. Exploitation techniques are good at finding local minima (or maxima). The GA uses crossover as an exploitation mechanism.
Exploration	The process of visiting entirely new regions of a search space, to see if anything promising may be found there. Unlike exploitation, exploration involves leaps into unknown regions. Random search is an example of exploration. Problems which have many local minima (or maxima) can sometimes only be solved using exploration techniques such as random search. The GA uses mutation as an exploration mechanism.
A genotype	A potential solution to a problem, and is basically the string of values chosen by the user, also called chromosome.
A phenotype	The meaning of a particular chromosome, defined externally by the user.

Table 12.2. (*Continued*)

Term	Definition
Genetic drift	The name given to the changes in gene/allele frequencies in a population over many generations, resulting from chance rather than from selection. It occurs most rapidly in small populations and can lead to some alleles to become extinct, thus reducing the genetic variability in the population.
A niche	A group of individuals which have similar fitness. Normally in multiobjective and multimodal optimization, a technique called **fitness sharing** is used to reduce the fitness of those individuals who are in the same niche, in order to prevent the population to converge to a single solution, so that stable sub-populations can be formed, each one corresponding to a different objective or peak (in a multimodal optimization problem) of the function [24].

```
generate initial population, G(0);
evaluate G(0);
t:=0;
repeat
   t:=t+1;
   generate G(t) using G(t-1) (applying genetic operators);
   evaluate G(t);
until the stop condition is reached
```

First, an initial population is randomly generated. The individuals of this population will be a set of chromosomes or strings of characters (letters and/or numbers) that represent all the possible solutions to the problem.

1	0	0	1	1	0	1

Fig. 12.1. Example of the binary encoding traditionally adopted with the genetic algorithm.

One aspect that has great importance in the case of the genetic algorithm is the encoding of solutions. Traditionally, a binary encoding has been adopted, regardless of the type of decision variables of the problem to be solved [45]. Holland provides some theoretical and biological arguments for using a binary encoding [57]. However, over the years, other types of encodings have been proposed, including the use of vectors of real numbers and permutations, which lend themselves as more "natural" encodings for certain types of optimization problems [69,81].

Table 12.3.   The basic selection schemes of genetic algorithms.

Scheme	Definition
Proportionate Selection	This term is used generically to describe several selection schemes that choose individuals for birth according to their objective function values f. In these schemes, the probability of selection p of an individual from the ith class in the tth generation is calculated as $$p_{i,t} = \frac{f_i}{\sum_{j=1}^{k} m_{j,t} f_j} \qquad (12.1)$$ where k classes exist and the total number of individuals sums to n. Several methods have been suggested for sampling this probability distribution, including Monte Carlo or *roulette wheel* selection [60], *stochastic remainder* selection [11,12], and *stochastic universal* selection [8,49].
Ranking Selection	In this scheme, proposed by Baker [7] the population is sorted from best to worst, and each individual is copied as many times as it can, according to a non-increasing assignment function, and then proportionate selection is performed according to that assignment.
Tournament Selection	The population is shuffled and then is divided into groups of k elements from which the best individual (i.e. the fittest) will be chosen. This process has to be repeated k times because on each iteration only m parents are selected, where $$m = \frac{population\ size}{k}$$ For example, if we use binary tournament selection ($k = 2$), then we have to shuffle the population twice, since at each stage half of the parents required will be selected. The interesting property of this selection scheme is that we can guarantee multiple copies of the fittest individual among the parents of the next generation.
Steady State Selection	This is the technique used in Genitor [92], which works individual by individual, choosing an offspring for birth according to linear ranking, and choosing the currently worst individual for replacement. In steady-state selection only a few individuals are replaced in each generation: usually a small number of the least fit individuals are replaced by offspring resulting from crossover and mutation of the fittest individuals. This selection scheme is normally used in evolving rule-based systems in which incremental learning (and remembering what has already been learned) is important and in which members of the population collectively (rather than individually) solve the problem at hand [71].

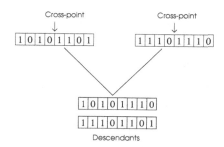

Fig. 12.2. Use of a single-point crossover between two chromosomes. Notice that each pair of chromosomes produces two descendants for the next generation. The cross-point may be located at the string boundaries, in which case the crossover has no effect and the parents remain intact for the next generation.

Once an appropriate encoding has been chosen, we apply a *fitness function* to each one of these chromosomes in order to measure the quality of the solution encoded by the chromosome. Knowing each chromosome's fitness, a *selection* process takes place to choose the individuals (presumably, the fittest) that will be the parents of the following generation. The most commonly used selection schemes are described in Table 12.3 [46].

After being selected, *crossover* takes place. During this stage, the genetic material of a pair of individuals is exchanged in order to create the population of the next generation. There are three main ways of performing crossover:

(1) *Single-point crossover.* A position of the chromosome is randomly selected as the crossover point as indicated in Fig. 12.2.
(2) *Two-point crossover.* Two positions of the chromosome are randomly selected as to exchange chromosomic material, as indicated in Fig. 12.3.
(3) *Uniform crossover.* This operator was proposed by Syswerda [87] and can be seen as a generalization of the two previous crossover techniques. In this case, for each bit in the first offspring it decides (with some probability p) which parent will contribute its value in that position. The second offspring would receive the bit from the other parent. See an example of 0.5-uniform crossover in Fig. 12.4. Although for some problems uniform crossover presents several advantages over other crossover techniques [87], in general, one-point crossover seems to be a bad choice, but there is no clear winner between two-point and uniform crossover [71,69].

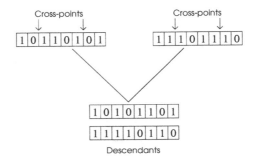

Fig. 12.3. Use of a two-point crossover between two chromosomes. In this case the genes at the extremes are kept, and those in the middle part are exchanged. If one of the two cross-points happens to be at the string boundaries, a single-point crossover will be performed, and if both are at the string boundaries, the parents remain intact for the next generation.

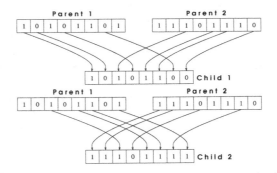

Fig. 12.4. Use of 0.5-uniform crossover (i.e. adopting a 50% probability of crossover) between two chromosomes. Notice how half of the genes of each parent go to each of the two children. First, the bits to be copied from each parent are selected randomly using the probability desired, and after the first child is generated, the same values are used to generate the second child, but inverting the source of procedence of the genes.

Mutation is another important genetic operator that randomly changes a gene of a chromosome. If we use a binary representation, a mutation changes a 0 to 1 and viceversa. An example of how mutation works is displayed in Fig. 12.5. This operator allows the introduction of new chromosomic material to the population and, from the theoretical perspective, it assures that — given any population — the entire search space is connected [13].

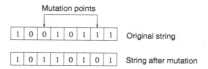

Fig. 12.5.   An example of mutation using binary representation.

If we knew in advance the final solution, it would be trivial to determine how to stop a genetic algorithm. However, as this is not normally the case, we have to use one of the two following criteria to stop the GA: either give a fixed number of generations in advance, or verify when the population has become homogeneous (i.e. all or most of the individuals have the same fitness). Traditionally,

> *Genetic algorithms do not have a self-adaptation mechanism. Therefore, one of their main drawbacks is that their parameters tend to be fine-tuned in an empirical manner (i.e., by trial-and-error).*

Some representative applications of genetic algorithms are the following [45,32,5]: data mining, optimization (structural, combinatorial, etc.), pattern recognition, and robot motion planning.

12.2.4. *Genetic Programming*

One of the original goals of artificial intelligence (AI) was the automatic generation of computer programs that could produce a desired task given a certain input. During several years, such a goal seemed too ambitious since the size of the search space increases exponentially as we extend the domain of a certain program and, consequently, any technique will tend to produce programs that are either invalid or highly inefficient.

Some early evolutionary algorithms were attempted in automatic programming tasks, but they were unsuccessful and were severely criticized by some AI researchers [35]. Over the years, researchers realized that the key issue for using evolutionary algorithms in automatic programming tasks was the encoding adopted. In this regard, John Koza [66] suggested the use of a genetic algorithm with a tree-based encoding. In order to simplify the implementation of such an approach, the original implementation of this sort of approach (which was called "genetic programming") was done under LISP, taking advantage of the fact that such programming language has a built-in parser.

The tree-encoding adopted by Koza obviously requires of different alphabets and specialized operators for evolving randomly generated programs until they become 100% valid. Note however, that the basic principles of this technique may be generalized to any other domain and, in fact, genetic programming has been used in a wide variety of applications [66]. The trees used in genetic programming consist of both functions and terminals. The functions normally adopted are the following [66]:

```
Arithmetic operations (e.g., +, -, ×, ÷ )
Mathematical functions (e.g., sine, cosine, logarithms, etc.)
Boolean Operations (e.g., AND, OR, NOT)
Conditionals (IF-THEN-ELSE)
Loops (DO-UNTIL)
Recursive Functions
Any other domain-specific function
```

Terminals are typically variables or constants, and can be seen as functions that take no arguments. An example of a chromosome that uses the functions F = {AND, OR, NOT} and the terminals T = {A0, A1} is shown in Fig. 12.6.

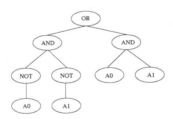

Fig. 12.6.   An example of a chromosome used in genetic programming.

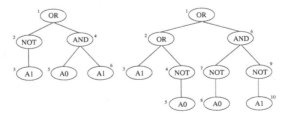

Fig. 12.7.   The tree nodes are numbered before applying the crossover operator.

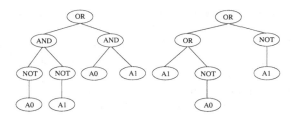

Fig. 12.8.   The two offspring generated after applying the crossover operator.

Crossover can be applied by numbering the tree nodes corresponding to the two parents chosen (see Fig. 12.7) and (randomly) selecting a point in each of them such that the subtrees below such point are exchanged (see Fig. 12.8, where we assume that the crossover point for the first parent is 2 and for the second is 6). Typically, the sizes of the two parent trees will be different as in the example previously shown. It is also worth noticing that if the crossover point is the root of one of the parent trees, then the whole chromosome will become a subtree of the other parent. This allows the incorporation of subroutines in a program. It is also possible that the roots of both parents are selected as crossover points. Should that be the case, the crossover operator will have no effect and the offspring will be identical to their parents.

Normally, genetic programming implementations impose a limit on the maximum depth that a tree can reach, as to avoid the generation (as a byproduct of crossover and mutation) of trees of very large size that could produce a memory overflow [9].

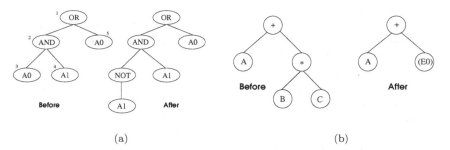

(a)   (b)

Fig. 12.9.   An example of mutation in genetic programming (a) and of encapsulation in genetic programming (b).

Mutation in genetic programming takes place through a (random) selection of a certain node tree. The subtree below the chosen node is replaced by another tree which is randomly generated. Fig. 12.9(a) shows an example of the use of this operator (the mutation point in this example is node 3).

In genetic programming is also possible to protect or "encapsulate" a certain subtree which we know to contain a good building block, as to avoid that it is destroyed by the genetic operators. The selected subtree is replaced by a symbolic name that points to the real location of the subtree. Such subtree is separately compiled and linked to the rest of the tree in an analogous way to the external classes of object oriented languages. Figure 12.9(b) shows an example of encapsulation in which the right subtree is replaced by the name (**E0**).

Finally, genetic programming also provides mechanisms to destroy a certain percentage of the population such that we can renovate the chromosomic material after a certain number of generations. This mechanism, called **execution**, is very useful in highly complex domains in which our population may not contain a single feasible individual even after a considerably large number of generations.

12.3. A More General View of Evolutionary Algorithms

Despite the obvious differences and motivations of each of the aforementioned paradigms, the trend in the last few years has been to decrease the differences among the paradigms and refer (in generic terms) simply to evolutionary algorithms when talking about any of them.

In Table 12.4, the basic components to implement an EA in order to solve a problem [69,32] are given. Also, it is shown that EAs differ from traditional search techniques in several ways [13].

It is worth indicating that the ever-growing popularity of evolutionary algorithms in a variety of application domains tends to be related to their good reputation as "optimizers" (either for single-objective or for multi-objective problems [75,16]). This is remarkable if we consider that some of them (namely, genetic algorithms) were not originally proposed for that type of application and that their use in optimization tasks has been questioned by some well-established researchers in the evolutionary computation community [61]. Apparently, the reported success of evolutionary algorithms has resulted sufficiently convincing for practitioners and therefore their popularity [6].

12.4. Some Applications in Biometrics

Since their very inception, several researchers have attempted to use evolutionary algorithms for tasks related to pattern recognition, image

Table 12.4. Basic components to implement an EA and particular features of an EA.

Requirements of an EA	Particular features of an EA
(1) A representation of the potential solutions to the problem. The selection of an appropriate encoding scheme tends to be crucial for the good performance of an EA [81].	
(2) A way to create an initial population of potential solutions (this is normally done randomly, but deterministic approaches can also be used).	• EAs do not require problem specific knowledge to carry out a search. However, if such knowledge is available, it can be easily incorporated as to make the search more efficient.
(3) An evaluation function that plays the role of the environment, rating solutions in terms of their "fitness". The definition of a good fitness function is also vital for having a good performance.	• EAs use stochastic instead of deterministic operators and appear to be robust in noisy environments. • EAs are conceptually simple and easy to implement.
(4) A selection procedure that chooses the parents that will reproduce.	• EAs have a wide applicability. • EAs are relatively simple to parallelize.
(5) Evolutionary operators that alter the composition of children (normally, crossover and mutation).	• EAs operate on a population of potential solutions at a time. Thus, they are less susceptible to false attractors (i.e. local optima).
(6) Values for various parameters that the evolutionary algorithm uses (population size, probabilities of applying evolutionary operators, etc.).	

processing, classification and machine learning [15,19,91,34,85,93,33,94]. However, most of this early work was mainly focused on the algorithmic development aspect rather than on an specific application. Thus, for some time, there was relatively little research on the development of evolutionary algorithms for specific biometric applications. This situation has changed in the last few years [84], although the use of other soft computing techniques such as neural networks is still more common than the use of evolutionary algorithms [58].

Next, we will review a few case studies in this area which aim to provide a general picture of the type of research being conducted nowadays. A summary of these case studies is provided in Table 12.5.

Table 12.5. Applications of evolutionary algorithms in biometrics.

Physiological biometrics	
Biometric	**Technique of EA**
Fingerprints	In [48], a coevolutionary genetic algorithm is proposed based on an *Enforced Sub-Populations* technique [47] and on a mathematical technique called *Lifting* to find wavelets that are specially adapted to a particular class of images. Other researchers have also worked on the use of evolutionary algorithms for solving other related problems such as optimizing the alignment of a pair of fingerprint images [3], estimation of the ridges in fingerprints [1] and fingerprint classification [77].
Face	In [53], a genetic algorithm-based approach has been proposed for facial modeling from an uncalibrated face image using a flexible generic parameterized facial model. Other researchers have also used evolutionary algorithms for related problems, such as human face detection [95], automated face recognition [89,59], and human posture estimation [80], among others.
Palmprints	In [65], a cooperative coevolutionary genetic algorithm is used for hand-based feature selection. The outcome of the approach is a number of good (i.e. tight and well-separated) clusters which exist in the smallest possible feature space.
Behavioral biometrics	
Signature	In [90], genetic programming has been adopted for handwritten character recognition. Fitness function consists of the weighted sum of errors produced by each of the programs generated (zero-argument functions, which consists of the features to be extracted from the images that are presented as test cases). Each program also receives a penalty when there is a lack of diversity. Other researchers have also used evolutionary algorithms for related problems, such as unsupervised learning in handwritten character recognition tasks [72], design of a voting system to select from among multiple classifiers adopted for cursive handwritten text recognition [50], and interpretation of characters for ebooks [68].
Keystroke dynamics	In [96], a combination of a support vector machine and a genetic algorithm is used for keystroke dynamics identity verification. The authors also propose the use of an ensemble model based on feature selection to alleviate the defficiencies of a small training data set.
Voice	In [43] a locally recurrent probabilistic neural network was adopted for speaker verification. The approach is based on a three-step training procedure, from which the third step was done with differential evolution (an evolutionary algorithm that is highly competitive for numerical optimization tasks in which the decision variables are real numbers).

12.4.1. *Fingerprint Compression*

Grasemann and Miikulainen [48] proposed a coevolutionary genetic algorithm based on a technique known as *Enforced Sub-Populations* [47] and on a mathematical technique called *Lifting* to find wavelets that are specially adapted to a particular class of images. This approach was tested in the fingerprint domain against standard wavelets (including a winner of a competition held by the FBI in the early 1990s) to find the best wavelet fingerprint compression. The evolutionary algorithm outperformed the algorithm used by the FBI, which attracted a lot of attention from the media and from the evolutionary computation community, because this was a clear example of how an evolutionary design can outperform a human design.

Enforced sub-populations refers to an approach in which a number of populations of individual neurons (from a neural network) are evolved in parallel. In the evaluation phase, this approach repeatedly selects one neuron from each subpopulation to form candidate neural networks. The fitness of a particular neuron is the average fitness of all the neural networks in which it participated. This same idea was used in this work, but evolving different wavelets instead. These wavelets are constructed using the so-called *lifting step* [86], which is a finite filter that can generate new filter pairs from existing pairs.

To validate the approach, the authors used 80 available images. The size of each image was 300 by 300 pixels, at 500 dpi resolution. Each of them was used once as a test image and 79 times as part of the training set. In the experiments performed, the best wavelet found after each generation was used to compress the test image. It turns out that after only 50 generations, the evolutionary algorithm was able to produce results that outperformed the wavelets used by the FBI. Overall, the evolutionary algorithm introduced between a 15% and a 20% decrease in the space requirements for the same image quality. This is very significant if we consider that the FBI has over 50 million fingerprint cards on file and has recently started to store them electronically (the FBI is currently digitizing and compressing between 30,000 and 50,000 new cards every day).

12.4.2. *Facial Modeling*

Ho and Huang [53] proposed a genetic algorithm-based approach for facial modeling from an uncalibrated face image using a flexible generic parameterized facial model (FGPFM). The authors adopted the so-called "Intelligent Genetic Algorithm" IGA [54] (which they had previously proposed) to tackle this problem. The approach of the authors consists in:

(a) Considering that the microstructural information can be expressed using the structural FGPFM with representative facial features that can be found in the image.

(b) The reconstruction procedure is considered as a block function of FGPFM where the input parameters are the 3D face-centered coordinates of a set of control points.

(c) Once these control points are given, the 3D facial model that we aim to find is determined based on the topological and geometrical descriptions of FGPFM.

So, the problem of reconstructing the 3D facial model is transformed into the problem of

How to acquire 3D control points that are sufficiently accurate as to provide the desired model.

This gives raise to a large parameter optimization problem that is solved using the IGA. The fitness function adopted considers two criteria for evaluating the quality of the facial model:

(a) The projection of the facial model from some viewpoint must coincide with the features in the given face image, and

(b) The facial model must adhere to the generic knowledge of human faces accepted by the human perception.

An interesting aspect of this work is that the authors adopt a coarse-to-fine approach to solve the problem. This refers to a solution process that consists of several stages and that combines a *global* (coarse) search, and a *local* (fine) search.

The approach was validated by performing several experiments in which synthetic face images and actual face images were analyzed. The results showed that the approach was robust and provided very good results with respect to other techniques based on genetic algorithms.

12.4.3. *Hand Image Classification*

Kharma et al. [65] used a genetic algorithm with a cooperative co-evolutionary scheme for dynamic clustering and feature selection. Cooperative coevolution refers to the simultaneous evolution of two populations whose fitnesses are coupled (i.e. the fitness of one population depends on the fitness of the other one). The authors use this sort of scheme to integrate dynamic clustering with hand-based feature selection. For the feature extraction task, the authors considered two types of features: (1) geometric (finger width, finger height, finger circumference, finger angle, finger base length and palm aspect ratio), and (2) statistical (central moments, Fourier descriptors and Zernike moments). The authors optimized three quantities: (1) the set of features used for clustering, (2) the cluster centers, and (3) the total number of clusters. In the scheme adopted, a number of cluster centers are initially proposed. Then, a point (input pattern) is assigned to the cluster whose center is closest to the point, and the points are re-assigned to the new clusters based on their distance from the new cluster centers. This process is repeated until the locations of the cluster centers stabilize. Two populations are adopted: one of cluster centers and another one of dimension selections). During the process, the less discriminatory features are eliminated, which leaves a more efficient subset for further use. So, the outcome of the approach is a number of good (i.e. tight and well-separated) clusters which exist in the smallest possible feature space.

In order to validate this approach, the authors used 100 hand images and 84 normalized features. The results were found to be quite promising, since the average hand misplacement rate was 0.0580 with a standard deviation of 2.044, and half of the original 84 features adopted were eliminated during the evolutionary process. However, results were not compared to any other type of approach.

12.4.4. *Handwritten Character Recognition*

Teredasai et al. [90] adopted genetic programming in handwritten character recognition. Traditionally, active pattern recognition involves the use of an active heuristic function (e.g. a neural network [29]) which adaptively determines: the *length* of the feature vector, and the *features* themselves used to classify an input pattern. The outputs of this stage are the *confidence* values and the *separability* values.

Then, in a postprocessing stage, decision making takes place, based on the output from the classification stage. This is followed by an iterative search process in which sub images of finer resolution are fed to the classification process until reaching a level of confidence that is satisfactory (or until we run out of time). Contrasting this traditional process, Teredasai et al. [90] proposed to eliminate the iterative search process and focused instead on searching for the areas in the image that have maximum separability information. They proposed two possible ways of achieving this goal:

(a) Use only those features from the feature set that provide maximum separability, and

(b) Generate a new set of features that represents the image data in a manner that provides a better classification accuracy.

These two tasks are tightly coupled and the authors propose to use genetic programming to decouple them. The most interesting aspect of the proposal is the fitness function adopted. Such fitness function consists of the weighted sum of errors produced by each of the programs generated. The "programs" are really zero-argument functions, which consist of the features to be extracted from the images that are presented as test cases. Each program also receives a penalty when there is a lack of diversity (i.e. poor accuracy of the classifiers is penalized). So, the authors were evolving features that provided maximum separability, but at the same time, were trying to achieve (through the penalty function adopted) a better classification accuracy.

The proposed approach was validated using several datasets which included the NIST handwritten digit sets, and the CEDAR-Digit data set (which has more noisy data). The NIST data set consists of 159,228 images in the training set, and 53,300 images in the test set. The CEDAR-digit data set consists of 7245 images in the training set and 2711 images in the test set. The CEDAR-digit data set is more challenging, because it contains more noisy data with images that were incorrectly recognized or rejected even by the current recognizers based on the K-nearest neighbor rule and neural networks. The proposed approach was compared with respect to a traditional passive classifier. The results indicated that the GP-based approach was able to outperform the traditional passive classifier, achieving an accuracy of 97.1% using a very low number of features.

12.4.5. *Keystroke Dynamics Identity Verification*

Yu and Cho [96] used a combination of a support vector machine (SVM) and a genetic algorithm (GA) for keystroke dynamics identity verification. It has been known at least since the 1980s that a user's keystroke pattern is highly repeatable and distinct from that of other users [44]. Thus, it is possible to collect a user's timing vectors and use them to build a model that discriminates the owner from possible imposters.

The authors adopted a SVM for novelty detection and a GA-wrapper feature selection approach to automatically select a relevant subset of features, while ignoring the rest. The authors argue that the use of a SVM only requires about 1/1000 of the training time that a neural network would require. The authors also propose the use of an ensemble model based on feature selection to alleviate the deficiencies of a small training data set. A wrapper approach tries many different sets of features by building a model using a learning algorithm, and then chooses the one with the best performance measure. Feature subset selection is basically an optimization problem, and the GA is used for that sake. On the other hand, the SVM is adopted as the base learner for the wrapper approach. The fitness function of the GA combines three different criteria: (1) the accuracy of the novelty detector, (2) the learning time and (3) the dimension reduction ratio. Thus, the GA deals with a population of SVMs which employ different feature subsets. At early stages of the evolutionary process, the candidate solutions or SVMs usually show a high level of diversity, but a low accuracy. At later stages of the search, the behavior is the opposite: there is less diversity and a higher accuracy. In order to deal with the problem of having insufficient data (something common within a keystroke dynamics identity verification system), the authors made a trade-off between diversity and accuracy by selecting diverse candidates from the population immediately before it converges. Such candidates are used to create an ensemble.

The proposed approach was validated using 21 different password-typing patterns and results were compared with respect to the use of neural networks, two other SVM-based techniques and a random feature ensemble approach. Although the proposed approach presented a similar performance as the others, its time requirements were considerably lower.

In [2], mouse dynamics is described as the characteristics of the actions received from the mouse input device for a specific user, while interacting with a certain graphical user interface. The authors used an artificial neural network for this study.

12.4.6.   *Voice Identification*

Ganchev et al. [43] used a locally recurrent probabilistic neural network for the process of speaker verification. The use of a locally recurrent probabilistic scheme aims to exploit the inter-frame correlation among the feature vectors extracted from successive speech frames. The approach is based on a three-step training procedure. The first training step creates the actual topology of the network. The second training step involves the computation of a smoothing parameter for each class. In the third step, the weights of the locally recurrent layer are computed. For this sake, a type of evolutionary algorithm called "differential evolution" [76] is adopted (this is done by minimizing an error function).

The proposed approach was validated using a text-independent speaker verification system previously developed by the same authors, which uses probabilistic neural networks [42]. Thus, the aim was to assess the improvements introduced by the new approach. The authors used data from POLYCOST (a telephone-speech database for speaker recognition) [52] for

Table 12.6. Possible research directions in the use of evolutionary algorithms in biometrics.

Direction	Strategy
Multiobjective optimization	Multiobjective optimization refers to the simultaneous optimization of two or more objective functions which are commonly in conflict with each other [16]. Due to the multiobjective nature of many real-world problems, it is very likely that the use of multiobjective optimization techniques becomes more popular in biometrics [72]. Among the possible techniques to solve multiobjective optimization problems, evolutionary algorithms present several advantages over traditional mathematical programming techniques [70,31]. For example, evolutionary algorithms have a population-based nature which allows them to generate several elements of the Pareto optimal set in one run. Additionally, evolutionary algorithms do not require an initial search point (as is the case of most mathematical programming techniques) and are less sensitive to the shape and continuity of the Pareto front [16,23].
Use of Genetic Programming	As we saw before, the use of tree-encodings in a genetic algorithm (the so-called "genetic programming") is a powerful aid for automated programming. Such type of encoding is also very useful for classification and data mining tasks [40,90]. Consequently, the use of genetic programming in biometrics, although not widely spread yet, is expected to considerably grow in the next few years. It is worth indicating, however, that the use of a more complex encoding, while allowing tackling more complex problems, also involves a higher computational cost (this applies to multiobjective optimization as well). Thus, the use of parallel computing seems an obvious choice in these cases [88].

their study. The results indicated that the proposed approach presented a relative reduction of the error rate of more than 28% with respect to the original scheme.

Despite the evident popularity of approaches such as statistical methods and other soft computing techniques (e.g. neural networks) in biometrics [67], the use of evolutionary algorithms has attracted a growing interest in the last few years. However, there are still many possible research directions that may be worth exploring in this area. A few of them are briefly discussed in Tables 12.6 and 12.7.

12.5.   Conclusions

This chapter has provided a short (and very general) introduction to evolutionary algorithms, including the historical roots of its main

Table 12.7. Possible research directions in the use of evolutionary algorithms in biometrics (continuation of Table 12.6).

Direction	Strategy
Use of alternative metaheuristics	In recent years, other biologically-inspired metaheuristics have become increasingly popular in a wide variety of applications [18]. It is expected that several of these approaches are eventually adopted in biometric applications. Representative examples of these new metaheuristics are the following: (a) **Particle Swarm Optimization**: Proposed by Kennedy and Eberhart [62,64], this metaheuristic simulates the movements of a group (or population) of birds which aim to find food. The approach can be seen as a distributed behavioural algorithm that performs (in its more general version) multidimensional search. In the simulation, the behavior of each individual is affected by either the best local (i.e. within a certain neighborhood) or the best global individual. The approach uses then the concept of population and a measure of performance similar to the fitness value adopted with evolutionary algorithms. The approach introduces the use of flying potential solutions through hyperspace (used to accelerate convergence) and allows individuals to benefit from their past experiences. This technique has been successfully used for both continuous nonlinear and discrete binary optimization [30,63,64] (b) **Artificial Immune Systems**: Computationally speaking, our immune system can be seen as a highly parallel intelligent system that is able to learn and retrieve previous knowledge (i.e. it has "memory") to solve recognition and classification tasks. Due to these interesting features, several researchers have developed computational models of the immune system and have used it for a variety of tasks, including classification and pattern recognition [21,73,74]. (c) **The Ant System**: This is a metaheuristic inspired by colonies of real ants, which deposit a chemical substance on the ground called *pheromone* [27,17,26,28]. This substance influences the behavior of the ants: they tend to take those paths where there is a larger amount of pheromone. Pheromone trails can thus be seen as an indirect communication mechanism among ants. From a computer science perspective, the ant system is a multi-agent system where low level interactions between single agents (i.e. artificial ants) result in a complex behavior of the entire ant colony. The ant system was originally proposed for the traveling salesman problem (TSP), and most of the current applications of the algorithm require the problem to be reformulated as one in which the goal is to find the optimal path of a graph. A way to measure the distances between nodes is also required in order to apply the algorithm [25]. Despite this limitation, this approach has been found to be very successful in a variety of combinatorial optimization problems [27,10,28].

paradigms, their original motivation and some of their applications. Then, we analyzed a few case studies on the use of evolutionary algorithms to solve problems in biometrics.

In the final part of the chapter, we provided some of the possible research paths that we consider that could be pursued within the next few years. A wide variety of tasks involved in biometrics tasks can be solved (and have been solved in the past) using evolutionary algorithms (e.g. classification, feature extraction and pattern recognition), which opens a lot of research paths that have been only scarcely explored. Thus, we believe that the use of evolutionary algorithms (perhaps in combination with other soft computing techniques) in biometrics will significantly increase within the next few years.

Acknowledgments

The author acknowledges support from CONACyT through project No. 45683-Y. The author would also like to thank Dr. V. P. Shmerko and Dr. S. N. Yanushkevich for their review and valuable comments on this work.

Bibliography

1. Ahmed, S. Abutaleb, and Kamel, M. (1999). A genetic algorithm for the estimation of ridges in fingerprints, *IEEE Trans. Image Processing*, **8**, 8, pp. 1134–1139.
2. Ahmed, A. A. E. and Traoré, I. (2006). A statistical model for biometric verification. This issue.
3. Ammar, H.T. and Tao, Y. (2000). Fingerprint registration using genetic algorithms, *Proceedings 3rd IEEE Symp. Application-Specific Systems and Software Engineering Technology*, pp. 148–154, Richardson, Texas.
4. Angeline, P. J., Fogel, D. B. and Fogel, L. J. (1996). A comparison of self-adaptation methods for finite state machines in a dynamic environment. In L.J. Fogel, P.J. Angeline, and T. Bäck, Eds., *Evolutionary Programming V*, pp. 441–449, Cambridge, Massachusetts, MIT Press.
5. Bäck, T. (1996). *Evolutionary Algorithms in Theory and Practice*. Oxford University Press, New York.
6. Bäck, T., Fogel, D.B., and Michalewicz, Z., editors (1997). *Handbook of Evolutionary Computation*. Institute of Physics Publishing and Oxford University Press, New York.
7. Baker, J.E. (1985). Adaptive Selection Methods for Genetic Algorithms. In J.J. Grefenstette, Ed., *Proc. 1st Int. Conf. Genetic Algorithms*, pp. 101–111. Lawrence Erlbaum Associates, Hillsdale, New Jersey.
8. Baker, J.E. (1987). Reducing Bias and Inefficiency in the Selection Algorithm.

In J.J. Grefenstette, editor, *Genetic Algorithms and Their Applications: Proc. 2nd Int. Conf. Genetic Algorithms*, pp. 14–22. Lawrence Erlbaum Associates, Hillsdale, New Jersey.

9. Banzhaf, W., Nordin, P., Keller, R.E. and Fancone, F.D. (1998) *Genetic Programming. An Introduction.* Morgan Kaufmann Publishers, San Francisco, California.

10. Bonabeau, E., Dorigo, M., and Theraulaz, G. (1999). *Swarm Intelligence. From Natural to Artificial Systems.* Oxford University Press, New York.

11. Booker, L.B. (1982). *Intelligent Behavior as an Adaptation to the Task Environment.* PhD thesis, Logic of Computers Group, University of Michigan, Ann Arbor, Michigan.

12. Brindle, A. (1981). *Genetic Algorithms for Function Optimization.* PhD thesis, Department of Computer Science, University of Alberta, Edmonton, Alberta.

13. Buckles, B.P. and Petry, F.E. editors (1992). *Genetic Algorithms.* Technology Series. IEEE Computer Society Press.

14. Cannon, W.D. *The Wisdom of the Body.* Norton and Company, New York.

15. Cavicchio, D.J. (1970). *Adaptive Search Using Simulated Evolution.* PhD thesis, University of Michigan, Ann Arbor, Michigan.

16. Coello Coello, C.A., Van Veldhuizen, D.A., and Lamont, G.B. (2002). *Evolutionary Algorithms for Solving Multi-Objective Problems.* Kluwer.

17. Colorni, A., Dorigo, M., and Maniezzo, V. (1992). Distributed Optimization by Ant Colonies. In F.J. Varela and P. Bourgine, Eds., *Proc. 1st European Conf. Artificial Life*, pp. 134–142. MIT Press, Cambridge, MA.

18. Corne, D., Dorigo, M. and Glover, F. editors. (1999). *New Ideas in Optimization.* McGraw-Hill.

19. Cornett, F.N. (1972). An Application of Evolutionary Programming to Pattern Recognition. Master's thesis, New Mexico State University, Las Cruces, New Mexico.

20. Darwin, C.R. (1882). *The Variation of Animals and Plants under Domestication.* Second edition. Murray, London.

21. Dasgupta, D., editor. (1999). *Artificial Immune Systems and Their Applications.* Springer.

22. Dawkins, R. (1990). *The Blind Watchmaker.* Gardners Books.

23. Deb, K. (2001). *Multi-Objective Optimization using Evolutionary Algorithms.* John Wiley & Sons, Chichester, UK.

24. Deb, K. and Goldberg, D.E. (1989). An Investigation of Niche and Species Formation in Genetic Function Optimization. In J.D. Schaffer, Ed., *Proc. 3rd Int. Conf. Genetic Algorithms*, pp. 42–50, Morgan Kaufmann Publishers, San Mateo, California.

25. Dorigo, M., Maniezzo, V. and Colorni, A. (1991). Positive Feedback as a Search Strategy. Technical Report 91-016, Dipartimento di Elettronica, Politecnico di Milano, Italy.

26. Dorigo, M., Maniezzo, V., and Colorni, A. (1996). The Ant System: Optimization by a colony of cooperating agents. *IEEE Trans. Systems, Man, and Cybernetics – Part B*, **26**(1):29–41.

27. Dorigo, M. and Di Caro, G. (1999). The Ant Colony Optimization Meta-Heuristic. In D. Corne, M. Dorigo, and F. Glover, Eds., *New Ideas in Optimization*, pp. 11–32, McGraw-Hill.

28. Dorigo, M. and Stützle, T. (2004). *Ant Colony Optimization*. The MIT Press, Cambridge, Massachusetts.

29. Duda, R. and Hart, P. (1973). *Pattern Classification and Scene Analysis*. Wiley.

30. Eberhart, R.C. and Shi, Y. (1998). Comparison between Genetic Algorithms and Particle Swarm Optimization. In V. W. Porto, N. Saravanan, D. Waagen, and A.E. Eibe, Eds., *Proc. 7th Annual Conf. Evolutionary Programming*, pp. 611–619. Springer.

31. Ehrgott, M. (2005). *Multicriteria Optimization*. 2nd edition. Springer.

32. Eiben, A.E. and Smith, J.E. (2003). *Introduction to Evolutionary Computing*. Springer.

33. Englander, A.C. (1985). Machine Learning of Visual Recognition Using Genetic Algorithms. In J.J. Grefenstette, Ed., *Proc. 1st Int. Conf. Genetic Algorithms and Their Applications*, pp. 197–201, Lawrence Erlbraum Associates, Hillsdale, New Jersey.

34. Fitzpatrick, J.M., Grefenstette, J.J. and Van Gutch, D. (1984). Image registration by genetic search. *Proc. IEEE Southeast Conf.*, pp. 460–464.

35. Fogel, D.B. (1995). *Evolutionary Computation. Toward a New Philosophy of Machine Intelligence*. The IEEE, New York.

36. Fogel, D.B., editor, (1998). *Evolutionary Computation. The Fossil Record. Selected Readings on the History of Evolutionary Algorithms*. The IEEE, New York.

37. Fogel, D.B., Fogel, L.J. and Atmar, J.W. (1991). Meta-evolutionary programming. In R.R. Chen, editor, *Proc. 25th IEEE Asimolar Conf. Signals, Systems, and Computers*, pp. 540–545.

38. Fogel, L.J. (1966). *Artificial Intelligence through Simulated Evolution*. John Wiley & Sons, New York.

39. Fogel, L.J. (1999). *Artificial Intelligence through Simulated Evolution. Forty Years of Evolutionary Programming*. John Wiley & Sons, New York.

40. Freitas, A.A. (1997). A Genetic Programming Framework for Two Data Mining Tasks: Classification and Generalized Rule Induction. In J.R. Koza, K. Deb, M. Dorigo, D.B. Fogel, M. Garzon, H. Iba, and R. L. Riolo, Eds., *Proc. 2nd Annual Conf. Genetic Programming*, pp. 96–101.

41. French, M. (1994). *Invention and Evolution. Design in Nature and Engineering*. Second edition. Cambridge University Press.

42. Ganchev, T., Fakotakis, N. and Kokkinakis, G. (2002). Text-Independent Speaker Verification Based on Probabilistic Neural Networks. *Proc. Acoustics Conf.*, pp. 159–166, Patras, Greece.

43. Ganchev, T., Tasoulis, D.K., Vrahatis, M.N. and Fakotakis, N. (2004). Locally Recurrent Probabilistic Neural Networks with Application to Speaker Verification. *GESTS Int. Trans. Speech Science and Engineering*, **1**(2), pp. 1–13.

44. Gaines, R., Lisowsky, W., Press, S., and Shapiro, N. (1980). Authentication

by Keystroke Timing: Some Preliminary Results. Technical Report No. R-2526-NSF, Rand Corporation, Santa Monica, California.

45. Goldberg, D.E. (1989). *Genetic Algorithms in Search, Optimization and Machine Learning.* Addison-Wesley, Reading, Massachusetts.

46. Goldberg, D.E. and Deb, K. (1991). A comparison of selection schemes used in genetic algorithms. In G.J.E. Rawlins, Ed., *Foundations of Genetic Algorithms,* pp. 69–93. Morgan Kaufmann, San Mateo, California.

47. Gomez, F. and Miikkulainen, R. (1999). Solving non-markovian control tasks with neuroevolution. *Proc. Int. Joint Conf. Artificial Intelligence,* pp. 1356–1361, AAAI, San Francisco, California.

48. Grasemann, U. and Miikkulainen, R. (2005). Effective image compression using evolved wavelets. In H.-G. Beyer et al., Ed., *Genetic and Evolutionary Computation Conf.,* vol. 2, pp. 1961–1968, ACM Press, New York.

49. Grefenstette, J.J. and Baker, J.E. (1989) How Genetic Algorithms work: A critical look at implicit parallelism. In J. D. Schaffer, Ed., *Proc. 3rd Int. Conf. Genetic Algorithms,* pp. 20–27, Morgan Kaufmann Publishers, San Mateo, California.

50. Günter, S. and Bunke, H. (2004). Optimization of Weights in a Multiple Classifier Handwritten Word Recognition System Using a Genetic Algorithm. *Electronic Letters on Computer Vision and Image Analysis,* 3(1):25–41.

51. Heitkoetter, J. and Beasley, D. (1995). The hitch-hiker's guide to evolutionary computation (faq in comp.ai.genetic). USENET. (Version 3.3).

52. Hennebert, J., Melin, H., Petrovska, D., and Genoud, D. (2000). POLYCOST: A Telephone-Speech Database for Speaker Recognition. *Speech Communication,* **31**, pp. 265–270.

53. Ho, S.-Y. and Huang, H.-L. (2001). Facial modeling from an uncalibrated face image using a coarse-to-fine genetic algorithm. *Pattern Recognition,* 34:1015–1031.

54. Ho, S.-Y., Shu, L.-S., and Chen, H.-M. (1999). Intelligent Genetic Algorithm with a New Intelligent Crossover Using Orthogonal Arrays. *Proc. Genetic and Evolutionary Computation Conf.,* Vol. 1, pp. 289–296, Morgan Kaufmann Publishers, San Francisco, California.

55. Holland, J.H. (1962). Concerning efficient adaptive systems. In M. C. Yovits, G. T. Jacobi, and G. D. Goldstein, editors, *Self-Organizing Systems—1962,* pp. 215–230. Spartan Books, Washington, D.C.

56. Holland, J.H. (1962a). Outline for a logical theory of adaptive systems. *J. Association for Computing Machinery,* **9**:297–314.

57. Holland, J.H. (1975). *Adaptation in Natural and Artificial Systems.* University of Michigan Press.

58. Huang, K. and Yan, H. (1997). Off-Line Signature Verification by a Neural Network Classifier. *Proc. 17th Australian Conf. Neural Networks,* pp. 190–194, Australian National University, Canberra, Australia.

59. Huang, R.J. (1998). *Detection Strategies for Face Recognition Using Learning and Evolution.* PhD thesis, Department of Computer Science, George Mason University, Fairfax, Virginia.

60. De Jong, K.A. (1975). *An Analysis of the Behavior of a Class of Genetic*

Adaptive Systems. PhD thesis, University of Michigan.

61. De Jong, K.A. (1993). Genetic Algorithms are NOT Function Optimizers. In L. D. Whitley, editor, *Foundations of Genetic Algorithms 2*, pp. 5–17. Morgan Kaufmann Publishers, San Mateo, California.

62. Kennedy, J. and Eberhart, R.C. (1995). Particle Swarm Optimization. *Proc. IEEE Int. Conf. Neural Networks*, pp. 1942–1948, IEEE Service Center, Piscataway, New Jersey.

63. Kennedy, J. and Eberhart, R.C. (1997). A Discrete Binary Version of the Particle Swarm Algorithm. *Proc. IEEE Conf. Systems, Man, and Cybernetics*, pp. 4104–4109, IEEE Service Center, Piscataway, New Jersey.

64. Kennedy, J. and Eberhart, R.C. (2001). *Swarm Intelligence.* Morgan Kaufmann Publishers, San Francisco, California.

65. Kharma, N., Suen, C.Y. and Guo, P.F. (2003). PalmPrints: A Novel Co-evolutionary Algorithm for Clustering Finger Images. In E. Cantú-Paz (editor) *2003 Genetic and Evolutionary Computation Conference*, pp. 322–331, Springer, Lecture Notes in Computer Science Vol. 2723, Chicago, Illinois.

66. Koza, J.R. (1992). *Genetic Programming. On the Programming of Computers by Means of Natural Selection.* The MIT Press, Cambridge, Massachusetts.

67. Kung, S.Y., Mak, W.M. and Lin, S.H. (2004). *Biometric Authentication: A Machine Learning Approach.* Prentice Hall.

68. Leung, K.F., Leung, F.H.F., Lam, H.K. and Ling, S.H. (2004). On interpretation of graffiti digits and characters for ebooks: neural-fuzzy network and genetic algorithm approach. *IEEE Trans. Industrial Electronics*, **51**, 2, pp. 464–471.

69. Michalewicz, Z. (1996). *Genetic Algorithms + Data Structures = Evolution Programs.* Third edition. Springer.

70. Miettinen, K.M. (1998). *Nonlinear Multiobjective Optimization.* Kluwer.

71. Mitchell, M. (1996). *An Introduction to Genetic Algorithms.* The MIT Press, Cambridge, Massachusetts.

72. Morita, M.E., Sabourin, R., Bortolozzi, F., and Suen, C.Y. (2003). Unsupervised feature selection using multi-objective genetic algorithm for handwritten word recognition. *Proc. 7th Int. Conf. Document Analysis and Recognition*, pp. 666–670, Edinburgh, Scotland.

73. Nunes de Castro, L. and Timmis, J. (2002). *Artificial Immnue System: A New Computational Intelligence Approach.* Springer.

74. Nunes de Castro, L. and Von Zuben, F.J. (2002). Learning and Optimization Using the Clonal Selection Principle. *IEEE Trans. Evolutionary Computation*, **6**, 3, pp. 239–251.

75. Osyczka, A. (2002). *Evolutionary Algorithms for Single and Multicriteria Design Optimization.* Physica Verlag.

76. Price, K.V. An Introduction to Differential Evolution. In D. Corne, M. Dorigo and F. Glover, Eds., *New Ideas in Optimization*, pp. 79–108, McGraw-Hill, London, UK.

77. Qi, Y., Tian, J. and Dai, R.-W. (1998). Fingerprint classification system with feedback mechanism based on genetic algorithm. *Proc. 14th Int. Conf. Pattern Recognition*, Vol. 1, pp. 163–165, IEEE.

78. Rao, S.S. (1996). *Engineering Optimization. Theory and Practice.* 3rd edition. John Wiley & Sons.
79. Rechenberg, I. (1973). *Evolutionsstrategie: Optimierung technischer Systeme nach Prinzipien der biologischen Evolution.* Frommann–Holzboog, Stuttgart, Germany.
80. Reinders, M.J.T., Sankur, B., and van der Lubbe, J.C.A. (1992). Tranformation of a general 3D facial model to actual scene face. *Proc. 11th IAPR Int. Conf. Pattern Recognition,* Vol. III, C: Image, Speech and Signal Analysis, pp. 75–78.
81. Rothlauf, F. (2002). *Representations for Genetic and Evolutionary Algorithms.* Physica-Verlag, New York.
82. Rudolph, G. (1994). Convergence Analysis of Canonical Genetic Algorithms. *IEEE Trans. Neural Networks,* **5**, pp. 96–101.
83. Schwefel, H.-P. (1981). *Numerical Optimization of Computer Models.* Wiley, Chichester, UK.
84. Soldek, J., Shmerko, V., Phillips, P., Kukharev, G., Rogers, W., and Yanushkevich, S. (1997). Image Analysis and Pattern Recognition in Biometric Technologies. *Proc. Int. Conf. on the Biometrics: Fraud Prevention, Enhanced Service,* pp. 270–286, Las Vegas, Nevada.
85. Stadnyk, I. (1987). Schema Recombination in a Pattern Recognition Problem. In J.J. Grefenstette, Ed., *Genetic Algorithms and Their Applications: Proc. 3nd Int. Conf. Genetic Algorithms,* pp. 27–35, Lawrence Erlbraum Associates, Hillsdale, New Jersey.
86. Sweldens, W. (1996). The lifting scheme: A custom-design construction of biorthogonal wavelets. *J. Applied and Computational Harmonic Analysis,* **3**, 2, pp. 186–120.
87. Syswerda, G. (1989). Uniform Crossover in Genetic Algorithms. In J.D. Schaffer, editor, *Proc. 3rd Int. Conf. on Genetic Algorithms,* pp. 2–9, Morgan Kaufmann Publishers, San Mateo, California.
88. Talay, A.C. (2005). An Approach for Eye Detection Using Parallel Genetic Algorithm. In V.S. Sunderam, G.D. van Albada, P.M.A. Sloot, and J.J. Dongarra, Eds., (2005). *Computational Science: Proc. 5th Int. Conf., Part III,* pp. 1004, Springer. Lecture Notes in Computer Science Vol. 3516, Atlanta, Georgia.
89. Teller, A. and Veloso, M. (1995). Algorithm Evolution for Face Recognition: What Makes a Picture Difficult. *Int. Conf. on Evolutionary Computation,* pp. 608–613, IEEE, Perth, Australia.
90. Teredasai, A., Park, J., and Govindaraju, V. (2001). Active Handwritten Character Recognition Using Genetic Programming. In J. Miller, M. Tomassini, P. L. Lanzi, C. Ryan, A.G.B. Tettamanzi, and W.B. Langdon, editors, (2001), *Genetic Programming. 4th European Conf.,* pp. 371–379, Springer, Lake Como, Italy.
91. Trellue, R.E. (1973). The Recognition of Handprinted Characters Through Evolutionary Programming. Master's thesis, New Mexico State University, Las Cruces, New Mexico.
92. Whitley, D. (1989). The GENITOR Algorithm and Selection Pressure: Why

Rank-Based Allocation of Reproductive Trials is Best. In J.D. Schaffer, editor, *Proc. 3rd Int. Conf. on Genetic Algorithms*, pp. 116–121. Morgan Kaufmann Publishers, San Mateo, California.

93. Wilson, S.W. (1985). Adaptive "Cortical" Pattern Recognition. In J.J. Grefenstette, Ed., *Proc. 1st Int. Conf. Genetic Algorithms and Their Applications*, pp. 188–196, Lawrence Erlbraum Associates, Hillsdale, New Jersey.

94. Yanushkevich, S. N., Stoica, A., Shmerko, V. P. and Popel, D. V. (2005). *Biometric Inverse Problems*, Taylor & Francis/CRC Press.

95. Yokoo, Y. and Hagiwara, M. (1996). Human faces detection method using genetic algorithm. *Proc. IEEE Int. Conf. Evolutionary Computation*, pp. 113–118.

96. Yu, E. and Cho, S. (2004). Keystroke dynamics identity verification-its problems and practical solutions. *Computers & Security*, **23**, 5, pp. 428–440.

97. Zhang, D.D. (2000). *Automated Biometrics: Technologies and Systems*. Kluwer.

Chapter 13

Some Concerns on the Measurement for Biometric Analysis and Applications

Patrick S. P. Wang

Image Processing Group,
College of Computer and Information Science,
Northeastern University, Boston, MA02115, USA
pwang@ccs.neu.edu
www.ccs.neu.edu/home/pwang

Some concerns of measurement for biometric analysis and synthesis are investigated. This research tries to reexamine the nature of the basic definition of "measurement" of distance between two objects or image patterns, which is essential for comparing the "similarity" of patterns. According to a recent International Workshop on Biometric Technologies: Modeling and Simulation at University of Calgary, June, 2004, Canada, biometric refers to the studies of analysis, synthesis, modeling and simulation of human behavior by computers, including mainly recognition of hand printed words, machines printed characters, handwriting, fingerprint, signature, facial expression, speech, voice, emotion and iris etc.

The key idea is the "measurement" that defines the similarity between different input data that can be represented by image data. This paper deals with the very fundamental phenomena of "measurement" of these studies and analysis. Preliminary findings and observations show that the concepts of "segmentation" and "disambiguation" are extremely important, which have been long ignored. Even while computer and information professionals and researchers have spent much effort, energy, and time, trying very hard and diligently to develop methods that may reach as high as 99.9999% accuracy rate for character and symbol recognition, a poorly or ill considered pre-designed board poster or input pattern could easily destroy its effectiveness and lower the overall performance accurate rate to less than 50%. The more data it handles, the worse the results. Its overall performance accuracy rate will be proportionally decreasing. Take road safety for example. If street direction signs are poorly designed, then even a most intelligent robot driver with perfect vision and 100% accurate rate of symbol recognition, still has at least 50% error rate. That is, it can achieve at most only 50% accuracy rate in determining which correct direction to follow. It's not just a matter of being slow or losing time.

However, in more serious and urgent life threatening situation like fire, accidents, terrorists attack or natural disaster, it's a matter of life and death. So the impact is enormous and widespread. The idea and concept of "ambiguity," "disambiguation," "measurement," and "learning," and their impacts on biometrics image pattern analysis and recognition, as well their applications are illustrated through several examples.

Contents

13.1. Overview (Introduction)

We first briefly review what's happening in biometric research and their results. According to [5], a majority of biometric studies fall into the categories of fingerprint, facial and symbol analysis and recognition, about 74%. It can be shown in the Fig. 13.1.

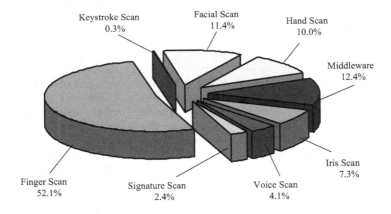

Fig. 13.1.   Percentage of biometric applications in real life [Int. biometric group (2003)], in which *biometrics* are defined as automated methods of recognizing a person based on the acquired physiological or behavioural characteristics.

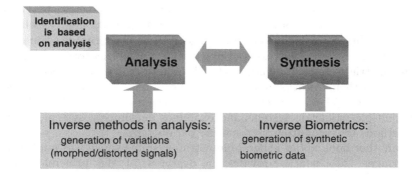

Fig. 13.2.   Analysis and synthesis of biometric images.

Inverse methods in analysis examples are shown in [10], [11], [12], [21] (Fig. 13.2):

(a) Modeling, or tampering (e.g. signature forgery in the task of signature recognition,
(b) Voice synthesis (telecommunication services), and
(c) Facial image synthesis (animation, "talking heads").

There are also some developments of generating synthetic biometric data, using computer graphics, computational geometry methodologies, with applications to the following fields, which are being implemented successfully [6], [7], [9], [13], [23], [24]:

> *Collecting large databases for testing the robustness of identification*
> *   systems,*
> *Training personnel on a system that deploys simulation of biometric*
> *   data,*
> *Cancelable biometrics,*
> *Biometric data hiding, and*
> *Biometric encryption.*

Further, according to a most recent book regarding advanced aspects of biometrics *Handbook of Multibiometrics* [14], consistent advances in biometrics help to address problems that plague traditional human recognition methods and offer significant promise for applications in security as well as general convenience. In particular, newly evolving systems can measure multiple physiological or behavioral traits and thereby increase overall reliability that much more. Multimodal Biometrics provides an accessible, focused examination of the science and technology behind multimodal human recognition systems, as well as their ramifications for security systems and other areas of application. After clearly introducing multibiometric systems, it demonstrates the noteworthy advantages of these systems over their traditional and unimodal counterparts. In addition, the work describes the various scenarios possible when consolidating evidence from multiple biometric systems and examines multimodal system design and methods for computing user-specific parameters.

In another book about general biometrics systems which was published by Springer [20], which provides an overview of the principles and methods needed to build reliable biometric systems. It covers 3 main topics: key biometric technologies, testing and management issues, and the legal and system considerations of biometric systems for personal verification/identification. It focuses on the four most widely used technologies — speech, fingerprint, iris and face recognition. It includes:

(a) In-depth coverage of the technical and practical obstacles which are often neglected by application developers and system integrators and which result in shortfalls between expected and actual performance,
(b) Detailed guidelines on biometric system evaluation, and
(c) Protocols and benchmarks which will allow developers to compare performance and track system improvements.

The book of Biometric Systems — Technology, Design and Performance Evaluation is intended as a reference book for anyone involved in the design, management or implementation of biometric systems.

According to a recent paper regarding metric learning for text documentation analysis and understanding [8], many algorithms in machine learning rely on being given a good distance metric over the input space. Rather than using a default metric such as the Euclidean metric, it is desirable to obtain a metric based on the provided data.

> *We consider the problem of learning a Riemannian metric associated with a given differentiable manifold and a set of points. Their approach to the problem involves choosing a metric from a parametric family that is based on maximizing the inverse volume of a given data set of points. From a statistical perspective, it is related to maximum likelihood under a model that assigns probabilities inversely proportional to the Riemannian volume element.*

We discuss in detail learning a metric on the multinomial simplex where the metric candidates are pull-back metrics of the Fisher information under a Lie group of transformations. When applied to text document classification the resulting geodesic distance resemble, but outperform, the *tfidf* cosine similarity measure.

In the literature, there is another recent book regarding some guidance to biometrics published in 2004 [1]. This complete, technical guide offers some rather detailed descriptions and analysis of the principles, methods, technologies, and core ideas used in biometric authentication systems. It explains the definition and measurement of performance and examines the factors involved in choosing between different biometrics. It also delves into practical applications and covers a number of topics critical for successful system integration. These include recognition accuracy, total cost of ownership, acquisition and processing speed, intrinsic and system security, privacy and legal requirements, and user acceptance.

From above investigations, it can be seen that most recent development and applications largely rely on the fundamental definition of pattern matching, which in turn depends on the measurement of distances between two input images. It has been observed that the concept of "ambiguity" plays a very important fundamental roles of all these distance measurement and image pattern processing in both learning (analysis) and recognition (synthesis), which is also the most difficult obstacles of essentially all recognition problems in the real daily life. Therefore in the next section, we are going to discuss these problems and how input images (patterns) can and should be "disambiguated" in dealing with real life problem applications.

13.2. The Problem of "Ambiguity," and Its Solutions

> *The main purpose of "Pattern Recognition" is trying to achieve high accuracy, fast speed, and saving memory and cost. It basically has two portions: learning (analysis) and recognition (synthesis), as can be shown in Fig. 13.3.*

Notice that here we adapt a general diagram that incorporate the capability to handle general 3D objects, including "articulated" objects, which is a compromise between two extreme categories of "rigid" objects such as bricks, of which the object has fixed shape and "deformable" objects such as newspaper and handkerchief, of which the shape can easily vary arbitrarily. Interesting enough, this category of "articulated" objects also include some human body movements such as hands, fingers, arms, elbows, arms stretch, legs running, body language, hand sign language for handicapped people gymnastic movements, and swimming diving etc [16], [17].

We also include "interactive" dynamics in which human beings are involved in the process interactively with the machine, such as making decision as for which input pattern should be included in the compiled dictionary in "learning" as well as "recognition stage" [15].

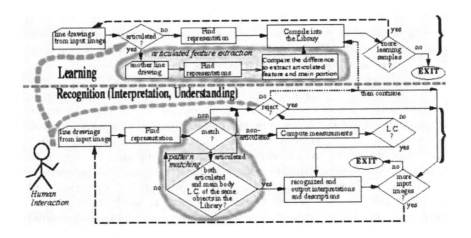

Fig. 13.3.   Learning/recognition general diagram.

Recently, through rapid advancements and developments of past years, there have been tremendous achievements in reaching these goals. For example, take character recognition, there have been methods that can reach as high as almost 99.99% accuracy rate [2]. However, even with 100% accuracy rate, there still might be more work need to be done to overcome serious difficulties in solving real life problems. This is related to "syntax," "semantics," and "pragmatics" problems. It is well known that "ambiguity" refers to the same "syntax" with multiple "meanings" or "interpretations," also known as "semantics".

Let's call this "1-st level of syntax-to-semantics ambiguity." Likewise, even with the same "semantics," there might be different interpretations, therefore, different implementations accordingly, also known as "pragmatics". Let's call this 2nd level *semantics-to-pragmatics ambiguity.* The 1st level "semantics ambiguity" is mainly due to the following three categories and their variations:

Syntax (pattern, shape) of the input images are very similar but with different semantics (interpretations, meanings), like English capital letter 'O' and numerical number '0'; English small letter 'l' and numerical number one '1' etc. Some hand printed English characters examples are shown in Fig. 13.4.

Segmentation difficulty, like the famous statement,

Read my lips, NO NEW TAXES!

(by former President George Bush), which can be interpreted in at least two different ways, namely:

NO NEW TAXES,

and

NO NEWT AXES!

If the word strings were segmented differently. And there is other example and commonly misinterpreted hand printed mathematical formula of $X = H + G$ as $X = 1 - 1 + CT$. It is particularly troublesome with those which are in the immediate neighborhood of "continuous transformation" [18], [19], for examples, O and C, C and $|$, H and A etc., as shown in Fig. 13.5.

Grammatical rule tree parsing, for example,

TIME FLIES LIKE AN ARROW,

which has at least three different interpretations, each corresponds a parsing tree. It also deals with "learning" experience. For example, let's take a look at the following scenario in a typical elementary school classroom:

```
Teacher:  Let's learn an English letter Oh "O", got it?
Students:  yes, (it is easy)
Teacher:  Now let's learn another numerical symbol zero
"0", got it?
Students:  yes, (it is easy)
Teacher:  Now that you have learned two symbols, let me
give you a quick quiz:  I'll write a circle on the
blackboard like "0", can you recognize what it is?
Students:  ?  ?  ?
```

I think by now, all students will pause for a while, because there are two possible answers.

Fig. 13.4.   Some hand printed characters samples of "A".

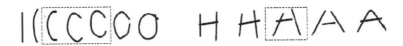

Fig. 13.5.   Some examples of "Nearmiss" in the immediate neighborhoods of "Continuos Transformation (CT) [Wang (1985)].

In general, the degree of "ambiguity" can be defined as the number of interpretations in a syntax within certain domain. In the above example, the degree of ambiguity of symbol 'O', in the domain of alpha-numerals (AL) denoted as

$$deg('O', AL) = 2,$$

because there are two interpretations. The higher the degree of ambiguity, the more difficult it is to be recognized accurately. Notice that it depends largely on the domain of the problem we are dealing with. If the domain of the above example is only in alphabet (A), then the degree

$$deg('O', A) = 1.$$

It has been observed that the accuracy rate of recognition of an input image is reversely proportional to it degree of ambiguity. That is, in general, the higher the degree of ambiguity, the lower its recognition accuracy, and vise versa. Some more details discussion and formulation can be found in [18].

Example 3. Different degrees of ambiguities are shown in Fig. 13.6 and Fig. 13.7.

Correspondingly, we can also establish a chemical-bond like structure, illustrating the continuous transformation (CT) of hand printed symbols (including all alpha-numerals, arithmetic symbols, and Greek letters), as

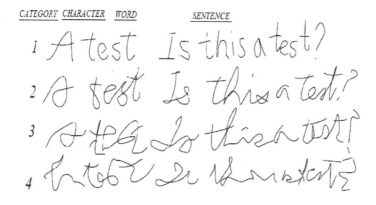

Fig. 13.6. Multiple 4-levels of ambiguities of English handwritings.

shown in Fig. 13.8. Notice that those which are "well behaved" patterns in the dashed rectangles are comparatively easy to distinguish and recognize, while those in the transition area from one symbol to another, known as "nearmiss" are normally ambiguous, are not easy to distinguish with other symbols, therefore are more challenging to be recognized accurately.

This illustrates the close relation between learning and recognition, and shows it is not always easy to learn implies easy to recognize, but sometimes it is easy to learn but hard to recognize and vise versa, as shown in Fig. 13.9.

Those problems and difficulties sometimes may be resolved by the means of contextual information, with a larger database of dictionary. However, many times, even though the 1st level "syntax-to-semantics ambiguity" can be resolved with contextual information and the surrounding environment known, it is still likely to make wrong decisions, therefore wrong execution and implementations (pragmatics), because there still needs to resolve a 2-nd level "semantics-to-pragmatics ambiguity" problems.

Example 4. Suppose a robot driver is designed according to human vision capability of 100% accuracy rate of symbol (alpha-numerals and graphics symbols like arrows) recognition. When this robot driver encounters a situation at the cross road of *Louvre* museum and *Notre Dam*, and if the direction sign is not designed properly, then the robot still has 50% of making wrong decision to turn to the wrong direction because of the 2-nd level ambiguity due to the wrong "semantics-to-pragmatics" segmentation. That is, it still has only 50% accuracy rate of turning into right direction at the "semantics-to-pragmatics" level, even though if it has 100% accuracy rate of character recognition at the "syntax-to-semantics" level (Fig. 13.10).

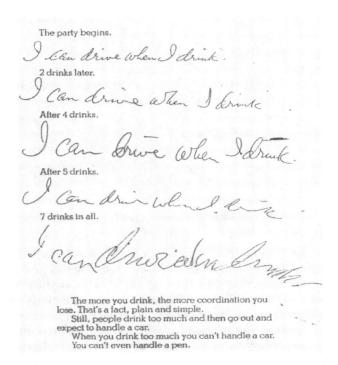

Fig. 13.7. Multiple 5-levels of ambiguities of English handwritings.

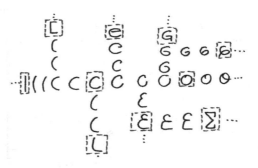

Fig. 13.8. Continuous Transformation (CT) of hand printed symbols (including all alpha-numerals, arithmetic symbols, and typesetting Greek letters).

Fig. 13.9.   Relation between Learning and Recognition.

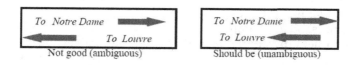

Fig. 13.10.   An example of bad (ambiguous) design and good (unambiguous) design of road signs.

Suppose there are 10 thousands tourists visiting Paris every day. Most of them are interested in touring the famous Louvre museum, and half of them take subway. But when they arrive at the cross section of *Louvre* and *Notre Dam*, they have 50% of chance to get to the wrong train, thereby wasting at least 1/2 an hour or 30 minutes of travel. In average, there will be a waste of

$$0.5 \times (1/2 \times 10,000) = 0.5 \times 5,000 = 2,500 \text{ hours}$$

of waste of tourists' time. This is just a conservative estimation.

There might be way more than just 10 thousands tourists visiting Paris per day. Suppose there are 100 thousands people visiting Paris each day, then the average waste of tourists time would be 25,000 hours per day total. The more visitors the more waste of their total time. I personally encountered such an experience when touring Paris some time ago. Since robotics basically emulates human behavior, and can not beat human vision capability in terms of accuracy rate at least for the time being, one can imagine the situation would be much worse if a robotic driver encounters the similar situation as to what "right" direction it should turn to at the cross section of an ill-designed road sign.

Example 5. A more serious example, as a matter of life and death, can also be seen at the regular houses or residential apartments or a company office and compound, or a school building, or a government building. An emergency switch is supposed to prevent faster in a very quick way in case of fire. But there are many improperly designed emergency switches that serve negative purpose. Again, it is due to improper "semantics-to-implementation" segmentation. At an emergency situation, one has only seconds to decide how to turn the switch to prevent fire. However, a wrongly segmented or improper designed switchboard might miss lead users to turn to the wrong direction, resulting, instead of preventing, a fire disaster. In this case, even though the accuracy rate of recognizing symbols may be as high as 100%, yet the accuracy rate of making right decision, therefore preventing fire disaster is only 50%. That is, there are still as high as 50% error rate at the pragmatics level. I have also personally seen such examples of ill-designed emergency switchboard in real life, in many apartments and houses, including my own.

There are similar examples in the design of hospital sign "H", supposedly to help drivers (or robotics drivers, ultimately) to get to the destination quickly, and "correctly" towards to the right direction to the nearest hospital. It is not only saving time, but more importantly, also to save lives. It is as important as a matter of life and death. Nowadays many countries, including U.S.A. are facing seriously the war against terrorism. One of the main key issues is how to save peoples' lives in case of terrorists attacks, or any natural and or man-made disastrous situations. The issues of designing proper road signs to avoid miss leading information and accurately leading to right direction becomes ever more important and urgent. Figure 13.11 illustrates even more serious situation, in which lives may be endangered, because of its misleading signs of directions. Even worse, there are more serious situation, as matter of life and death, say, in an emergency situation of fire. If an emergency switch is designed improperly, then it might mislead people to turn to wrong direction, leading to disaster, rather than preventing fire as originally desired. Notice that, in an emergency situation such as fire or driving a car to hospital and encounter a cross road, people has only second to decide what to do or which direction to turn to. It is true for human beings, and it is certainly even more so for a "robot driver with vision or eyes".

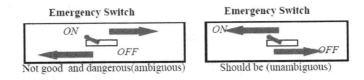

Fig. 13.11. Another example of bad (Dangerous) versus good (safer) design of emergency signs.

Notice that, in both cases (Figs. 13.10 and 13.11) there are two possible interpretations in each example, therefore the degree of ambiguity of each sign is two, i.e.

$$deg(\text{``}Sign\text{''}, \texttt{SymbolsSet}) = 2.$$

These illustrations show that even though the symbol (alpha-numerals plus graphics symbols including the arrow signs) accuracy rate is as high as 100%, the overall effective recognition rate could be greatly reduced to 50% or lower, due to improper design of the input image (road sign poster or label).

Fig. 13.12. A university campus directory sign poster: Which is which?

Example 6. There are wide spread examples falling into this category, as can be seen in the following Fig. 13.12 of a university campus directory sign poster. In a quick look, can anybody easily figure out which is which? that is, which direction leads to West Village A, B, C, G, E, G, or H? Or more specifically, if a human being can not figure it out, wouldn't it be even more difficult for a robot driver to figure it out, even this robotic has vision capability with recognition rate as high as 100%. Figure 13.13 shows a city road sign poster, which is ambiguous, and the degree of ambiguity here is two, because there are two possible ways of interpreting the meaning of its directions associated with the word annotations.

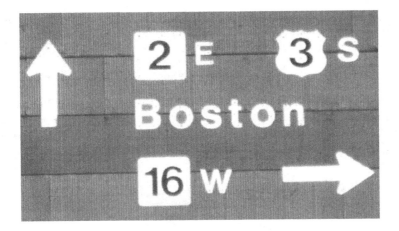

Fig. 13.13. An ambiguous road sign poster.

Fig. 13.14. Another example of (perhaps less ambiguous) road sign poster (will be even more ambiguous without the darker and lighter background green color).

Finally, last but not the least, let's take a look at the following example.

Example 7. Consider the scenario below:

```
Father to his toddler son:  "Will you.take out the garbage?"
Son:  "You are closer."
Then the father measures the distance from son's nose to the
garbage.
Father:  "Actually, I believe you're closer."
Then, the son went to his mom, and asked: Son:  "Do you measure
garbage from feet or noses?"
```

The child intuitively feels his father is "closer" to the garbage. Then his father uses a ruler to "measure" the distance between the garbage and the kid, comparing the distance between the garbage and the father himself, from their "noses", not from their "feet". Since the father is much taller than the kid, therefore the father reaches a conclusion that the kid is closer to the garbage, and should take out the garbage, even though actually the father is indeed closer to the garbage, according to the regular sense of measurement of distance from feet (Fig. 13.15).

ROSE IS ROSE by Pat Brady

Fig. 13.15.   Illustrations showing the importance of "measurement" and disambiguation.

Although it is just a simple cartoon comic, yet, it demonstrates how important it is to have a proper definition of "measurement" at the first place for pattern matching and pattern recognition of input images.

13.3.   Discussions and Conclusions

We have observed some basic phenomena of "ambiguity" in learning and recognition in a biometric environment. The key idea here is the concept of "measurement" which can define the distance between two images. The illustrative examples given in previous sections are just a few selected to demonstrate how important it is to treat the problems of "ambiguity" and "segmentation" seriously and cautiously. In our daily life of essentially every field one can imagine of, there are numerous, hundreds of thousands of examples that fall into this category.

For the future research, one shall pay more attention to its segmentation and ambiguity problems from the beginning, and make sure that the definition of "measurement," which is the foundation of analysis and recognition of essentially all kinds of image patterns, including all types of images or data in the biometric study and research as well as applications, including, but not limited to the following problems of analysis and recognition of hand printed characters, fingerprints, signatures, facial expressions, and iris, among other things. These difficulties can be resolved from global semantic to pragmatics point of view as discussed in previous section. It will reduce a lot of problems and troubles in our daily life, and make our living environment safer. On the other hand, if we continue to ignore the importance of it, we definitely will face more and more troubles in solving our real life problems.

Bibliography

1. Bolle, R., Connell, J., Pankanti, S. Ratha, N. and Senior, A. (2004). *Guide to Biometrics* Springer.
2. Chen, C. and Wang, P. Eds. (2005). *Handbook of Pattern Recognition and Computer Vision,* World Scientific Publishers.
3. Cook, P. (2002). *Real Sound Synthesis for Interactive Applications,* A K Peters, Natick, MA.
4. Fua, P. and Miccio, C. (1999). Animated heads from ordinary images: a least-squares approach, *Computer Vision and Image Understanding,* **75**, 3, pp. 247–259.
5. Int. biometric group, www.biometricgroup.com/press_releases_2003.html.
6. Jain, A. K., Bolle, R. and Pankanti, S. (2005). Biometrics: Personal Identification in Networked Society, *Springer.*
7. Jain A. and Uludag, U. (2003). Hiding biometric data. *IEEE Trans. Pattern Analysis and Machine Intelligence,* **25**, 11, pp. 1494–1498.
8. Lebanon, G. (2006). Metric learning for text documents, *IEEE Trans. Pattern Analysis and Machine Intelligence,* **28**, 4, pp. 497–508.
9. Ma, Y., Schuckers, M. and Cukic, B. (2005). Guidelines for appropriate use of simulated data for bio-authentication research *Proc. 4th IEEE Workshop Automatic Identification Advanced Technologies,* pp. 251–256.
10. Parker, J. R. (2006). Composite systems for handwritten Ssignature recognition, this issue
11. Plamondon, R. and Srihari, S. (2000). On-line and off-line handwriting recognition: A comprehensive survey, *IEEE Trans. Pattern Analysis and Machine Intelligence,* **22**, , pp. 63–84.
12. Popel, D. (2006). Signature analysis, verification and synthesis in pervasive environments, This issue.
13. Ratha, N., Connell, J. and Bolle, R. (2001). Enhancing security and privacy in biometrics-based authentication systems, *IBM Systems J.* **40**, 3, pp. 614–634.
14. Ross, A., Nandakumar, K. and Jain, A. (2006). *Handbook of Multibiometrics,* International Series on Biometrics Springer, **6**.
15. Wang, P. (2004). Some new development of 3D object visualization and recognition using graphic representations, and applications, *Proc. Int.*

Workshop on Biometric Technologies: Modeling and Simulation, University of Calgary, Canada, pp. 69–79.

16. Wang, P. (2001). 3D articulated object understanding, learning, and recognition from 2D images, In J. Shen, P. Wang and T. Zhang), Eds, *Proc. Multispectral Image Processing and Pattern Recognition Conf.,* pp. 5–16.

17. Wang, P. (2001). Articuated object recognition using Web technology, *Proc. Web Imaging Conf.* Maebashi, Japan, pp. 61–74.

18. Wang, P. (1994). Learning, representation, understanding and recognition of words – an intelligent approach, In S. Impedovo, Ed., *Fund. in Handwriting Recog.,* Springer, pp. 81–112.

19. Wang, P. (1985). A new character recognition scheme with lower ambiguity and higher recognizability, *J. Pattern Recognition Letters,* **3**, pp. 431–436

20. Wayman, J., Jain, A., Maltoni, D. and Maio, D. Eds. (2005). Biometric Systems Technology, Design and Performance Evaluation, *Spinger*

21. Yanushkevich, S. N., Stoica, A., Shmerko, V. P. and Popel, D. V. (2005). *Biometric Inverse Problems.* CRC Press/Taylor & Francis Group, Boca Raton, FL.

22. Yanushkevich, S. N., Stoica, A. and Shmerko, V. P. (2006). Fundamentals of biometric-based training system design, this issue.

23. Zhang, Y., Prakash, E. and Sung, E. (2004). Face understanding and recognition *J. Visible Languages and Computing,* **15**, pp.125–160.

24. Zhang, Y., Sung, E. and Prakash, E. (2001). A physically-based model for real-time facial expression animation *Proc. 3-rd IEEE Int. Conf. 3D Digital Imaging and Modeling,* pp. 399–406.

Chapter 14

Issues Involving the Human Biometric Sensor Interface

Stephen J. Elliott*, Eric P. Kukula, Shimon K. Modi

Biometrics Standards, Performance, and Assurance Laboratory,
Knoy Hall Room 372, 401 N. Grant Street,
West Lafayette, IN- 47906, USA
* elliottpurdue.edu

This chapter outlines the role of Human Biometric Sensor Interaction (HBSI) which examines topics such as ergonomics, the environment, biometric sample quality, and device selection, and how these factors influence the successful implementation of a biometric system. Research conducted at Purdue University's Biometric Standards, Performance, and Assurance Laboratory has shown that the interaction of these factors has a significant impact on the performance of a biometric system. This chapter examines the impact on biometric system performance of a number of modalities including two-dimensional and three-dimensional face recognition, fingerprint recognition, and dynamic signature verification. By applying ergonomic principles to device design as well as understanding the environment in which the biometric sensor will be placed, can positively impact, the quality of a biometric sample, thus resulting in increased performance.

Contents

Glossary

FRVT — The Face Recognition Vendor Test
HBSI — Human Biometric Sensor Interaction
IEA — The International Ergonomics Association
NIST — The National Institute of Standards and Technology

14.1. Introduction

Biometric technologies are defined as the automated recognition of behavioural and biological characteristics of individuals. Using various techniques of pattern recognition, specific behavioural and biological traits of individuals are collected, extracted, and matched to produce a result, which will either validate the claim of identity or reject it. Biometric technologies are used currently for a multitude of applications for authenticating an individual either as a stand alone application, or as part of a multi-factor authentication system (i.e. combined with other possessions such as an ID card, or knowledge such as a PIN). When deciding whether to implement a biometric system, many factors have to be examined to evaluate the potential performance of the system against other non-biometric solutions. One important factor in the evaluation of a biometric system is to understand how the target population will react to that system, and what their resulting performance will be.

As biometrics become more pervasive in society, understanding the interaction between the human and the biometric sensor becomes imperative, and each biometric modality have their strengths and weaknesses depending on the application and target population. Biometrics can be split into two different types, commonly termed

Behavioural and
Biological (also sometimes referred to as *physiological*).

Behavioural biometrics include signature and voice, and biological or physiological biometrics include face, finger, hand geometry and iris. Although to some extent these categories overlap as some modalities are functions of both behavioural and biological voice, face, and signature can have influences from both behavioural and physiological.

Biometric modalities can also be classified by five properties outlined by [9]:

(a) *Universality,*
(b) *Uniqueness,*

(c) *Permanence,*
(d) *Collectibility,* and
(e) *Acceptability.*

Herein lies one of the challenges associated with large scale deployments of biometrics and the topic of the paper the majority of biometrics are challenged by these five categories.

Example 8. Not everybody has a particular biometric trait or that individuals biometric trait maybe significantly different from the common expected trait for example in hand geometry an individual may have a missing digit which may prevent them from providing an image to the sensor.

Another issue is that a biometric characteristic should be *invariant* over time, but in the case of fingerprints, the scarring of the print over time, may affect its character (a measure of image quality) which will subsequently affect the performance of the biometric matcher. And as biometrics become deployed in a variety of locations, the positioning of the sensor, and its surrounding environment become important.

As biometric technologies become more pervasive in society, research must be undertaken to assess the impact of those subjects who have non-normal biometric traits, in order to predict the impact of such samples on the biometric system performance, and with such research emerges the field of Human Biometric Sensor Interaction (HBSI). The research described below examines some of the problems associated with biometric authentication, namely with the environment, population and devices. These research areas are fundamental, because the widespread adoption of biometric technologies requires an understanding of how each of these factors can affect the performance of the device. Both government and industry need to understand

(a) *How biometric samples perform over time,*
(b) *How changes to biometric samples affect system performance, and*
(c) *How the image quality of a biometric sample changes with age.*

The Face Recognition Vendor Test (FRVT), designed by the National Institute of Standards and Technology (NIST) and conducted in 2000 and 2002 in support of the *U.S.A. Patriot Act,* assessed if advancements in face recognition algorithms actually improved performance. Ten vendors participated in this test in 2002; test results proved that variation in outdoor lighting conditions, even for images collected on the same day, drastically reduced system performance.

Example 9. The verification performance for the best face recognition systems drops from 95 percent to 54 percent when the test environment is moved from indoors to outdoors [36,38].

This poses the fundamental question:

Is the photograph of an individual who enrolls indoors at a drivers license bureau represent an environment of controlled illumination?

(a) If controlled lighting is used, then facial recognition systems will verify that particular individual 95 percent of the time in that same environment.
(b) If verification is attempted outdoors, then the facial recognition system performance would fail about 46 percent of the time, even on the same day as enrollment.

This scenario of failure in different environments but on the same day as enrollment eliminates the factor of template aging. In most real-world scenarios, an individual will not be identified

(a) In the same environment or
(b) On the same day as enrollment.

Therefore, more research must be conducted to assess system performance in a variety of conditions.

It is important to consider the effect of *poor quality biometric samples* on the performance of the overall system.

Image quality for fingerprints of the elderly population has been shown to have poor ridge definition and lower moisture content, resulting in lower image quality [44]. Fingerprint recognition systems that will be used by an elderly population need to take into account these factors so that impact on performance is minimal.

Example 10. A research study was conducted to examine image qualities of younger group (18–25 years old) and elderly group (62 years and above). The research compared the number of minutiae and image quality from the two different age groups, and the research showed that there was a significant difference between the image qualities of the fingerprints from the two age groups.

How biometric samples differ when collected on different devices. Another important consideration involves how biometric samples differ when collected on different devices [51].

Example 11. In [12], more than 15,000 dynamic signatures were collected on different mobile computing devices and the resulting signatures analyzed to see which variables are stable among different digitizers. The purpose of this research was to test signature verification software on both traditional digitizer tables and on wireless/mobile computing devices, in order to assess how the dynamics of the signature signing process change on such devices.

Example 12. The research examined the differences on table-based digitizers (Wacom Technologies Intuos and Interlink Electronics, Inc.'s ePad electronic signature devices) and mobile computing devices (Palm, Inc.'s Palm IIIxeTM, Symbol 1500 and 1740 digitizers). The research showed significant differences in specific variables across different digitizers.

Therefore, when deciding to implement dynamic signature verification using different digitizers, the device type and identity needs to be embedded in the signature data. Therefore, the study of the environment, population, device variables, and how the human interacts with the sensor is important to the field of biometrics, and will be discussed in this chapter.

14.2. Ergonomics and an Introduction to the Human Biometric Sensor Interaction

Traditionally, biometric hardware and software research and development have focused on increasing performance, increasing throughput, and decreasing the size of the sensor or hardware device. However, understanding how system design impacts both the user interaction and system performance is of utmost importance and is one of the objectives of Ergonomics.

Ergonomics is a derivative of the Greek words "ergon," or work, and "nomos," meaning laws. In 2000, the International Ergonomics Association (IEA) defined ergonomics or human factors as:

```
The scientific discipline concerned with the understanding of
interactions among humans and other elements of a system, and the
profession that applies theory, principles, data and methods to
design in order to optimize human well-being and overall system
performance [18].
```

In 2003, leading biometric experts met under the sponsorship of the National Science Foundation to "develop a rational and practical description of crucial scholarly research to support the development of biometric systems" [41]. Within the proposed research agenda was an item on ergonomic design of the capture system and usability studies to evaluate the effect on system performance [41]. And at the time of writing, the authors' have found minimal literature or research in the fusion of biometrics and ergonomics.

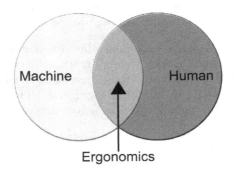

Fig. 14.1.   The fit of ergonomics in the human and machine interaction [8].

There is potential for merging the disciplines of ergonomics and biometrics, creating an interdisciplinary research opportunity called the HBSI. In design, ergonomics attempts to achieve an optimal relationship between humans and machines in a particular environment. The goal of ergonomics according to [48] is:

> *The goal of ergonomics is to fit, or adapt, the work to individuals, as opposed to fitting the individuals to the work or device.*

Figure 14.1 presents the model proposed by [48] to show where ergonomics fits in the human-machine interaction. If we apply *Tayyari and Smith's model* for biometrics — which includes the human-machine (sensor) interaction (Fig. 14.1), and add ergonomics to the model, the overlap of all three areas is defined as the HBSI, which can be seen in Fig. 14.2.

The HBSI applies ergonomic principles in the design of a biometric device or system to minimize unneeded stressors on the human body to simplify the way individuals interact with the sensor, system, etc. This paper examines the assertion that by improving the HBSI overall system performance could be improved through a higher rate of repeatability of samples by humans using the system. This is important as atypical users such as those with musculoskeletal disorders, disabilities such as missing digits, or the elderly for example may have problems either using the device to produce an image/sample or have problems producing repeatable samples or images.

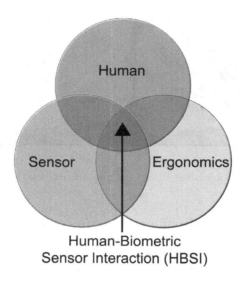

Fig. 14.2. Interaction of the human, biometric sensor/device, and ergonomics to form the HBSI.

Example 13. An example of a "normal" and "problematic" user using a hand geometry reader can be seen in Fig. 14.3. The "problematic" user has difficulty aligning their hand on the platen as the ring finger does not completely reach the guide pin, which causes highly variable placements, ultimately affecting the ability for this individual to be recognized by the system. The flat shape of the platen, wrist extension angles, and motion of the fingers may also cause discomfort or awkwardness, resulting in users not being able to use the device, or even worse individuals not being able to repeat hand placements in the same manner over time. The impact of non-repeatable hand placements is sizeable, as this challenges the algorithm and system to match hand images from users with varying characteristics and attributes.

Thus:

(a) By incorporating ergonomic principles into biometric sensor/device research and development an optimal relationship between the device and users can potentially be reached.
(b) Through optimization, system performance benefits a greater number of users that could successfully use the device and produce repeatable samples.

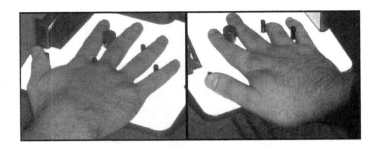

Fig. 14.3. Example of a normal user (left) and a problematic user with a partial missing digit (right).

14.3. Influence of Environmental Factors on Face Recognition

Another factor that can impact the human biometric interface is environment. The discussion in this section involves the impact of illumination on both 2D and 3D face. It must be noted the studies cannot be directly compared as they were conducted at different times, on different populations, using different cameras, different light levels, but did follow similar testing protocols. Thus the two stories are presented here to give specific examples of how one environmental factor illumination impacted the human biometric interface and the system performance.

14.3.1. *2D Face Recognition*

Numerous experts, reports and papers have stated that illumination has a significant impact on facial recognition performance. There is, however, only anecdotal evidence to support this claim [1], [32], [37], [38], [40], [42], [45], [46], [47]. In [44] evaluated the performance of a commercially available 2D facial recognition algorithm across three illumination levels. The illumination levels were determined by:

(a) Polling levels in two locations, with the third light level (medium light) determined by calculating the mean of the high and
(b) Low light levels.

Figure 14.4 shows sample images illustrating the differences between the three illumination levels.

The study consisted of:

(a) Thirty individuals, of which 73 percent were males, Caucasian and between the ages of 20 and 29.

Fig. 14.4.   2D face images captured at varying light levels [17].

(b) Other ethnicities represented in the study included Hispanic and Asian/Pacific Islander.

There were:

1. Six enrollment failures out of ninety-six attempts, representing an overall failure to enroll (FTE) rate of 6.25 percent.
2. The failure to acquire (FTA) rates for low light (7–12 lux), medium light (407–412 lux) and high light (800–815 lux) were 0.92%, 0.65 %, and 0% , respectively [23].

The 2D evaluation compared the face recognition algorithm's ability to verify each enrollment illumination condition template to the sample verification images taken at each illumination level.

Example 14. Sample verification images (3 attempts at each illumination condition in Fig. 14.4: low (left), medium (middle), and high (right)) were compared to the low illumination enrollment template (Fig. 14.4, leftmost image).   The process was then repeated for the other two enrollment conditions and took place over three visits.

The statistical analysis revealed that when the high illumination enrollment template was used, the illumination of verification attempts was not statistically significant, based on the three tested illumination levels. So the results of this study reveal that when illumination conditions during verification are expected to be variable, the illumination level at enrollment should be as high as practically possible.   For the low and medium enrollment templates, the illumination used for the verification attempts was statistically significant, meaning that enrollments in a darker environment do not perform well when environmental lighting conditions for verification vary. In summary, the lessons learned from [23] show that the enrollment illumination level is a better indicator of performance than the illumination level of the verification attempts. Thus, stringent quality metrics should be in place during enrollment, so that the template created is a representation that can be classified by the biometric system at a later time.

14.3.2. *3D Face Recognition*

To further evaluate face recognition and the effects illumination conditions has on performance, the authors evaluated a three dimensional face recognition algorithm in [24,25]. According to 3D face manufacturers the technology has advantages over 2D face as it compares the 3D shape of the face which is invariant in different lighting conditions and pose, as opposed to points, features, or measured distances that are evaluated in 2D face recognition systems.

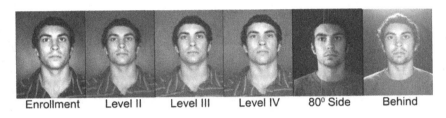

Enrollment   Level II   Level III   Level IV   80⁰ Side   Behind

Fig. 14.5.   Six light levels and directions used for the 3D face evaluation.

The evaluation consisted of:

(a) Four frontal illumination levels, one side light condition, and
(b) One behind the head light condition.

The six illumination conditions can be seen in Fig. 14.5. To eliminate external effects on the experiment and to emphasize the sole effect of lighting on the performance of the system, the subjects position, facial pose, and face covering artifacts were defined by the test protocol, which followed ANSI INCITS Face Recognition Format for Data Interchange standard [4]. Each image reflects the specifications for captured images as shown in Fig. 14.6. The resulting 3D model created from two 2D face images is shown in Fig. 14.7.

Like the 2D face evaluation, the 3D evaluation consisted of 30 individuals, of which 75% were male. The hypotheses were set up to establish whether there was a significant difference in the performance of the algorithm using the enrollment level illumination 3D template and verification attempts using all illumination levels shown in Fig. 14.5. A statistical analysis revealed that there was no statistically significant

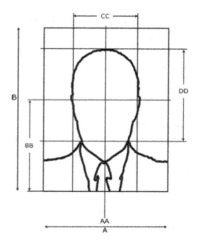

Fig. 14.6.   INCITS face recognition data format image requirement [20].

Fig. 14.7.   3D face recognition model.

difference in the performance of the algorithm when the light level was measured

(a) Between light level 1 (220–225 lux), and
(b) The other front illumination levels (320–325 lux; 650–655 lux; 1020–1140 lux), side light, and behind light conditions.

So the conclusions in [24,25] showed an increase performance over prior 2D tests, but encountered a tradeoff of speed for accuracy, as an average verification (1:1) attempt took roughly 10 seconds, which was significantly higher than the 2D evaluation in [23].

To summarize the effects of environmental factors on biometric system performance, the system designer must be:

(a) Cognizant of the application,
(b) Fit the correct biometric system into that environment.
(c) Quality metrics should be established during enrollment to ensure the stored enrollment template is usable in the conditions where the biometric system is implemented.

To continue the discussion on quality, the next section of this chapter will focus on fingerprint image quality across different age groups.

14.4. Fingerprint Image Quality

Fingerprint image quality issues have been widely discussed in the biometrics literature [44,5,21]. Variations in the quality and attributes of the fingerprints from the same finger is a common issue for fingerprint recognition systems [20]. The inconsistencies in biometric sample quality lead to an overall degraded system performance. The usefulness of quality metrics is acknowledged in the biometrics field, but there is no standard that provides guidelines for objective scoring of biometric samples. The current standardization effort refers to three different connotations of quality:

(a) *Character* that represents inherent features of the source from which the biometric sample is derived.
(b) *Fidelity* that represents how closely the samples from the same individual match, and
(c) *Utility* that represents the contribution of a biometric sample to the overall performance of a system.

The utility of a sample is affected by the character and fidelity of the sample, and performance of a system is affected by the utility of a biometric sample.

Non-uniform and irreproducible contact between the fingerprint and the platen of a fingerprint sensor can result in an image with poor utility quality. Non-uniform contact can result when the presented fingerprint is too dry, or too wet, as shown in Fig. 14.8. Irreproducible contact occurs when the fingerprint ridges are semi-permanently or permanently changed due to manual labor, injuries, disease, scars or other circumstances such as loose skin [19].

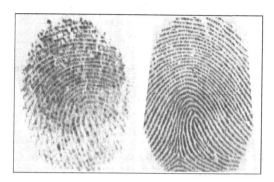

Fig. 14.8. Image of a low quality dry fingerprint (left). Image of a normal good quality fingerprint (right). Both images were acquired using an optical sensor.

These contact issues can affect the sample provided to the fingerprint sensor when an elderly user presents a fingerprint to the fingerprint device [19]. Due to effect of ageing, the skin becomes drier, the skin sags from loss of collagen, and the skin becomes thinner and loses fat as a direct result of elastin fibers. All of these decrease the firmness of the skin, which affects the ability of the sensor to capture a high quality image [2]. The skin of elderly individuals is likely to have incurred some sort of damage to the skin over life of the individual. Medical conditions like arthritis affect the ability of the user to interact with the fingerprint sensor. The data collection and signal processing sub-systems are directly affected by non-uniform contact, irreproducible contact and inability of the subject to interact with the system.

Interaction between the user and the system. The problem of interaction between the user and the system affects the sub-category of presentation within the data collection module. *Irreproducible* and *non-uniform* contacts affect the sub-categories of feature extraction within the signal-processing module. Low quality images affect the feature extraction process, and results in extraction of inaccurate and spurious features. The overall utility of the fingerprint image is reduced because it affects the performance of the fingerprint recognition system.

Example 15. A research study was performed to analyze fingerprint image quality from a younger group (18–25 years old) and elderly group (62 years and above). The fingerprint images were collected using an optical sensor.

Fingerprints from 60 elderly subjects and 79 young subjects were collected. The NIST Fingerprint Image Software 2 (NFIS2) was used for the analysis of the fingerprint images. The mindtct process was used to extract the minutiae count for each fingerprint, and the nfiq process was used to compute the quality score for each fingerprint.

A comparison of average minutiae count for the younger group and average minutiae count for the elderly group showed that there was a difference between the two. Figures 14.9 and 14.10 shows the frequency count of the minutiae count for the young and the elderly groups.

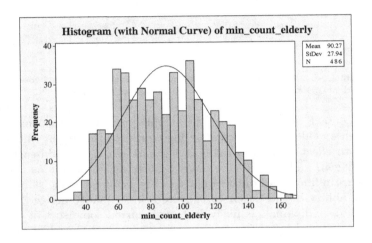

Fig. 14.9.   Minutiae count for the elderly group.

The younger group had an average minutiae count of 54.79 and the elderly group had an average minutiae count of 90.27. Previous research has shown that good quality fingerprints have between 40–100 minutiae [20]. The minutiae count analysis for fingerprints from the elderly group indicates that spurious minutiae were introduced during the fingerprint capture process. The spurious minutiae could be introduced by any number of factors mentioned previously in this section.

The quality analysis showed that:

(a) The fingerprints from the young group had an average quality score of 1.81 and the elderly group had an average quality score of 4.

(b) The quality score has a range of 1 to 5, where 1 indicates the highest quality score possible and 5 indicates the lowest quality score possible.

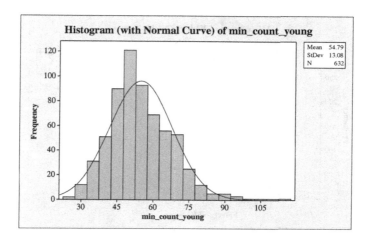

Fig. 14.10. Minutiae count for the young group.

(c) Statistical analysis of the fingerprint images from the young and elderly group showed that both the groups were significantly different in their quality scores.

Figures 14.11 and 14.12 show the frequency count of quality scores for the young and elderly group. The lower quality scores for fingerprint images from the elderly group indicated that aging and drying issues can introduce spurious minutiae and affect the quality scores for fingerprints. Figure 14.13 shows the correlation between age and fingerprint image quality. The study found that the elderly group had lower quality fingerprint images and a higher minutiae count compared to the younger group.

For fingerprint recognition systems that use minutiae matching algorithms, the introduction of spurious minutiae points affects the overall performance of the recognition systems. It is crucial for the system to be designed keeping these considerations in mind. During enrollment and identification/verification phase,

(a) Low quality fingerprints should be rejected so that overall performance does not degrade.
(b) Awareness about the general demographics of the target users can help lower the False Accept Rates and False Reject Rates for the system.
(c) For large scale systems like the US-VISIT program quality issues related to fingerprint images of the elderly can have a significant impact on the performance of the system.

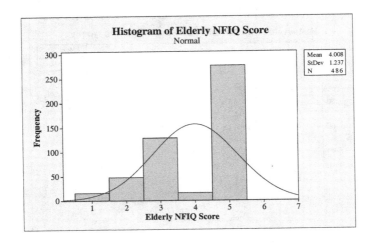

Fig. 14.11.   Quality scores for the elderly group.

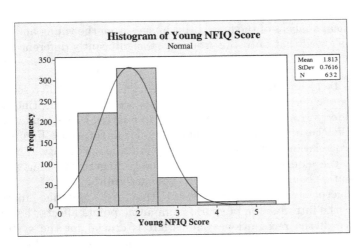

Fig. 14.12.   Quality scores for the young group.

14.5.   Digital Signature Verification

The use of written signatures to seal business and personal transactions has a long history. Before one can implement a signature verification system and collect data from it, one must decide on the method that will be used to acquire the signature signal. On-line or dynamic signature verification

Quality

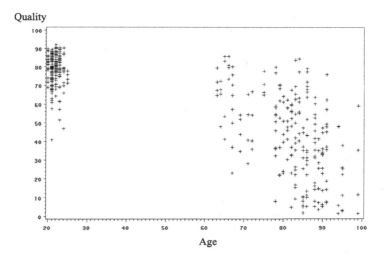

Fig. 14.13. Pearson correlation of image quality vs. age.

has various methods of acquiring data. Each of the various devices used in studies demonstrate different physical and measurement characteristics. Image and template quality are also important for system performance.

Many vendors publish specification sheets that outline some of the features of the device, including resolution. Other factors that will affect the quality of the image presented to the sensor will be:

(a) The cleanliness of the sensor,
(b) Ergonomics (size of the sensor and how it interacts with the body), and
(c) The feedback to the user (in signature verification, some devices display the signature under the pen in the same way as in a pen/paper interaction).

Detailed device specification is also importantdynamic signature verification uses a digitizer tablet, the quality of which will vary. Although some academic research papers do not list the specifications of the digitizer, it is of fundamental importance, affecting the quality and performance of the device within an e-commerce transaction environment. As technology improves, there is now no "standard" digitizer, or hardware, to collect DSV variables. There have been a number of dynamic signature verification studies that have centered on capturing these traits as a pen moves across a digitizer tablet [3,49]. These devices include both table-based digitizers and mobile computer (PDA type) digitizers, both of which are becoming more sophisticated, capturing an increasing number of measures such as x, y coordinates, velocity, and pressure. Furthermore, the evolution of mobile electronic commerce will increase the requirement to sign documents online with wireless mobile computing devices.

As hardware progresses towards providing solutions to mobile and electronic commerce, there may be a shift away from table-based digitizers. Hardware properties can affect the variables collected in the data acquisition process, and therefore the quality and performance of the device. Different digitizers and software enable the capture of different parameters of the signature, such as differing resolutions and speed.

According to [26], there is no consensus within the research community on the best method of capturing this information, although the table-based digitizer tablet is by far the most popular. As hardware manufacturers develop other solutions to mobile and electronic commerce, there may be a shift away from these table-based digitizers. Hardware properties can affect the variables collected in the data acquisition process; as a result, the quality and performance of the device may also be affected. Different digitizers and software enable the capture of different parameters of the signature, at differing resolutions and speed. Typical acquisition features include:

- Equivalent time interval samples,
- X and Y co-ordinates of pen-tip movements,
- Pen pressure,
- Pen altitude (the height of the pen from the digitizer),
- Pen azimuth (the angle between the reference line and the line connecting a point with the origin) [50].

Example 16. The central focus of [11] was to examine whether there are statistically significant differences in the measurable variables across devices. The volunteer crew was consisted of 203 individuals whose demographics tended toward the 19- to 26-year-old age groups, due to the location of the testing environment. Although not representative of the U.S. population except for gender, it is representative of the population found in college and university environments. Males accounted for 66% of the volunteer group. Females accounted for 35% of the volunteer group and were typically older than the males. Right-handed individuals accounted for 91% of the volunteer group, and left-handed individuals accounted for 9%.

The system in this study accepted all the signatures, and the failure to enroll rate was 0.0%. In [24] and [25] they define the "failure to acquire" rate as the "proportion of attempts for which the system is unable to capture or locate an image of sufficient quality." Typically, this is the case when an individual is unable to present a biometric feature to the device due to either an injury or to insufficient image quality. As such, this does not apply to dynamic signature verification because the biometric feature (signature) is presented by a stylus rather than direct interface with the biometric sensor. Therefore, a paradox arises between a signature that is repeatedly signed and acquired by the sensor yet provides an image different to a normal paper based signature. One subject did have difficulty in providing a signature that is similar to a pen and paper version; although the system acquired the image, the signature was thrown out, as it was a statistical outlier with over 15,000 segments as opposed to 2 to 19 segments exhibited by other crew members. The failure to acquire rate in this study was 0.0%.

When examining area of the signature across two different digitizers, the number of pen-up, pen-down sequences should not change significantly when the signing space is restricted. However, in [11] the uptime variable was significantly different at an $\alpha = 0.01$ level. Durations of the signature, i.e. the time for individual 0.01 mm groupings, were also significantly different across these two devices. When the variables associated with the graphical nature of the signature were examined, there was no significant difference in the number of variables relating to the signing area of the device. Area was not a factor in [11], as it was held constant on both devices; and those variables measuring area, such as distance (p-**value** = 0.181), net area (p-**value** = 0.676), and rubber area (p-**value** = 0.344) show that there is no statistically significant difference in the signing areas of the two devices. When two other digitizers were chosen (one a PDA, and the other a table based digitizer), there was a statistically significant difference between the two devices for 61 variables, including uptime, distance, and event variables. Even though the signing area between the two devices was constant, and the variables related to area measurement such as measuring:

Area,
Distance (p-value = 0.000),
Net area (p-value = 0.000), and
Rubber area (p-value = 0.000),

showed significant differences, so too did:

Speed of the signature (p-value = 0.002),
Speed variation (p-value = 0.006),
Average velocity of the signing (p-value = 0.000),
Up-speed (the speed when the pen is off the digitizer) (p-value = 0.000),
Net speed (p-value = 0.000), and
Average uptime (p-value = 0.000)

---

were all significantly different.

When signing restriction was removed in [11], out of the 91 variables, 53 were significantly different at $a = 0.01$. The signing space on the Wacom digitizer was 8×6 inches, and the signing space on the PDA was 2.3×2.3 inches. From this difference, the signing area variables, including the bounding area of the signature, net area, and the ratio of signature ink length to area, were all significantly different. This result would validate the assumption that people sign with a larger signature when given a larger signing space. Again, the variables that differed on the SPT 1500 compared to the template signature on the Wacom digitizer were the local maximum or minimum ratios. Other variables that were statistically significant were velocity, the time the pen was on the digitizer, the average speed that the pen was on the digitizer, and the net speed of the signature.

The final hypothesis was to examine whether there were any changes on one particular digitizer when an individual was sitting and then standing. There were no restrictions on the signing space. The motivation for this hypothesis was that in many environments where this digitizer was used, the user would sign both seated and standing.

Example 17. In many retail environments, customers stand when signing a credit card receipt at a checkout; but in cases where the Interlink E-Pad is currently being used, for example in banking and brokerage, the user may also be seated. No differences were statistically significant when a t-test was performed at $\alpha = 0.01$, showing that when a member of the volunteer crew stands and enrolls and then sits to sign, the variables of the signature do not differ statistically at the $\alpha = 0.01$ level.

Based upon the research, the major conclusion that can be drawn from the study is that

> *There are significant differences in signature trait variables across devices, yet these variables are not significantly different within device families.*

When these devices are grouped together, these variable differences continue to be significant. The fact that the signature variables differ across devices is significant. Dynamic signature verification is used within the realms of electronic document management and contracts and will typically be verified only if the document validity is questioned. Therefore when deploying dynamic signature verification applications, it is important to note that the design of the device can have an impact on the variables collected, and depending on the weighting of the variables within a specific algorithm, the HBSI may have an impact on the system performance. Furthermore, for audit control purposes, the type of device needs to be embedded in some way to the signature so that document examiners can compare signatures captured on the same type of device. Based upon the data, the major conclusion that can be drawn from the study is that there are significant differences in the variables across devices, yet these variables are not significantly different within device families from Wacom Technologies, Palm and Interlink Technologies.

14.6. Summary

Ergonomics, environment, biometric sample quality, and device selection have a direct impact on analysis of biometric data. The interaction of these factors has a significant impact on the performance of a biometric system, as shown by the different studies mentioned throughout this chapter. The research in this chapter indicates that a better understanding of ergonomics, environmental issues, sample quality, and device interoperability is crucial if biometric technologies are to become more pervasive. All the issues mentioned, along with HBSI, are open research topics which need to be addressed to optimize the analysis of biometric data.

Acknowledgments

This research has been supported by Purdue University Department of Industrial Technology, College of Technology; the Purdue University Discovery Park e-Enterprise Center, and the Center for Education, Research, and Information Assurance and Security at Purdue University.

Bibliography

1. Alim, O., Elboughdadly, N., El Morchedi, C., Safawat, E., Abouelseoud, G., El Badry, N., Mohsen, N. (2000). Identity Verification using Audio-Visual Features. *Proc. 17th National Radio Science Conf.*, Egypt, pp. C12/1–C12/8.
2. American Academy of Dermatology. (2002). Mature Skin. Retrieved Nov. 14, 2002, http://www.aad.org/NR/exeres/
3. Ansano, H., Sumi, A., Ramzan, Z., and Zhu, I. (2000). Wireless Electronic Commerce Security. Retrieved March 22, 2000, http://wwvv.nokia.com
4. ANSI INCITS. (2004). Information Technology Face Recognition Format for Data Interchange: American National Standards Institute - International Committee for Information Technology Standards. No. 385-2004. 66p.
5. Behrens, G. (2002). Assessing the Stability Problems of Biometric Features. *Proc. Intl. Biometrics Conf. and Exhibition*, Amsterdam, NL.
6. Bioscrypt. (2003). Bioscrypt's Pattern-Based Algorithm. Retrieved June 20, 2004, http://www.bioscrypt.com/assets/bioscrypt_algorithm.pdf
7. Blackburn, D., Bone, M. and Phillips, J. (2001). Facial Recognition Vendor Test 2000 Evaluation Report. 61p.
8. Buettner, D. (2001). A Large-Scale Biometric Identification System at the Point of Sale. Retrieved Sept. 29, 2002, http://www.itl.nist.gov/div895/isis/bc2001/FINAL
9. Clarke, R. (1994). Human identification in information systems: Management challenges and public policy issues, *Information Technology and People*, 7, pp. 6–37
10. Dullink, H., Van Daalen, B., Nijhuis, J., Spaanenburg, L., and Zuidhof, H. (1995). Implementing a DSP Kernel for Online Dynamic Handwritten Signature Verification Using the TMS320DSP Family (SPRA304), France: EFRIE, 26p.
11. Elliott, S. (2001). A comparison of on-line Dynamic Signature Trait Variables across different computing devices, Ph.D. Thesis, Department of Industrial Technology, Purdue University, 215p.
12. Elliott, S. (2002). A Comparison of On-Line Dynamic Signature Trait Variables vis--vis Mobile Computing Devices and Table-Based Digitizers, *3rd IEEE Workshop Automatic Identification Advanced Technologies*, pp. 121–125
13. Greiss, F. (2000). On-Line Signature Verification, M.S. Thesis, Department of Computer Science, Michigan State University, 88p.
14. Hamilton, D., Whelan, J., McLaren, A., and MacIntyre, I. (1995). Low Cost Dynamic Signature Verification System, *Proc. of European Convention on Security and Detention*, pp. 202–206
15. Holmes, J., Wright, L., and Maxwell, R. (1991). A Performance Evaluation of Biometric Identification Devices, Albuquerque: Sandia National Labs No. SANDIA91-0276.
16. INCITS (2003). Finger Pattern Based Interchange Format (Version 7). Washington, DC: INCITS.

17. Interlink. (2001). E-Pad Data Sheet. Retrieved July 4, 2001, http://www.interlinkelectronics.com
18. International Ergonomics Association (IEA). (2006). The Discipline of Ergonomics. Retrieved Jan. 27, 2006, http://www.iea.cc/ergonomics/
19. Jain, A., Hong, L., Pankanti, S., and Bolle, R. (1997). An Identity-Authentication System Using Fingerprints, *Proceedings of the IEEE*, 85(9), 1365–1388.
20. Jain, L., Halici, U., Hayashi, I., Lee, S., and Tsutsui, S. (Eds.). (1999). *Intelligent Biometric Techniques in Fingerprint and Face Recognition.* CRC Press, Boca Raton, FL.
21. Jiang, X., and Ser, W. (2002). Online Fingerprint Template Improvement. *Proc. Pattern Analysis and Machine Intelligence*, 24 (8), pp. 1121–1126.
22. Komiya, Y., and Matsumoto, T. (1999). On-Line Pen Input Signature Verification PPI. *Proc. 6th Int. Conf. Neural Information Processing*, 3, pp. 1236–1240.
23. Kukula, E. (2004). The Effects of Varying Illumination Levels on FRS Algorithm Performance, M.S. Thesis, Department of Industrial Technology, Purdue University, 98p.
24. Kukula, E. (2004). Effects of Light Direction on the Performance of Geometrix FaceVision 3D Face Recognition System, *Proc. Biometric Consortium*, Retrieved Oct. 21, 2004, http://www.biometrics.org/bc2004/program.htm.
25. Kukula, E., and Elliott, S. (2004). Effects of Illumination Changes on the Performance of Geometrix FaceVision 3D FRS, *Proc. 38th Annual Int. Carnahan Conf. Security Technology (ICCST)*, pp. 331–337.
26. Leclerc, F., and Plamondon, R. (1994). Automatic Signature Verification, *Int. J. of Pattern Recognition and Artificial Intelligence*, 8(3), pp. 673–660
27. Lee, H. and Gaensslen R. (2001). *Advances in Fingerprint Technology (2nd ed.)* Elsevier, New York.
28. Lee, L., Berger, T., and Aviczer, E. (1996). Reliable On-Line Human Signature Verification Systems, *IEEE Trans. Pattern Analysis and Machine Intelligence*, 18(6), pp. 643–647.
29. Lee, W., Mohankrishnan, N., and Paulik, M. (1998). Improved Segmentation through Dynamic Time Warping for Signature Verification Using a Neural Network Classifier, *Proc. Int. Conf. Image Processing*, pp. 929–933.
30. Li, X., Parizeau, M., and Plamondon, R. (1998). Segmentation and Reconstruction of On-Line Handwritten Scripts, *Pattern Recognition*, 31(6), pp. 675–684.
31. Mansfield, T., Kelly, G., Chandler, D., and Kane, J. (2001). Biometric Product Testing Final Report. Retrieved March 17, 2005, www.cesg.gov.uk/site/ast/biometrics/media/BiometricTestReportpt1.pdf
32. Mansfield, A., and Wayman, J. (2002). Best Practices in Testing and Reporting Performances of Biometric Devices: Biometric Working Group. 32p.
33. Martens, R., and Claesen, L. (1997). On-Line Signature Verification: Discrimination Emphasised, *Proc. 4th Int. Conf. on Document Analysis and*

Recognition, pp. 657–660.

34. Mingming, M., and Wijesoma, W. (2000). Automatic On-Line Signature Verification Based on Multiple Models, *Proc. IEEE/IAFE/INFORMS Conf. on Computational Intelligence for Financial Engineering*, pp. 30–33.

35. Narayanaswamy, S., Hu, J., and Kashi, R. (1999). User Interface for a PCS Smart Phone, *Proc. IEEE Int. Conf. Multimedia Computing and Systems*, pp. 777–781.

36. Phillips, P., Grother, P., Bone, M., Micheals, R., Blackburn, D., and Tabassi, E. (2003). Face Recognition Vendor Test 2002 Evalulation Report. 56p.

37. Phillips, P., Martin, A., Przybocki, M., and Wilson, C. (2000). An Introduction to Evaluating Biometric Systems, *Computer*, 33, pp. 56–63.

38. Phillips, P., Rauss, P., and Der, S. (1996). FERET (Face Recognition Technology) Recognition Algorithm Development and Test Report. 73p.

39. Plamondon, R., and Parizeau, M. (1988). Signature Verification from Position, Velocity and Acceleration Signals: A Comparative Study, *Proc. 9th Int. Conf. Pattern Recognition*, pp. 260–265.

40. Podio, F. (2002). Personal Authentication Through Biometric Technologies.*Proc. 4th Int. Workshop on Networked Appliances*, pp. 57–66.

41. Rood, E., and Jain, A. (2003). Biometric Research Agenda: Report of the NSF Workshop. 17p.

42. Sanderson, S., and Erbetta, J. (2000). Authentication for Secure Environments Based On Iris Scanning Technology, *Proc. IEEE Colloquium Visual Biometrics*, 8, pp. 1-7.

43. Schmidt, C., and Kraiss, K. (1997). Establishment of Personalized Templates for Automatic Signature Verification, *Proc. 4th Int. Conf. Document Analysis and Recognition*, pp. 263–267.

44. Sickler, N. (2003). An Evaluation of Fingerprint Quality across an Elderly Population vis--vis 18-25 Year Olds, M.S. Thesis, Department of Industrial Technology, Purdue University.

45. Sims, D. (1994). Biometric Recognition: Our Hands, Eyes, and Faces Give Us Away, *Proc. IEEE Computer Graphics and Applications*, 14(5), pp. 14–15.

46. Starkey, R., and Aleksander, I. (1990). Facial Recognition for Police Purposes Using Computer Graphics and Neural Network, *Proc. IEEE Colloquium on Electronic Images and Image Processing in Security and Forensic Science*, pp. 1–2.

47. Sutherland, K., Renshaw, D., and Denyer, P. (1992). Automatic Face Recognition, *Proc. 1st Int. Conf. Intelligent Systems Engineering*, pp. 29–34.

48. Tayyari, F., and Smith, J. (2003). *Occupational Ergonomics: Principles and Applications*. Kluwer, Norwell.

49. Wacom Technology Co. (2000). Wacom Product Specification. Retrieved November 20, 2000, from http://www.wacom.com

50. Yamazaki, Y., Mizutani, Y., and Komatsu, N. (1999). Extraction of Personal Features from Stroke Shape, Writing Pressure, and Pen Inclination

in Ordinary Characters. *Proc. 5th Int. Conf. Document Analysis and Recognition*, pp. 426–429.

51. Yanushkevich, S.N., Stoica, A. and Shmerko, V. P. (2006). Fundamentals of Biometric-Based Training System Design, this issue.

Chapter 15

Fundamentals of Biometric-Based Training System Design

S. N. Yanushkevich*, A. Stoica[†], V. P. Shmerko[‡]

Biometric Technology Laboratory: Modeling and Simulation,
University of Calgary, Canada

Humanoid Robotics Laboratory, California Institute of Technology,
NASA Jet Propulsion Laboratory, California Institute of Technology,
Pasadena, CA, USA, and ECSIS - European Center for Secure
Information and Systems, Iasi, Romania

Biometric Technology Laboratory: Modeling and Simulation,
University of Calgary, Canada

* syanshk@ucalgary.ca
[†] adrian.stoica@jpl.nasa.gov
[‡] vshmerko@ucalgary.ca

This chapter describes the fundamental concepts of biometric-based training system design and training methodology for users of biometric-based access control systems. Such systems utilizes the information provided by various measurements of an individual's biometrics. The goal of training is to develop the user's decision making skills based on two types of information about the individual: biometric information collected during pre-screening or surveillance, and information collected during the authorization check itself. The training system directly benefits users of security systems covering a broad spectrum of social activities, including airport and seaport surveillance and control, immigration service, border control, important public events, hospitals, banking, etc.

Contents

Glossary

BBN	—	Bayesian Belief Network
HQP	—	High Quality Personnel
IR	—	Infrared Image (band, spectrum)
PASS	—	Physical Access Security System
SPA	—	Security Personnel Assistant
T-PASS	—	Training Physical Access Security System

15.1.   Introduction

This paper presents a summary of theoretical results and design experience obtained in investigations conducted at the Biometric Technology Laboratory, University of Calgary, Canada, in collaboration with the Humanoid Robotics Laboratory at the NASA Jet Propulsion Laboratory, California Institute of Technology, in Pasadena, California [52]. These investigations focus on using biometrics as early warning information rather than data for authentication. Initial studies led to the idea of designing and prototyping a biometric-based Physical Access Security System (PASS) using an early warning paradigm and integrated knowledge-intensive tools at the global level with distributed functions.

Many research and industrial applications of training systems exist in such diverse domains as flight simulation [13,24], driving simulation [2], environment simulation [30], and surgical simulation [27].

The military has long experience in simulation and training. The rapid pace of development of new military hardware requires simulators that are flexible, upgradeable, and less expensive. For example, there is a need to replace aircraft components on flight line trainers with simulated units designed specifically for this application.

Another trend is remote simulation through networking, eliminating the need to transport trainees to the simulator site. Networking is also requires in team simulations [26].

Surgical simulation systems have been developed for training medical personnel and for testing new technologies which might prove unreliable for direct use in a real clinical intervention. Image-guided surgery systems use virtual surgical tools in a 3D virtual scene. These systems provide the surgeon with a spatial image formation which he/she previously would have had to reconstruct mentally. Finally, robotic surgery strives for the minimization of human error during surgery.

Effective utilization of all available information, and the intelligence to process it, is a key objective of a biometric-based PASS [51,52]. Binding all sources of information into this objective can provide reliable identification of an individual. A Guidance Package Biometrics for Airport Access Control [4] provides criteria that airport operators need to know in order to integrate biometric devices into their access control systems and to revise their security plans. However, PASS is a semi-automated system; that is, the final decision must be made by the highly qualified personnel (HQP) human users. The HQP must be trained to use these tools in practice. The fundamental premise of our approach is that the training of HQP can be carried out on the existing PASS itself, expanded with knowledge-based analytical and modeling tools.

Two directions in access control system design can be discern today [1,8,45]. The first direction includes approaches that are based on the expansion of sources of biometric data and, as a result, the amount of professional skills required of the officer increases. In these approaches, the mathematical problems of supporting the officer are critically simplified, allowing a cost-efficient, but in practice not very suitable, solution. The second direction aims at high-level automated system design, where the human factor is critically reduced. Such systems are often considered cognitive systems. However, these automated systems are not acceptable in applications where high confidentiality and reliable decision making are critical.

Most of today's physical access control systems extensively utilize the information collected during the authorization check itself, but make limited or no use of the information gained in the conversation with individual during the authorizing procedure. The officer must make a decision in a limited time based on authorization documents, personal data from the databases, and his/her own experience. This may not be sufficient to minimize the probability of mistakes in decision making.

Consider, for example, the Defense Advanced Research Projects Agency (DARPA) research program, HumanID, which is aimed at the detection, recognition and identification of humans at a distance in early warning support systems for force protection and homeland defense [45]. Existing screening systems exclusively utilize visual appearance to compare against "lookout checklists" or suspected activity, and do not effectively use the time slot before/during registration to collect biometric information (body temperature, surgical changes) about an individual. In addition, they do not arm the officer with decision-making support (such as modeling of the aging face, artificial facial accessories, and other features, for further comparison against the information available in databases.

The job of a customs or security officer becomes complicated if submitted documents do not match the appearance of an individual. In order to draw firm conclusions, additional information is needed. This information is available in the next-generation PASS system, which uses both *pre-screening* and *check-point* biometric information. The skilled PASS user possesses an ability to manipulate and efficiently utilize various sources of information in the decision making process. Hence, a training system is needed to implement the PASS in practice [51,52]. However, personnel training cannot keep pace with rapid changes in technology for physical access control systems. This is the driving force behind developing novel training paradigms to meet the requirements of security policy.

Traditionally, training is implemented on a specifically desqigned training system. For example, training methodologies are well developed for pilots [13], astronauts, surgical applications [27], and in military systems. These are expensive professional simulation systems, which are difficult to modify or extend, since they are unique in architecture and functions. For example, the traditional flight simulator is a large domelike structure placed on a motion platform and housing an identical replica of an airplane cockpit. Such training is very realistic (including effects such as engine sounds, airflow turbulence, and others) but expensive. An example of the newer generation of simulators is the networked simulation environment in [7]. Mission team level training is based on networked nodes of several fighter simulators, air warning system (AWACS) simulators, and synthetic computer-aided designed enemy forces. The whole simulation is taped, allowing postmission debriefing either at the local or global level. Typically, pilots fly two 1-hour missions each day, with a debriefing after each mission. The debriefing analyzes the mission mistakes, which pilots can learn to avoid in a repeat simulation. The pilot's performance is assessed by instructors using monitoring stations.

This methodology is partially acceptable for training PASS users. In our approach, the design of a expensive training system is replaced by an inexpensive extension of the PASS, already deployed at the place of application. In this way, an important effect is achieved: modeling is replaced with real-world conditions. Furthermore, long term training is replaced by periodically repeated short-time intensive computer-aided training.

> *The PASS and T-PASS implement the concept of multi-target platforms, that is, the PASS can be easy reconfigured into the T-PASS and vice versa.*

We propose a training paradigm utilizing a combination of various biometrics, including visual-band, IR, and acoustic acquisition data for identification of both physical appearance (including natural factors such as aging, and intentional (surgical) changes), physiological characteristics (temperature, blood flow rate, etc.), and behavioral features (voice). Other biometrics can be used in pre-screening and at check-points: gait biometrics are used between the points of the appearance of an individual and near-distance, non-contact and contact biometrics at the check-point. Note that gait can be thought of as learned motor activity of the lower extremities, which when performed, results in the movement of the subject in the environment. By changing various parameters of a gait (stride length, velocity, heel and toe strike, timing, etc.) one can perform an infinite variety of gaits [15,31].

In complex systems, such as PASS and its user training system — T-PASS, several local levels and a global level of hierarchy are specified. In our approach, biometric devices are distributed at the local levels.

> *The role of each biometric device is twofold:*
> *(a) Their primary function is to extract biometric data from an individual and*
> *(b) their secondary function is to "provoke" question(s).*

For example, if high temperature is detected, the question should be formulated as follows: "Do you need any medical assistance?" The key of the proposed concept of decision making support in PASS is the process of generating questions initiated by information from biometric devices. In this way, the system assists the officer in access authorization. Additionally, by using an appropriate questionnaire technique, some temporary errors, and the temporary unreliability of biometric data, can be alleviated. The proposed approach can be classified as a particular case of *image interpretation* technique, deploying *case-based reasoning* using questionnaire strategies for acquisition and evaluation of knowledge [34].

In our approach, experience gained from dialog system design is utilized. In particular, PASS and T-PASS are based on architectural principles similar to SmartKom [19], i.e. sensor specific input processing, modality specific analysis and fusion, and interaction management. The differences follow from the target function: PASS and T-PASS aim to support an officer in *human-human* interaction, while SmartKom aims to provide *human-machine* interaction.

Training techniques and training system design use domain specific terminology. An example of this terminology is given in Table 15.1 These terms will be used in various combinations. For example, training scenarios can be considered for real-world conditions, where hybrid biometric data are used — that is, real and synthetic (simulated) data. Knowledge and signal domains are considered in the context of the space of available information and the space of decision making.

Our results are presented as follows. Section 15.2 formulates the requirements for a training methodology using PASS and T-PASS. The focus of Section 15.3 is a technique for estimation of training, including the skills of personnel and the effectiveness of the training system. The basic approaches to the design of the system are given in Section 15.4. The results we obtained from prototyping the proposed system are given in Section 15.5 We summarize the results obtained and discuss open problems for further research in Section 15.6.

15.2. Basics of Training Security System Design

In this section, we focus on fundamental requirements for training systems, modeling of the biometric-based physical access control system, and decision making support for security personnel.

Table 15.1. Terminology of training systems.

Term	Comments
Simulation	Generation of artificial (synthetic) biometric data and/or virtual environment (Figs. 15.4 and 15.13)
Training scenario	System configuration protocols that provide for the particular goals of training (Section 15.4.3)
Real-world training	Training of personnel in a real-world environment
Early warning paradigm	A scenario that provides data using distance surveillance
Knowledge and signal domain	Data representation that is acceptable for human and machine communication, respectively (Fig. 15.1)
Space of decision making	A set of biometric features acceptable for decision making (Fig. 15.5)
Space of available information	The set of biometric sources of information in the system (Fig. 15.5)
Degradation, training, and evolution functions	Functions that characterize loss of skill, efficiency of training strategy, and improving decision making by extending the space of available information, respectively (Section 15.3)

15.2.1. *Requirements of Training Systems*

Flexibility

The development of PASS must be synchronized with the training of its users. PASS is a dynamically developing system because of its reconfigurable architecture and the capability of integration of new devices. Hence, the training methodology and architectural properties of the training system, T-PASS, must *reflect the dynamic nature* of PASS. This property is called *flexibility*, that is, the capability of the training system for extension, modification, and accommodation to changes in tasks and environments beyond those first specified.

Learnability

The use of each system requires specific skills. The T-PASS is a complex system that requires certain skills of its personnel, therefore the personnel need time for learning the system. *Learnability* is measured by the time and effort required to reach a specified level of user performance. This amount should be minimized. In our approach,

> *The experience of the PASS user can be used in the T-PASS because they both run on the same hardware and software platform, which minimizes learning time.*

Short-term, periodically repeated, and intensive training

Traditionally, training is a long-term and costly process and particular skills (often not required of the trainee in practice) degrade over time. In contrast,

> *The concept of the mobile PASS requires short-term training in a new environment.*

Hence, the training methodology should be *short-term, periodically repeated*, and *intensive*. In our approach, the security system can be used as a training system (with minimal extension of tools) without changing of the place of deployment. In this way we fulfill the criterion of cost-efficiency and satisfy the above requirements.

Training for extreme situations

Simulation of extreme scenarios is aimed at developing the particular skills of the personnel. The modeling of extreme situations requires developing specific training methodologies and techniques, including virtual environments.

Example 18. Severe Acute Respiratory Syndrome (SARS) is an illness with a high infection and a high fatality rate [47]. At the early stage of this disease, it appears as a high temperature spot in the IR image. Early detection and immediate isolation of those infected are the keys to preventing another outbreak of such an epidemic.

Example 19. Monitoring of the breathing function of a pre-screening individual and one who is being screened can be used to predict heart attacks, in particular. The breathing rate during the dialog is considered as the individual's response to questions, i.e. differences in breathing rates correspond to different questions [36].

Confidentiality

Confidentiality of data must be implemented at many levels of the security system, including local biometric devices and databases, and information from the global database. Processing biometric data such as IR images and their interpretation in semantic form must addresses privacy and ethical issues.

Interpretability

This issue is related to biometric-based security systems. Interpretability of the biometric patterns encoded in biometric devices is often more important than the predictive accuracy of these devices.

Example 20. The interpretation of data from medical devices is especially important in medical applications, where not only the correct decision is desired, but also an explanation of why that particular decision has to be made. This is why systems which provide human-comprehensible results are considered to be more acceptable than "black box" biometric devices. The decision process should be easily followed and understood.

Ergonomic issues

The objective of ergonomics is to maximize an operator's efficiency and reliability of performance, to make a task easier, and to increase the feeling of comfort and satisfaction. The work of the officer is very intensive: he/she must minimize both the time of service and the risk of a wrong authorization. Improvements to PASS during authorization could be made, for example, using footpedals, a footmouse, several monitors, or a speech interface. Some ergonomic problems are discussed in [11].

15.2.2. *Model of the Biometric-Based Access Control*

PASS is a next generation security system.

> *PASS combines several basic approaches: early warning data fusion, multimode sources of information, and authorization techniques.*

The generic model of the PASS is given in Fig. 15.1, and details are provided in Table 15.1. Two processes occur in the system simultaneously: pre-screening of an individual-A (collection of information) and screening of an individual-B (analysis of screened information). *Early warning data* is the result of pre-screening analysis. This data belongs to the one of three clusters:

Alarming cluster (extreme scenario),
Warning cluster (authorization with caution), and
Regular cluster (regular authorization).

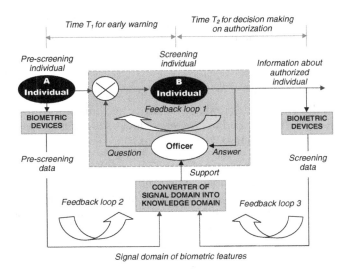

Fig. 15.1.   Model of the PASS: a semi-automatic system with three feedback loops.

The PASS consists of the following modules: *cameras* in the visible and IR bands placed at two levels of observation, *processors* of preliminary information and online data, the *knowledge domain converter* for interpretation of data from the signal domain and manipulation of data in semantic form, and a *personal file* generating module. Two-level surveillance is used in the system: pre-screening and the screening of an individual. The protocol of a personal file includes (a) preliminary information using surveillance during pre-screening of individuals in a line/gate in the visible and IR bands, and (b) information extracted from observation, conversation, and available additional sources during screening itself, or authorization procedure.

Note that biometrics such as iris and retinal patterns, ear and face recognition require high resolution images. A typical surveillance camera often captures poor quality, low resolution images. From this point of view, gait biometrics prove convenient.

In PASS, each of the biometric devices automatically generates a local decision making outcome corresponding to local levels of the hierarchy of decision making. These decisions are stored and processed by knowledge-intensive techniques to support the officer. This biometric information should be simulated in the training system. To this end, so-called *hybrid* data are used; that is, real, synthetic, and mixed biometric data [48,50].

Two groups of biometric devices are used in the system: devices for the collection and processing of early warning biometric data about the individual-A, and devices for processing of biometric data from the individual-B. These data are represented in the signal domain. The converter transforms biometric features from the signal domain to the knowledge domain and generates supporting questions for the officer. The officer follows his/her own scenario of authorization (question/answer). The questions generated by the convertor are defined as supporting data, and can be used or not by the officer in the dialogue.

Three feedback loops address the three main data flows: the main flow (loop 1), where the professional skills of personnel play the key role, and supporting flows (loops 2 and 3), where biometric data are processed and decisions on detection or recognition of various patterns are automated.

Table 15.2. Functions of feedback loops in the PASS model.

LOOP	DESIGN AND IMPLEMENTATION
1	Experience of the officer in interviewing supported by knowledge-based technology
2	Non-contact biometrics in visible and IR bands, representation of pattern in semantic form.
3	Non-contact and contact biometrics; representation of patterns in semantic form.

The following properties of the screening procedure of service are utilized in the PASS and in the training system (Fig. 15.2):

Property 1. Two phase service, where the interval of service T can be divided into two subintervals T_1 and T_2. The first phase T_1 (pre-screening) is suitable for obtaining early warning information using surveillance. At the second phase, T_2, available sources of information are used to identify and authorize an individual.

Property 2. There is a distance between the individual and the registration desk. This distance can be used to obtain extra information using video and IR surveillance, and, thus, facial, IR, and gait biometrics.

The above model can be interpreted from various perspectives in complex system design, in particular, from learning knowledge-based systems, adaptive knowledge-based systems, and cognitive systems [14,10].

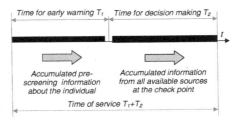

Fig. 15.2. Screening during time T: time of service includes of early warning time T_1 (pre-screening) and check-point time T_2.

15.2.3. *Biometrics*

The most important feature of the PASS is the *hierarchical distribution of biometrics* in the system; that is, various types of biometric devices are embedded into PASS with respect to architectural, topological, and functional properties. This is a necessary condition for implementation of the early warning paradigm in the access control system.

Biometrics provide many sources of information that can be classified with respect to various criteria; in particular, spectral criteria (visible, IR, and acoustic), behavioural and inherent criteria, contact and non-contact, correlation factors between various biometrics (multi-biometric) [22,23,42,53,54], computing indirect characteristics and parameters such as drug and alcohol consumption, blood pressure, breathing function, and temperature using the IR band [16,33,46], emotion [44], psychological features and gender [15,29], and implementation. In T-PASS, these criteria are used to satisfy the topological and architectural requirements of the system, machine-human interaction, computing a correlation between biometric patterns.

We also utilize the correlation between various biometrics. Traditionally, the correlation is calculated in the signal domain, for example, between facial images in the visible and IR bands [9,43,44]. The correlation between biometrics can be represented in semantic form (knowledge domain). For this, biometric features generated by different biometric devices in the signal domain are converted into the knowledge domain, where their correlation in semantic form is evaluated.

Example 21. The correlation between handwriting, signature, stroke dynamics, facial features, and gait can be represented in semantic form as some special conditions (for example, drunkenness or drug affection, physical condition or disability, emotional state, gender, and others) (Fig. 15.3).   The advantages to this are: acceptability for humans, adaptation for different biometrics, and the possibility of compensating for the unreliability of biometric data by using fusion from several sources combined with high-level description.

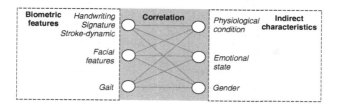

Fig. 15.3. Correlation of biometric feature are described in semantic form using knowledge representations and indirect computing.

Acoustic domain

In the acoustic domain (Fig. 15.4), the pitch, loudness and timbre of an individual's voice are the main relevant parameters. For analysis and synthesis, the most important frequency range is 70 to 200 Hz for men, 150 to 400 Hz for women, and 200 to 600 Hz for children. Several levels of speech description are used: representations in the *acoustic*, *phonetic*, *phonological* (speech is described in linguistic units, phonemes), *morphological* (phonemes combine to form words), *syntactic* (restricted by the rules of syntax), and *semantic* (semantic features are attached to words giving the sentence meaning) domains. Each of these levels can be used to extract behavioural information from speech. Voice carries emotional information by *intonation*. The *voice tension* technology utilizes varying degrees of vibration in the voice to detect emotional response in dialogue (certain languages require exceptional treatment, as for example, in Chinese, intonation carries semantic information: four tones distinguish otherwise identical words with different meanings from each other).

Infrared band

Early warning and detection using IR imaging is considered to be a key to mitigating service time and incidents. Techniques of IR imagery have been studied with respect to many applications, including tracking faces [12], and detecting inflamed regions [16,38,44]. It is well documented that the human body generates waves over a wide range of the electromagnetic spectrum, from 10^{-1} Hz (resonance radiation) to 10^9 Hz (human organs) (Fig. 15.4). The human face and body emit IR light both in the mid-IR (3–5 μm) and

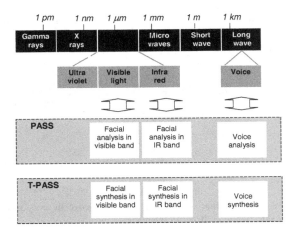

Fig. 15.4. Sources of information for supporting decision making and simulation of them in the training system.

the far-IR range (8–12 μm). Most light is emitted in the longer-wavelength, lower energy far-IR band.

The human body radiates invisible IR light in proportion to its temperature. Human body temperature varies from $T = 300$ K to 315 K. Mid-IR provides the best spectral range for human temperature measurements. Below the mid-IR, at about 1.4 μm, human skin reflects light [17,40]. This is the basis for skin detection [12,18] by the thermogram; skin detection precedes temperature evaluation. A skin disease can be recognized in an IR image.

Blood flow rate measures

In [16], heat-conduction formulas at the skin surface are derived. The thermodynamic relation between the blood flow rate V_S at the skin level, blood temperature at the body core T_{blood}, and the skin temperature T_{skin} is used to convert IR intensity to temperature $\frac{dV_s}{dt} = f(T_{\text{blood}}, T_{\text{skin}})$. The solution of this equation is the blood flow rate for every point in the image.

Non-direct measures

In healthy individuals, heat energy is released from the cellular metabolism and transferred into the environment through the skin by sweating and vasodilatation. Sweating occurs during exercise, and its evaporation is the most important mechanism in maintaining the core temperature as close to 37°C as possible. The thermoregulatory centre in the hypothalamus controls all processes. Emotional sweating is a paradoxical response in contrast to the thermal sweating of thermoregulation. Emotional stress elicits vasoconstriction in the hands and feet combined with profuse sweat secretion on the palmar and plantar skin surfaces.

Gait biometrics

Using gait, it is possible to determine the gender of a walker [15,29,31] (male walkers tend to swing their shoulders more compared with female walkers, who tend to swing their hips). However, physical condition can affect gait; for example, pregnancy, affliction of the legs or feet, and drunkenness.

Sensor fusion

The distribution of biometrics in a system brings up many issues of environment, human-machine cooperation and human-human dialogue, social, ethical, and psychological issues, the effectiveness of biometric information, the implementation of biometric devices, as well as training skills for manipulating various streams of information. We focus on the last aspect, that is, the skills of personnel for using available data to make a decision.

Figure 15.5 shows the available sources of information in a system which includes distance (A) and contact biometrics (B), and other data from local and global databases (C). From these sources, a space of decision making is formed with respect to a given authorization scenario.

15.2.4. Cooperation Between the Officer and the Screened Individual

The security personnel and the individual must *cooperate* or *collaborate* to perform an access authorization procedure. In the training distributed environment, the personnel cooperate and collaborate with each other and with a virtual individual. Two or more officers are said to *cooperate* if they simultaneously interact with a virtual individual performing a given simulation task. Conversely, the personnel *collaborate* if only one officer interacts with the virtual individual.

This problem relates to human-human dialog, human-machine interaction, and machine-assisted dialogue. Human-machine interaction is a discipline concerned with the design, evaluation and implementation of interactive computing systems for human use and with the study of the major phenomena surrounding them.

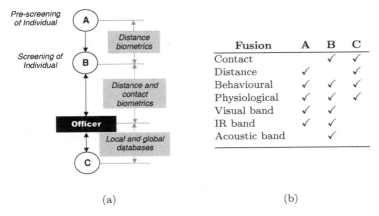

(a)   (b)

Fig. 15.5.   Space of available information and space of decision making are specified by three characteristic points, *A*, *B*, and *C* in the topology of the PASS (a); type of biometrics for these points are specified using classification of biometrics (b).

The term "assisted" in PASS and T-PASS implies that the operator (officer), is still responsible for the outcome and overall control of the authentication. The system is there as an aid; thus, the officer has to decide as to whether the information provided by the system is reliable or not.

There are two approaches to decision support system design:

The data-driven approach, based on on-line analytical processing and data mining applications, using databases (interpretation of data, identification of specific or unknown patterns, etc.), and

The model-driven approach, which uses quantitative and/or qualitative models of data; users can perform

What-If

analysis to gain more insight into the problem and its potential solution.

These models can be used in various modifications, for example, for vehicle routing [25]. Our classification of state-of-the-art human-machine cooperation is given in Table 15.3 The first two types of cooperations, i.e. machine-machine and human-human, are trivial. The differences lie in the form of representation of information: the signal domain for machines and the knowledge domain for humans.

Example 22. (a) Machine-machine cooperation can result in the correlation of biometric data, for example, the correlation of facial images in the visible and IR bands [43]. This result is represented in the signal domain. (b) Human-human, i.e. officer-individual cooperation can result in the authorization of an individual.

For collaboration between a machine and a human, a converter from the signal domain to the knowledge domain (and vice versa) is needed.

Example 23. Human-machine cooperation results in the representation of correlation between biometric data in semantic form. There are two approaches: (a) data from each device are converted into the knowledge domain where the correlation is derived, and/or (b) the correlation from the signal domain is converted to the knowledge domain.

Table 15.3. Classifications of cooperation between humans and machines.

Cooperation	Specification	Representation
Machine-Machine interaction	The request of the machine M_2 (signal domain) activates the machine M_1 (signal domain). The response of M_1 shifts M_2 to another state	
Human-Human interaction (dialogue)	The question of human H_1 (knowledge domain) is the request to the human H_2 (knowledge domain) and vice versa, the response of H_1 is the answer to H_2	
Human-Machine interaction	The question of the human H (knowledge domain) is converted into a request to machine M (signal domain), and vice versa, the response of M is converted to an answer for H	
Machine-Human interaction	The request of the machine M (signal domain) is converted into a question to human H (knowledge domain), and vice versa, the answer of H is converted to the response of M.	

In PASS, three agents collaborate: the officer, the individual, and the machine. This is a *dialogue-based interaction*. Based on the constraint that the officer has priority in decision making at the highest level of the hierarchy of a system, the role of the machine is defined as assistance or support of the officer. Three possible supporting strategies are introduced in Table 15.4. The first one satisfies the requirements of PASS and T-PASS. The other strategies can be considered in the further development of systems.

Table 15.4.   Support for cooperation between humans and machines.

Cooperation	Specification	Representation
Supporting questionnaire	The questionnaire technique of the human H_2 is supported by machine M over weak cooperation with humans H_1 and H_2	
Supporting answering	The answering technique of human H_1 is supported by machine M over weak cooperation with humans H_1 and H_2	
Supporting dialogue	Questionnaire and answering techniques of humans H_1 and H_2 are supported by machine M over weak cooperation with humans H_1 and H_2	

15.2.5.   *Decision Making Support*

Human error often focuses on cognitive errors (e.g. human made a poor decision) rather than behavioral errors (e.g. hand slipped off the control). Poor performance in decision making means that we need to help the person in order to make him or her a better decision-maker.

The obvious approach is to improve the skills of the human. However, training will be inefficient if the information for improving decision making is insufficient. Hence, a change in the information provided by the external environment is needed to support better decision making. Explicit help for decision makers can take many forms.

A set of biometric devices at the local level of PASS generates patterns and makes decisions automatically. These patterns and decisions are generated by means of some probability. Hence, the risk of a wrong decision always exists. Knowledge-based support partially alleviates this problem, as follows:

(a) It analyzes all decisions with respect to sources of information; weights are assigned to sources of information based on the criteria of the reliability of biometric information and contribution to the final decision.
(b) Specifies the potential uncertainty, and
(c) Describes it in the form of questions for the individual requesting authorization.

The generated questions proposed to the officer are intended to clarify the situation.

Example 24. An expert system is a software tool designed to capture one or more experts' knowledge and to provide answers in the form of recommendations. Expert systems take situational cues as input and provide an analysis and suggested action as an output.

Cognitive support is the more successful approach. *Decision support* is implemented by an interactive system that is specifically designed to improve the decision making capability of its users by extending the user's cognitive decision making abilities. The user delegates a task to the communication assistant. The interface recognizes the user intentions and goals, and involves the user in feedback and collaboration. This interactive style is partially employed within the T-PASS.

Consider the role of the officer in the authorization process. The result of an individual's data verification is a decision made by the officer. There are three sources of faults that influence decision making (Fig. 15.6):

(a) Temporal faults in biometric devices,
(b) Insufficiency of information, and
(c) Professional skills of personnel.

All these faults are not exactly predictable, i.e. they are random, and not manifested, for example, as the failure of a device. These faults implicitly influence decision making. The situation becomes unpredictable if several of these faults occur in the system. In Table 15.5, these faults are specified.

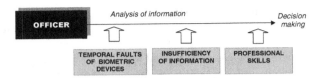

Fig. 15.6. The decision making model. Minimization of the risk of a wrong decision can be achieved by using reliable biometrics, multi-biometrics and better trained personnel.

Table 15.5. Three sources of faults that influence decision making.

Source	Comments
Temporal faults and errors	*Intermittent* (soft) faults are presented occasionally due to the shortcomings of algorithms, their program implementations, and the nature of biometrics. *Transient* faults are caused by temporary environmental conditions.
Insufficiency of information	The conditions of uncertainty in decision making. The solution is based on integrating additional sources of biometric data
The skills of security personnel	Traditionally, the officer's experience is accumulated over years of practice. To minimize this time, intensive training is needed, accomplished by assisting the officer in developing his/her skills through computer aided training using modeling of various scenarios, including extreme situations.

Example 25. Often, biometric devices are unable to extract data under non-perfect conditions. Figure 15.7 illustrates such a situation, which can be recognized as a temporal fault of a biometric device.

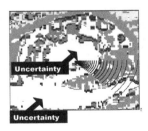

Fig. 15.7. Glints on the iris image significantly decrease the probability of correct identification.

Example 26. It is impossible to recognize, through visual observation of the individual's appearance and at the desk, any changes due to plastic surgery; or in other words, to determine whether an individual is trying to pass him/herself off as someone else. Hence, this situation is characterized as *insufficiency* of information.

Three spaces for representation and manipulation of biometric data can be identified in PASS:

1. Space for the *total* biometric sources,
2. Space for *available* biometric sources, and
3. Space for *decision making*.

Example 27. Consider decision making in PASS using six available sources of biometric information:

(a) Distance facial biometrics in the visible and IR bands,
(b) Behavioural biometrics of the voice, signature, and gait, and
(c) Measures of the size of the pupils.

The contributions of these automatic devices in authorization are different and depend on the scenario. Suppose that drugs and alcohol intoxication are detected in an IR image at the lowest level. In this case, an additional source of data can be used, i.e. the size of pupil.

The usefulness and uncertainty of the information is estimated using statistical criteria: ranging, weighting, and rating. Based on these estimations, the integrated knowledge-intensive system forms a space for a global solution. This space includes useful and uncertain information that is defined by statistical rules and used to provoke a set of specific questions. These questions are generated by the system and aim to minimize uncertainty in decision making. However, decisions are always made under some degree of uncertainty (perhaps very small).

15.3. Training Estimation Techniques

No comprehensive model of human behavior exists, owing to its large individual variability. The validity of a simulation is qualitative and can be quantified mathematically only fragmentarily. The more human parameters are involved in the model, the more difficult it is to have a valid interpretation of the human-human and human-machine interactions. Human factor studies have to adhere to a well-documented experimental protocol (trials, sessions, etc.).

> Training is a discrete random process which can be estimated using statistical methods. The objectives of this estimation should be formulated in terms of the quality of personnel (ability to make decisions in regular and extreme situations), the effects of various innovations and the installation of new devices, and different training strategies with respect to scenarios.

Training characteristics should provide reliable estimations and predict possible effects. Prediction is one of the most important features of a training system, which models real world processes and estimates the effects using controlled events.

15.3.1. *Bayesian Statistics*

Most decision making algorithms are based on Bayesian statistics [1,55]. Bayesian statistics utilize current information and update the training process with new data. The Bayesian equation is

$$\underbrace{P(\text{Training effect})}_{Prior-to-data\ probability} \times \underbrace{P(\text{Data}|\text{Training effect})}_{Revised\ probability} = \underbrace{P(\text{Training effect}|\text{Data})}_{Updated\ probability}.$$

That is, the prior-to-data probability of a training effect, revised with the probability model of additional data, gives the updated probability of a training effect. Hence, the Bayesian approach involves two steps that can be iteratively repeated. In training, the first step is the formation of a training paradigm based on modeling of available information. At the second step, the likelihood of the training effect is revised or updated with new data.

Using the Bayesian approach, the following estimations of training can be obtained: What kinds of skills degrade, and how fast? How efficiently can training improve or recover skills? To answer these and other questions, we use the following functions to evaluate the effects of HQP training:

1. *Degradation* function, $D(t)$
2. *Training* functions, $T(t)$, and
3. *Evolution* function, $I(t)$

The unit for these functions is the probability of making a correct decision. More precisely, various statistical methods can be used, such as Bayesian methods and maximum likelihood estimation, as well as confidence and prediction intervals.

15.3.2. *Degradation Function*

The measure of qualification of personnel is the *degradation* function $D(t)$. It characterizes losses of skills as a function of time t. This function should be understood as follows: the personnel obtain experience through regular duties and daily routine, but their skills for making decisions in extreme situations degrade. In terms of Bayesian statistics, degradation can be expressed as follows

$$P(\text{Degradation}) \times P(\text{Data}|\text{Degradation}) = P(\text{Degradation}|\text{Data}),$$

> The prior-to-data probability of degradation revised with the probability model of additional data gives the updated probability of degradation.

Example 28. In Fig. 15.8, the skills of the officer have degraded significantly to the critical level C during the period T_3. Re-training that takes a time of $t_3 \ll T_3$, can improve or recover his/her lost skills that should be in the confidence interval of degradation. The degradation function $D(t)$ should be evaluated periodically by the training system.

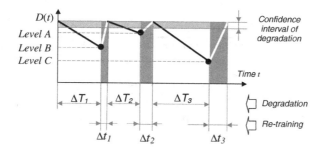

Fig. 15.8. Degradation function $D(t)$ of personnel quality for the periods T_1, T_2, and T_3, and of recovering the skills by re-training t_1, t_2, and t_3, respectively.

15.3.3. *Training Function*

The *training* function $T(t)$ is useful for the estimation of the efficiency of various strategies of training. The function $T(t)$ is a non-decreasing function of time t.

Example 29. In Fig. 15.9, strategies A and B are compared and the effect is calculated as a difference in time to achieve a given level of qualification (confidence interval for training). Note that the difference $\Delta t = t_2 - t_1$ corresponds to the cost-function of training.

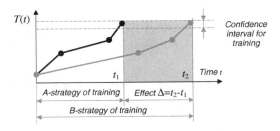

Fig. 15.9.   Training function $T(t)$ for estimation of A and B strategies of training.

The training function $T(t)$ is useful for the estimation of the impact of various modifications and installations of new devices and systems on the personnel's skills. In terms of Bayesian statistics, the effect of a new device installation can be expressed as

$$P(Skills) \times P(New\ device|Skills) = P(Skills|New\ device),$$

that is,

> The prior-to-data probability of the HQP skills revised with the probability model of additional devices gives the updated probability of skills.

The training function $T(t)$ is useful for the estimation of the impact of various modifications and installations of new devices on the quality of personnel.

Example 30. Assume that two new devices have been installed in t_1 and t_2 (Fig. 15.10). Without training, personnel are not able to use these devices, and the training function drops.

15.3.4.   *Evolution Function*

A function that shows how additional information can improve decision making is called an *evolution* function $I(t)$ in our approach. This is a non-decreasing function of time t. In terms of Bayesian statistics, the efficiency of installation of new biometric devices can be expressed as

$$P(New\ device) \times P(Data|New\ device) = P(New\ device|Data),$$

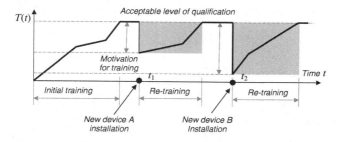

Fig. 15.10. Installation of the new devices A and B at t_1 and t_2, respectively, require re-training of personnel.

> The prior-to-data probability of the new biometric devices revised with the probability model of any additional device gives the updated probability of the new devices.

Example 31. The evolution function $I(t)$ is interpreted in Fig. 15.11 as a function of the number of newly installed biometric devices. They provide additional information about the authorized individual. The effect of installing new devices is an increase in the probability of making the correct decision.

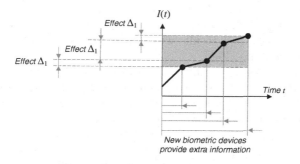

Fig. 15.11. The function of insufficiency of information $I(t)$ vs. time of installation for new biometric devices.

Example 32. The IR image indicates that there is no drug or alcohol consumption. However, through conversation with a person, the officer suspects that drugs and alcohol have been consumed. His decision may include testing of this individual's iris, or a pupil analyzer. The evolution function is also useful for predictive analysis both in the PASS and the training system.

15.4. Training System Design

Prototyping a training system is based on *functional* analysis, which includes *design analysis* and *system synthesis.* Design analysis aims at establishing requirements, specifications, and design evaluation criteria for the designed system. System synthesis is the translation of system requirements and functions into hardware and software specifications.

15.4.1. *Desired Characteristics*

T-PASS is distinguished from existing training systems by many features, which can be grouped with respect to architectural, topological, functional, communication, cost, and other criteria. To minimize cost, T-PASS utilizes PASS architectural principles and related characteristics and functionalities. Below, a sample of components and desired characteristics of T-PASS are given:

Screening, which is widely used in applications such as airport checkpoints, facilities access, etc.

Decision making support. A knowledge intensive technique can be useful. However, alternative techniques can be applied to achieve the same but more cost-efficient result (we used multi-valued logic and decision diagram techniques [49])

Real-world conditions of training, which is accomplished by the reconfiguration of PASS into the training system using minimal extra tools.

Flexibility, the extent to which the system can accommodate changes to the tasks and environments beyond those first specified.

Expandability and adaptibility. In the proposed training system, the basic configuration includes tools for generation of early warning data using visible and IR images, and tools for generation of other synthetic biometrics. The first will model the pre-screening process, and the second will model the authorization process at the check-desk.

Early warning, that is, the ability to get preliminary information about an individual, and in this way to reserve extra time for analysis of data before authorization.

Multiprocessor and distributed configuration. Three main information streams should be processing in real time: pre-screening data, screening data, and data from local and global databases.

Mobile configuration. T-PASS inherits this property from PASS which can be deployed at any public place where authorization is needed (important public events, exhibitions, and other social meetings or high profile settings where VIP security is required on a temporary basis).

These characteristics were the desired aims at the various phases of the system design.

15.4.2. *Architecture*

In the development of the training system T-PASS, most of the features of PASS are inherited. In Fig. 15.12, the topological and architectural properties of both of these systems are shown. It follows that the T-PASS architecture uses the PASS platform. However, T-PASS is different from PASS in the following features:

(a) Biometrics of pre-screened and screened individuals are simulated by the system,
(b) The appearance of biometric data can be modeled by the user using modeling tools,
(c) Local and global databases are simulated.

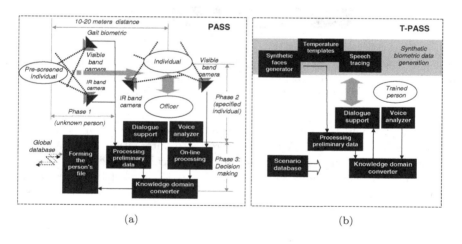

(a)   (b)

Fig. 15.12.   The PASS (a) and T-PASS (b) architecture.

The following distribution of biometric devices is related to the screening discipline (Fig. 15.13):

Distance, or non-contact biometrics (facial image in visible and IR bands), used to provide early warning information about a pre-screened individual. Gait biometrics is a good candidate for obtaining the additional information when individuals move from pre-screened to screened positions.

Contact biometrics, used to provide information about an individual (fingerprints, palmprint, hand geometry, signature, stroke dynamic, etc.) Correlation between various biometrics in semantic form provide a useful interpretation of biometric patterns.

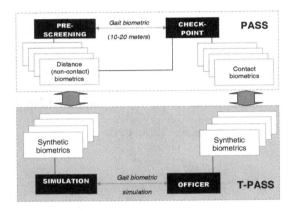

Fig. 15.13. The topological properties of PASS and the distribution of various types of biometric: distance (non-contact) biometrics are used to obtain early warning information and near distance and contact biometrics are used in real time authorization.

Our approach to T-PASS design is related to virtual modeling techniques [5]. Virtual environments refer to interactive systems that provide the user with a sense of direct physical presence and direct experience. These systems include visual and feedback techniques, and allow natural interaction by manipulation of computer-generated objects. In T-PASS, synthetic biometric data are used that can be understood as virtual biometric data.

Example 33. Surgeons and security personnel do not prefer immersion-based training systems since it makes them feel isolated from the real operation scene.

15.4.3. *Training Scenarios*

Humans form mental models through experience, training and instruction. Important human factors that are studied through the mental model are the prioritization of events and their sequence of occurrence. In PASS, scenarios are developed using architectural and topological characteristics. Possible scenarios are divided into three groups: regular, non-standard, and extreme. Most scenarios in each group are generated automatically using computer-aided design tools.

Scenarios based on distance IR images

Let us consider an example of a training scenario, in which the system generates the following data about the screened person:

```
Time 00.00.00:
Protocol of pre-screening person 45
Warning, level 04
Specification:  Drug or alcohol consumption, level 03
Possible action:
1.  Direct special inspection
2.  Register with caution
```

It follows from this protocol that the system evaluates the fourth level of warning using automatically measured drug or alcohol consumption for the screened individual. A knowledge-based sub-system evaluates the risks and generates two possible solutions. The officer can, in addition to an automatic analysis, analyze the acquired images in the visible and IR spectra.

Scenarios corresponding to the results of matching

We distinguish scenarios that correspond to the results of the matching of an individual's data with data in local and global databases. This process is fully automated for some stationary conditions, for example:

```
Time 00.00.00:
Protocol of screening person 45
Warning, level 04
Specification:  Drug or alcohol consumption, level 03
Local database matching:  positive
Possible action:
1.  Register
2.  Clarify and register
```

Note that data may not always be available in a database — this is the worst case scenario, and knowledge-based support is vital in this case.

Scenarios based on the analysis of behavioral biometric data

The example below introduces scenarios based on the analysis of behavioural biometric data.

```
Time 00.00.00:
Protocol of screening person 45
Warning, level 04
Specification:  Drug or alcohol consumption, level 03
Local database matching:  positive
Proposed dialog:
Question 1:  Do you need medical assistance?
Question 2:  Any service problems during the flight?
Question 3:  Do you plan to rent a car?
Question 4:  Did you meet friends on board?
Question 5:  Do you prefer red wine or whisky?
Question 6:  Do you have drugs in your luggage?
```

The results of automatically analyzed behavioral information are presented to the officer:

```
Time 00.00.00:
Protocol of screening person 45
Warning, level 04
Specification:  Drug or alcohol consumption, level 03
Local database matching:  positive
Level of trustworthiness of Question 6 is 03:  Do you have drugs in
your luggage?
Possible action:
1.  Direct to special inspection
2.  Continue clarification by questions
```

15.4.4. *Decision Making Support*

In the training system, decisions are generated by special tools and various scenarios in PASS, instead of the local biometric device response (Fig. 15.14).

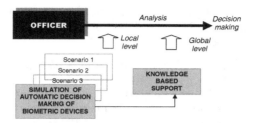

Fig. 15.14. Automatic generation of decisions at local levels is replaced in the training system by a simulator.

15.5. Training System Prototype

Prototyping of T-PASS is aimed at proving the basic concept of training system design. In this section, we introduce some details of the platform, data flows and their representations, computing of indirect characteristics, conversion patterns from signal to knowledge domain, and estimations.

15.5.1. *Platform*

In prototyping we used an *extended* platform, that is, we implemented the hardware and software tools without constraints on the cost and criteria of optimal configuration. The following facilities were utilized in prototyping [52]:

Hardware platform: a network of four computers connected with cameras in the IR and visible bands,

Software platform: about a dozen packages on biometric data processing and simulation,

Local and *global* communication resources, and

Biometric devices: about a dozen contact and non-contact devices for the acquisition and recognition of behavioral and physical biometrics.

15.5.2. *Data Flow*

In prototyping, we distinguish several levels of data flows. First of all, the information flows are divided into the following streams:

Static document record and biometric data from local and global databases.

Dynamic biometric data obtained from surveillance facilities and conversation with an individual at the desk, including behavioural biometric data (voice, facial expressions, signature).

We distinguish four forms of the individual's information:

(a) File data from databases which are well structured and standard;
(b) Appearance, or visual data, acquired mainly in real-time;
(c) Other appearance information, such as IR images, which requires specific skills for understanding and analysis;
(d) Various forms of behavioural biometrics (facial expressions, voice, signature) that are acquired in real-time (during pre-screening or check-in procedure).

15.5.3. *Indirect Characteristic Computing*

In our experimental study, we focus on various aspects of indirect characteristic computing, in particular (Fig. 15.15): interpretation of biometric patterns in semantic form, initialization of questionnaires, reliability of biometric data, and their improvement using correlation in the knowledge domain.

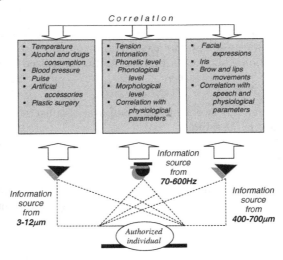

Fig. 15.15.   Prototyping indirect biometric characteristics using correlation technique.

A fragment of this design is given in Fig. 15.16. The officer manipulates and efficiently uses biometric data to make a decision according to various scenarios. Many factors render decision making a complex procedure. To simplify this procedure, decision making is divided into several simple steps with respect to time and levels.

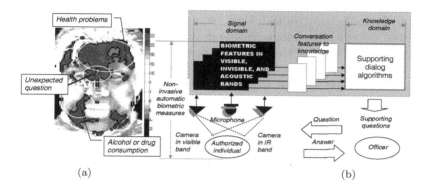

(a) (b)

Fig. 15.16. Prototyping of T-PASS: interpretation of temperature regions in an IR image (a) and generation of supporting questions in PASS (b).

15.5.4. *Conversion of Data*

The key component of a system is the *converter* from the signal domain to the knowledge domain. The efficiency of the converter is critical in decision making support. The converter implements a knowledge-based technology. In the development of the converter and its prototyping, we utilize properties of the system such as

(a) Limited sets of supported questions (biometric patterns of a given class initialize 3–5 questions),
(b) Limited types of biometrics (about 10), and
(c) A limited number of possible correlation (about 20).

The example below explains our approach.

Example 34. Let us assume that there are three classes of samples assigned to "Disability," "Alcohol intoxication," and "Normal," and denoted as values of the three-valued logic variable {0}, {1}, and {2}, respectively. Let us also assume that feature patterns from facial IR image and gait correspond to the above classes. Using a logic function representation and decision diagram technique [35,49], the following semantic constructions can be generated by the system

```
Construction 1:  I do not know, but something is not normal.
Construction 2:  Possible alcohol intoxication.
Construction 3:  An individual with a disability.
```

In this example, a 3-valued logic is used. Given m classes of construction; an m-valued logic function can be applied. Generally, psychologists must be consulted about the report formulation.

Note that this approach is a case of data mining, that is, a process of nontrivial extraction of implicit and potentially useful information [34].

15.5.5. *Modularization*

The modularity principle is implemented of all levels of the architecture. In particular, the following modules, or Security Personnel Assistants (SPAs) correspond to the specific biometric devices:

SPA for distance temperature measure,
SPA for distance blood pressure and pulse measure,
SPA for distance drug and alcohol intoxication estimation,
SPA for distance artificial accessory detection,
SPA for distance plastic surgery detection.

15.5.6. *Estimations*

The prototyping is also aimed at rendering the estimation techniques introduced in Section 15.3. This can be accomplished by using a Bayesian Belief Network (BBN) which describes the probability distribution of a set of variables by specifying a set of conditional independence assumptions together with a set of causal relationships among variables and their related joint probabilities. The BBN is modelled by a direct acyclic graph, where each node represents a variable to which a group of conditional probabilities are attached. The BBN can be thought of as a probabilistic inference engine that can answer queries, or What-If questions, about the variables in the network. These queries include calculation of conditional probabilities, and BBNs have the potential to make inferences under uncertain conditions.

Let $X_1, \ldots, X_n$ denote a set of n variables. The joint probability distribution of variables in a direct acyclic graph is defined as $p(X_1 = x_1, \ldots, X_n = x_n) = \prod_{i=1}^{n} p(x_i|x_i^*)$, where x_i^* is defined from the topology of the graph (each node interacts with a bounded set of neighbors).

The following examples illustrate our approach to measurement of the efficiency of analysis and processing data at various levels of a system.

Example 35. The effects of pre-screening can be estimated as the difference of time, $\Delta T = T_2 - T_1$, and the difference of risks used to make a decision, $\Delta I = I_2 - I_1$ (Fig. 15.2(b)). The distance between the pre-screening area and the authorization check can be used for obtaining additional information (Fig. 15.17).

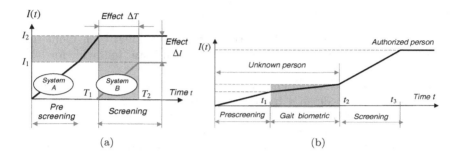

(a)   (b)

Fig. 15.17.   Screening T: effects of pre-screening can be estimated as the differences $\Delta T = T_2 - T_1$ and $\Delta I = I_2 - I_1$ (a), and, utilizing the topological and gait biometrics in the training system (b).

Example 36. Figure 15.18 shows how the efficiency of questionnaire techniques can be estimated using BBN.

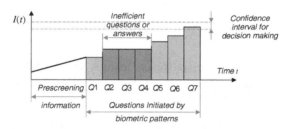

Fig. 15.18.   Estimation of knowledge-based support using a function of insufficiency $I(t)$.

15.6.   Discussion and Open Problems

We have developed a uniform and cost-effective platform of hardware and software tools for PASS and for training users of PASS. We have introduced our experience and study in prototyping the main components of the training system based on the PASS platform.

15.6.1. *Novel Approaches*

To achieve the study's goals, a number of novel approaches were used throughout design and prototyping, in particular:

Extension of the context of biometric data; we consider each biometric device in a system as a smart sensor and an initiator for the questionnaire.

Two-domain processing of biometric data; in addition to the signal domain where biometric devices traditionally operate, we use the knowledge domains, where biometric patterns are represented in semantic form.

Questionnaire techniques based on possible sources of information (biometric data, personal data from databases, and the experience of personnel)

Evaluation techniques are based on three main characteristics: degradation, training, and evolution functions. These functions can be transformed into each other and are well suited to the requirements of keeping personnel skills at a high level of professionalism, due to their ability to accept various strategies for improving the system.

Improving the reliability of biometric data. To alleviate the sensitivity of biometrics to the environment, we have developed two approaches. The first is based on a questionnaire technique that allows us to "locate uncertainty" in processing data. The second approach is based on correlation between biometric data in the signal and knowledge domains.

Interpretability of biometric patterns insures human-friendly interface when manipulating data, for example, derivations of the correlation between biometrics.

In addition, we intensively used throughout our design the principle of *formidable resemblance*, that is, the similarity of a set of algorithms based on the scheme:

$$\text{Data acquisition} \rightarrow \text{Information extraction}$$
$$\rightarrow \text{Knowledge representation} \rightarrow \text{Action}$$

15.6.2. *Open Problems*

Several problems were identified through the prototyping of the system.

Problem 1. Extension of the pattern recognition problem for various biometrics using a knowledge domain representation requires more sophisticated tools. This is known as the *integrated knowledge intensive approach.*

Example 37. One can observe the strong correlation between the facial IR image and gait analysis patterns in semantic form:

```
Possible alcohol or drug intoxication.
Warning:  level XX.
```

and

```
Gait:  Possible alcohol or drug intoxication.
Warning:  level 04.
```

The correlation is well identified in semantic form and can be used for initiation of supporting questions as indicated in the protocol in subsection 15.4.3. This technique should be improved.

Problem 2. Alternative solutions to the *questionnaire technique,* the basic component of decision making support technology, must be investigated. In the prototyping system, logic equations for describing the relationship between biometric patterns and sets of questions are currently in use.

Example 38. The pattern "Temperature between 37 and 38" is assigned to a value "1" of a four valued logic variable and corresponds to the semantic construction "Do you need medical assistance?" However, the pattern "Extremely high temperature" does not initiate a question, but rather an action that corresponds to the early warning manifestation of the "Alarm" scenario.

We have chosen this approach to knowledge representation and question generation for several reasons. First of all, automatic minimization of uncertainty and inconsistences is an approach that is still risky in practice and costly to develop and implement [14,27]. Specifically, defined inconsistencies were viewed through logical description and manifested to the personnel to be resolved. Secondly, a questionnaire technique can be adopted from well-studied polygraph systems [36].

Problem 3. *Synthetic biometric data generation.* It is remarkable that the next generation of security and training systems will be multi-biometric and will use synthetic biometrics. However, the development of solutions to inverse biometric problems (generating synthetic biometric data) and techniques for virtual environment design demand more effort. Our prototyping is based on synthetic biometric data automatically generated to "imitate" real data as described in [3,28,39,50,48].

Example 39. Models can be generated from real acquired data and can simulate age, accessories, and other attributes of the human face. This environment, or personnel with virtual reality modeling, as well as it can be used in decision making support.

Note that synthesis of biometric data brings up the classic problem of computer graphics and computer vision: the generation of novel images corresponds to an appropriate set of control parameters. In particular, in [3], learning techniques were used given "pose" and "expression" parameters for face synthesis.

We have not yet developed training scenarios using biometrics such as synthetic iris [41,48,50], ear topology [20,21], and signature [32,37]. Additional study is needed in the application of knowledge-based techniques [6].

15.7. Conclusion

This study contributes to the development of a methodology for training personnel in specific applications based on biometrics and security screening. This research confirms that:

(a) Screening is a technique already adopted in social infrastructure (for example, document-based checks, fingerprints, and mug-shot acquisition),

(b) Screening can be efficiently combined with analytical, knowledge intensive techniques,

(c) Screening is the basis for the implementation of the early warning paradigm in authorization techniques, and

(d) Training systems can be developed based on the hardware and software platform of the access control system.

Usage of various biometrics, including early detection and warning devices in the system, must involve sociological and economic considerations. The proposed decision making support system is considered to be a unique area to combine and prototype various methodologies, concepts, and techniques from many areas of the natural and social sciences, in particular: image processing and pattern recognition, virtual environment design and synthetic biometric data generation, distributed and multi-agent system design, human-machine interaction, integrated knowledge-intensive system design, communication, and psychology.

Acknowledgments

This Project is partially supported by the Natural Sciences and Engineering Research Council of Canada (NSERC), the Canadian Foundation for Innovations (CFI), the Government of the Province of Alberta, and ECSIS — the European Center for Secure Information and Systems, Iasi, Romania. A part of the project has been implemented as an initiative within the JPLs Humanoid Robotics Laboratory.

The authors are grateful to many researchers in biomedical engineering, knowledge-based decision making, security, and psychology, who provided their valuable remarks and suggestions on this Project. Several commercial companies provided their equipment and tools to prototype the described system. The authors acknowledge the valuable suggestions and remarks of Eur. Ing. Mr. Phil Phillips (Department of Information Security, Office of National Statistics, UK), Dr. M. S. Nixon (School of Electronics and Computer Science at the University of Southampton, UK), Dr. D. J. Hurley (Analytical Engines Ltd., UK), Dr. P. Perner (Institute of Computer Vision and Applied Computer Science, Leipzig, Germany), and Dr. R. Elio (University of Alberta, Canada).

Bibliography

1. Anderson, D. N., Thompson, S. E., Wilhelm, C. E. and Wogman, N. A. (2004). Integrating Intelligence for Border Security, *J. of Homeland Security,* Jan., www.homelandsecurity.org
2. Bayarri, S., Fernandez, M. and Perez, M. (1996). Virtual reality for driving simulation, *Communication of the ACM,* **39**, 5, pp. 72–76.
3. Beymer, D., Shashua, A. and Poggio, T. (1993) Examples Based Image Analysis and Synthesis, Technical Report 1431, *Massachusets Institute of Technology,* 20 p.
4. Biometrics for Airport Access Control Guidance Package. Response to Section 4011(a)(5), Version release Sept. 30, 2005, www.tsa.gov/interweb/ anetlibrary/Biometric
5. Burdea, G. C. and Coiffet, P. *Virtual Reality Technology* (2004). John Wiley & Sons.
6. Coello Coello, C. A. (2006). Evolutionary algorithms: basic concepts and applications in biometrics, this issue.
7. Crane, P. (1999). Implementing distributed mission training, *Communication of the ACM,* **42**, 9, pp. 90–94.
8. Davis, J. and Prosnitz, D. (2003). Technical and Policy Issues of Counterterrorism - A Primer for Physicists, *Physics Today,* **56**, 4, pp. 39–45, April.
9. Dowdall, J., Pavlidis, I. and Bebis, G. (2003). Face detection in the near-IR spectrum, *Image and Vision Computing,* **21**, 7, pp. 565–578.

10. Edwards, W. (1987). Decision making. In: G. Salvendy, Ed. *Handbook of human factors*, Wiley, NY, pp. 1061–1104.

11. Elliott, S. J., Kukula, E. P. and Modi, S. K. (2006). Issues involving the human biometric sensor interface, this issue

12. Eveland, C. K., Socolinsky, D. A. and Wolff, L. B. (2003). Tracking human faces in infrared Vvideo, *Image and Vision Computing*, **21**, pp. 579–590.

13. Flight simulation/simulators (1985), *Society of Automotive Engineers*, Warrendale, PA.

14. Forsythe, C., Bernard, M. L., Goldsmith, T. E. Eds, *Cognitive Systems: Human Cognitive Models in Systems Design*, (2006). Lawrence Erlbaum Associates Publishers, New Jersey

15. Foster, J. P., Nixon, M. S. and Prüugel-Bennett, A. (2003). Automatic gait recognition using area-based metrics Pattern, *Recognition Letters* **24**, pp. 2489-2497.

16. Fujimasa, I., Chinzei, T. and Saito, I. (2000). Converting far infrared image information to other physiological data, *IEEE Engineering in Medicine and Biology Magazine*, **10**, 3, pp. 71–76.

17. Gaussorgues, G. (1994). *Infrared Thermography*, Microwave Technology Series 5, Chapman and Hall, London.

18. Van Gemert, M. J. C., Jacques, S. L., Sterenborg, H. J. C. M. and Star, W. M. (1989). Skin optics, *IEEE Trans. Biomedical Engineering*, **36**, 12, pp. 1146–1154.

19. Herzog, G. and Reithiger, N. (2006). The SmartKom: Foundations of multimodal dialogue systems. In: W. Wahlster, Ed., Springer, Heidelberg, pp. 43–58.

20. Hurley, D. J., Nixon, S. M. and Carter, J. N. (2005). Force field feature extraction for ear biometrics, *Computer Vision and Image Understanding* **98**, pp. 491-512.

21. Hurley, D. J. (2006). Force field feature extraction for ear biometrics, this issue

22. Jain, A., Pankanti, S. and Bolle, R. Eds. (1999). *Biometrics: Personal Identification in Networked Society*, Kluwer, Dordrecht.

23. Jain, A. K., Ross, A. and Prabhakar, S. (2004). An introduction to biometric recognition, *IEEE Trans. on Circuits and Systems for Video Technology, Special Issue on Image- and Video- Based Biometrics*, **14**, 1.

24. Johannsen, G. and Rouse, W. B. (1983). Studies of planning behavior of aicraft pilots in normal, abnormal, and emergency sumulations, *IEEE Trans. Systems, Man, and Cybernetics,* **13**, 3, pp. 267-278.

25. Keenan, P. (1998). Spatial decision support system for vehicle routing. *Decision Support Systems*, **22**, 1 pp. 65–71.

26. Knerr, B., Breaux, R., Goldberg, S., Thrurman, R. (2002). National defence. In: K. Stanney, Ed. *The Handbook of Virtual Environments Technology*, Erlbaum, Mahwah, NJ, pp. 857–872.

27. Lapeer, R. J., Chios, P., Linney, A. D., Alusi, G. and Wright A. (2001). HCI: The next step towards optimization of computer-assisted surgical planning, intervention and training (CASPIT), In: Human Computer Interaction:

Issues and Challenges, Chen Q. Ed., *Idea Group Publishing*, Hershey, PA, pp. 232–246.

28. Ma, Y., Schuckers, M. and Cukic, B. (2005). Guidelines for Appropriate Use of Simulated Data for Bio-Authentication Research, *Proc. 4th IEEE Workshop Automatic Identification Advanced Technologies*, Buffalo, NY, pp. 251–256.

29. Mather, G. and Murdoch, L. (1994). Gender discrimination in biological motion displays based on dynamic cues. *Proc. Royal Society*, **258**, pp. 273–279.

30. Method to Testing Objects for Random Actions Thereon and Digital Simulator-Computer System for Effecting Same, Patent U.S.A. 4205383, Bakanovich, E. A., Shmerko, V. P. at al.

31. Nixon, M. S., Carter, J. N. Cunado, D., Huang, P. S. and Stevenage, S. V. (1999). Automatic gait recognition. In: A. Jain, R. Bolle, and S. Pankanti, Eds, *Biometrics: Personal identification in a networked society*, Kluwer, Dordrecht, pp. 231–250.

32. Parker, J. R. (2006). Composite systems for handwritten signature recognition, this issue

33. Pavlidis, I. and Levine, J. (2002). Thermal image analysis for polygraph testing, *IEEE Engineering in Medicine and Biology Magazine*, **21**, 6, pp. 56–64.

34. Perner, P. (2005). Case-based reasoning for image analysis and interpretation, In: C. Chen and P. S. P. Wang, Eds. *Handbook on Pattern Recognition and Computer Vision,* 3rd Edition, World Scientific Publisher, pp. 95–114.

35. Pleban, M., Niewiadomski, H., Buclak, P. Sapiecha, P. Yanushkevich, S. and Shmerko, V. (2002). Argument Reduction Algorithms for Multi-Valued Relations, *Proc. Int. Conf. Artificial Intelligence and Soft Computing,* Banff, Canada, pp. 609–614.

36. *The Polygraph and Lie Detection,* (2003). The National Academies Press, Washington, DC, 2003

37. Popel, D. V. (2006). Signature analysis, verification and synthesis in pervasive environments, this issue

38. Prokoski, F. J. and Riedel, R. B. (1999). Infrared identification of faces and body parts, In: A. Jain, R. Bolle, and S. Pankanti, Eds., *Biometrics: Personal Identification in Networked Society*, Kluwer, Dordrecht, pp. 191–212.

39. Ratha, N., Connell, J., and Bolle, R. (2001). Enhancing security and privacy in biometrics-based authentication systems, *IBM Systems J.*, **40**, 3, pp. 614–634.

40. Richards, A. (2001). *Alien Vision: Exploring the Electromagnetic Spectrum with Imagin Technology,* SPIE.

41. Samavati, F. F., Bartels, R. H. and Olsen, L. (2006). Local B-spline Mmultiresolution with example in iris synthesis and volumetric rendering, this issue.

42. Shmerko, V, Perkowski, M., Rogers, W., Dueck, G., and Yanushkevich, S. (1998). Bio-Technologies in Computing: The Promises and the Reality,

Proc. Int. Conf. on Computational Intelligence & Multimedia Applications, Australia, pp. 396–409.

43. Srivastava, A. and Liu, X. (2003). Statistical hypothesis pruning for identifying faces from infrared images, *Image and Vision Computing*, **21**, pp. 651–661.

44. Sugimoto, Y., Yoshitomi, Y. and Tomita, S. (2000). A Method for detecting transitions of emotional states using a thermal facial image based on a synthesis of facial expressions, *Robotics and Autonomous Systems*, **31**, pp. 147–160.

45. Total Information Awareness DAPRA's Research Program, (2003). *Information and Security*, **10**, pp. 105–109.

46. Tsumura, N. Ojima, N. Sato, K., Shiraishi, M. Shimizu, H., Nabeshima, H., Akazaki, S., Hori K. and Miyake, Y. (2003). Image-based skin color and texture analysis/synthesis by extracting hemoglobin and melanin information in the skin, *ACM Trans. on Graphics*, **22**, 3, pp. 770–779.

47. Wang, W., Zeng, Y., Ma, D., Jin, Z., Wu, H., Yuan, C., Yuan, Y., Lin, Z., Wang, C. and Qi, H. (2004). Clinical study on using thermal texture maps in SARS diagnosis, *Proc. 26th Annual Int. Conf. of the IEEE EMBS*, San Francisco, CA, pp. 5258–5264.

48. Yanushkevich, S. N., Stoica, A., Shmerko, V. P. and Popel, D. V. (2005). *Biometric Inverse Problems*. CRC Press/Taylor & Francis Group, Boca Raton, FL.

49. Yanushkevich, S. N., Miller, D. M., Shmerko, V. P. and Stanković, R. S. (2006). *Decision Diagram Techniques for Micro- and Nanoelectronic Design*, Handbook, Taylor & Francis / CRC Press, Boca Raton, FL.

50. Yanushkevich, S. N. (2006). Synthetic biometrics: A survey, *Proc. World Congress on Computational Intelligence*, Vancouver, Canada.

51. Yanushkevich, S. N., Stoica, A. and Shmerko, V. P. (2006). Experience of Design and Prototyping of a Multi-Biometric, Physical Access, Early Warning Security Systems Based on Screening Discipline of Service (PASS) and Training System T-PASS, *Proc. 32nd Annual IEEE Industrial Electronics Society Conf.*, Paris.

52. Yanushkevich, S. N. and Stoica, A. (2006). Supervisors, *Design and Prototyping of a Multi-Biometric Physical Access Screening Early Warning Security Systems (PASS) and Training System T-PASS*, Technical Report ECE2006.BT.R04, Biometric Technology Laboratory, University of Calgary (Canada) and Humanoid Robotics Laboratory, California Institute of Technology, NASA Jet Propulsion Laboratory, California Institute of Technology Pasadena, CA (USA).

53. Zhang, D. (2000). *Automated Biometrics: Technologies and Systems*, Kluwer, Dordrecht.

54. Zhang, D. (2004). *Palmprint Authentication*, Kluwer, Dordrecht.

55. Zellner, A. (1988). Optimal information processing and Bayes' theorem, *American Statistician*, **42**, 4, pp. 278–284.

About the Authors and Contributors

Ahmed Awad E. Ahmed is a research associate and a Ph.D. candidate at the Electrical and Computer Engineering Department, University of Victoria. His Ph.D. thesis presents a number of new trends in security monitoring through human computer interaction devices. He is a member of the Security and Object Technology (ISOT) Research Group and the principal investigator of Biotracker, an intrusion detection system based on  biometrics. Mr. Ahmed worked as a technology specialist and a quality assurance consultant in a number of leading software firms. He obtained his B.Sc. and M.Sc from Ain Shams University, Electrical and Computer Department on 1992, and 1997 respectively.

Richard Harold Bartels born in Ann Arbor, Michigan, in 1939, Bartels Bartels earned his Bachelor and Master degrees from the University of Michigan in 1961 and 1963, respectively. He earned his PhD from Stanford University in Computer Science with a specialization in numerical analysis in 1968. He worked for the Computer Science Department and the Center for Numerical Analysis at the University of Texas and Austin from 1968 through  1974, then for the Mathematical Sciences Department of The Johns Hopkins University in Baltimore, Maryland, until 1979, and finally for the Computer Science Department of the University of Waterloo in Ontario, Canada, until his retirement in 1999. He has authored papers in numerical linear algebra, numerical optimization, and computer-aided geometric modeling. He is the co-author of a book on the use of splines in computer graphics.

Carlos A. Coello Coello received the B.Sc. degree in civil engineering from the Universidad Autónoma de Chiapas, México, and the M.Sc. and the PhD degrees in computer science from Tulane University, USA, in 1991, 1993, and 1996, respectively. He is currently a professor (CINVESTAV-3D Researcher) at the Electrical Engineering Department of CINVESTAV-IPN, in Mexico City, México. He also chairs the Task Force on Multi-Objective Evolutionary Algorithms of the IEEE Neural Networks Society. He is a member of the IEEE, the ACM and the Mexican Academy of Sciences. His major research interests are: evolutionary multiobjective optimization, constraint-handling techniques for evolutionary algorithms, and evolvable hardware. Dr. Coello has authored and co-authored over 120 technical papers and several book chapters. He has also co-authored the book "Evolutionary Algorithms for Solving Multi-Objective Problems" (Kluwer Academic Publishers, 2002).

Stephen J. Elliott is involved in a number of activities related to biometrics and security. He is actively involved in biometric standards, acting as Secretary of the INCITS M1, the U.S. standards committee for Biometrics. Dr. Elliott is also involved in international standards activities including Testing and Evaluation. From 2003-2006 he was Vice Chair of the International Committee for Information Technology Standards. Recently, Dr. Elliott has been involved in testing and evaluating commercially available devices as well as giving lectures on biometric testing and product evaluation. Dr. Elliott has also given numerous conference lectures on distance education, biometric course development, and the results of the research at the Biometric Standards, Performance, and Assurance Laboratory. He is a member of Purdue University's e-Enterprise Center and the Center for Education, Research, Information Assurance and Security (CERIAS).

Marina L. Gavrilova received Diploma with Honors from Lomonosov Moscow State University in 1993 and Ph.D. degree from the University of Calgary in 1999. Associate Professor in the Department of Computer Science, University of Calgary, Dr. Marina Gavrilova conducts research in the area of computational geometry, image processing, optimization, spatial and biometric modeling. Prof. Gavrilova is co-founder and director of two innovative research labs: the Biometric Technologies Laboratory: Modeling and Simulation, sponsored by CFI and the SPARCS Laboratory for Spatial Analysis in Computational Sciences, sponsored by GEOIDE. Prof. Gavrilova publication list includes over ninety journal and conference papers, edited special issues, books and book chapters. Prof. Gavrilova acted as a Co-Chair of the International Workshop on Biometric Technologies BT 2004, Calgary, Alberta, Canada, June 21-23 2004. In 2001, Prof.. Gavrilova founded the annual International Workshop on Computational Geometry and Applications series, and in 2003 she Co-Founded the ICCSA International Conference on Computational Sciences and its Applications. Prof. Gavrilova is on the Editorial Board of the International Journal of Computational Sciences and Engineering and currently serves as a Guest Editor for International Journal of Computational Geometry and Applications. Her greatest accomplishment, in her own words, is finding a delicate balance between professional and personal life while striving to give her best to both. Together with her husband, Dr. Dmitri Gavrilov, she is a proud parent of two wonderful boys, Andrei and Artemy.

David J. Hurley founded Analytical Engines in the early 1980s in order to exploit newly emerging electronics and microcomputer technologies. The company was named after the ingenious mechanical computing invention of Charles Babbage, an eccentric 19th century English professor who held the same chair at Cambridge as Isaac Newton. Analytical Engines played a role in the uptake of the C programming language in the UK by marketing the first low cost C compiler, initially for the CP/M operating system, and later for MSDOS. This was before Microsoft and Borland had even entered the C market. He believes that computer vision and its related technologies are destined to be amongst the most influential technologies of the emerging century, especially as computer vision begins to take advantage of developments in high speed electronics. He has both Master of Engineering and Doctor of Philosophy degrees from the University of Southampton, where he demonstrated the viability of the human ear as a biometric for the first time, although it had been proposed more than a century before by the French biometrics pioneer Alphonse Bertillion. David hopes to see machine vision evolve to match and even overtake its biological counterpart during our lifetimes, and continues to work towards this goal in collaboration with researchers and industrial partners around the world.

Wai-Kin Kong received the BSc degree in mathematics from Hong Kong Baptist University with first class honors and obtained the MPhil degree from The Hong Kong Polytechnic University. During his study, he received several awards and scholarships from both universities, including scholastic awards and a two year Tuition Scholarship for Research Postgraduate Studies. Currently, he is studying at University of Waterloo, Canada for his PhD. His major research interests include pattern recognition, image processing and biometric security. He is a member of the IEEE.

Eric P. Kukula received his B.S. in Industrial Distribution and M.S. in Technology with a Specialization in Information Security from Purdue University. Currently, Eric is pursuing his Ph.D. in Technology with specialization in Computational Science and Engineering at Purdue University. His research interests include the HBSI, focusing on the interaction of ergonomic principles and biometric performance. He has received numerous awards including a graduate student teaching award, a summer research grant, and a graduate student scholastic award from the College of Technology. Eric is a member of INCITS M1, IEEE, and NAIT.

Qingmin M. Liao received the B.S. degree in radio technology from the University of Electronic Science and Technology of China, Chengdu, China, in 1984, and the M.S. and Ph.D. degrees in signal processing and telecommunications from the University of Rennes 1, Rennes, France, in 1990 and 1994, respectively. Since 1995, he has joined in Tsinghua University, Beijing, China. Now he is a professor in the  Department of Electronic Engineering of Tsinghua University, and is the directors of Visual Information Processing laboratory and Tsinghua-PolyU Biometrics Joint laboratory in the Graduate School at Shenzhen, Tsinghua University. From 2001 to 2003, he served as the Invited Professor at University of Caen, France. His research interests include image/video processing, transmission and analysis; biometrics and their applications.

Guangming Lu graduated from the Harbin Institute of Technology (HIT, China) in Electrical Engineering in 1998, where he also obtained his master degree in Control Theory and Control Engineering in 2000. Then he joined the Department of Computer Science and Technology of HIT and got the PH.D degree in 2005. He currently is a postdoctor in Visual Information Processing Lab, Graduate School at Shenzhen, Tsinghua University. His  research interests include pattern recognition, image processing, and biometrics technologies and applications.

Shimon Modi received the B.S. degree in Computer Science from Purdue University in 2003. He is a member of the Biometrics Standards, Performance and Assurance Laboratory in the Department of Industrial Technology. He is currently pursuing a Ph.D. in Technology with specialization in Computational Science and Engineering at Purdue University. He received a M.S. in Technology with specialization in Information Security at Purdue's Center for Education and Research in Information Assurance and Security (CERIAS) at Purdue University in 2005. He is a recipient of the Ross Fellowship awarded at Purdue University. His research interests include impact of biometric sample quality on performance, and use of biometrics in electronic authentication.

Mark Nixon is the Professor in Computer Vision in the ISIS research group at the School of Electronics and Computer Science at the University of Southampton, UK. His research interests are in image processing and computer vision. He has helped to develop new techniques for static and moving shape extraction (both parametric and non-parametric) which have found application in automatic face and automatic gait recognition and in medical image analysis. His team were early workers in face recognition, later came to pioneer gait recognition and more recently joined the pioneers of ear biometrics. Amongst previous research contracts, he was Principal Investigator in Southampton's part of the Fuzzy Land Information from Environmental Remote Sensing (FLIERS) project. and he was Principal Investigator with John Carter on the DARPA supported project Automatic Gait Recognition for Human ID at a Distance. Currently, his team is part of the General Dynamics Defence Technology Centre's program on data fusion (biometrics, naturally). He chaired the 9th British Machine Vision Conference BMVC'98 held at Southampton in September '98 (an issue of Image and Vision Computing containing some of the most highly rated conference papers was published as Volume 18 Number 9). The BMVC'98 Electronic Conference Proceedings remain online via the British Machine Vision Association. Apart from being a programme member/reviewer for many other conferences, he co-chaired the IAPR International Conference Audio Visual Biometric Person Authentication (AVBPA 2003) at Surrey, UK (the conference has now transmuted to the International Conference on Biometrics), he was Publications Chair for the International Conference on Pattern Recognition (ICPR 2004) at Cambridge UK and with Josef Kittler, he co-chaired the IEEE

7th International Conference on Face and Gesture Recognition FG2006 held at Southampton, UK in 2006. He has given many invited talks and presentations, most recently at IEEE Face and Gesture 2004, EUSIPCO 2004 and MMUA 2006. He has written over 250 publications, including papers, authored and edited books. His first book, Introductory Digital Design — A Programmable Approach, was published by MacMillan, UK, July 1995 and is still in print. His vision book, co-written with Dr. Alberto Aguado from the University of Surrey, entitled Feature Extraction and Image Processing was published in Jan. 2002 by Butterworth Heinmann/ Newnes, now Elsevier. It's been reprinted many times and a new Edition is in preparation. With Tieniu Tan and Rama Chellappa, his new book Human ID based on Gait is part of the New Springer Series on Biometrics, and was published late 2005.

Luke Olsen received his Bachelor's degree in Computer Science from the University of Calgary in April 2004. He is currently completing a Master's degree at the University of Calgary under the supervision of Dr. Faramarz Samavati and Dr. Mario Costa Sousa. His thesis topic is multiresolution techniques for arbitrary-topology surfaces, and research interests include sketch-based modeling, real-time rendering, and computer vision.

Jim R. Parker is a professor of Computer Science, specializing in pattern recognition, machine perception, and computer games and multimedia. He is the author of books on computer vision and game development, as well as over a hundred technical articles. He has worked on biometrics since 1992. Dr. Parker received his Ph.D. from the State University of Gent, Belgium with greatest distinction. He has been a member of the IEEE, the ACM,

and the International Game Developer's Association. He is interested in teaching, and among other things created, designed, and implemented the first course in computer game programming at a Canadian University — the Computer Science 585 course in Calgary, along with Radical Entertainment of Vancouver BC. Jim is the director of the Digital Media Laboratory at the University of Calgary, and works also in the Sport Technology Research lab in that school's world renowned Kinesiology faculty.

Eur. Ing. Phil Phillips, C.Eng. MIEE, FBCS Formerly National Information Security Adviser, NHS Scotland (2001–2004), now Head of Information Security for a UK Government Department. Phil has spent over 40 years in IT, Communications security and general security as a risk manager and consultant for various Government Departments and other business sectors. As President of the BCS Information Security Specialist Group, he has endeavoured to encourage closer security cooperation across the public sector and utilities in support of CNI and the E-Government programme and fosters a strong link between academia, industry and IT Security.

Denis V. Popel, PhD, Member IEEE, is an Associate Professor in the Computer Science Department, Baker University, KS, USA. He has a Master's degree with Honors in Computer Engineering (contextual support in document understanding systems, and biometrics) from the State University of Informatics and Radio- electronics, Minsk, Belarus, and a PhD with Honors in Computer Science (information theory, logic design, and decision making) from the Technical University, Szczecin, Poland. Dr. Popel held teaching positions at the University of Wollongong, Australia-Dubai campus, and at the Technical University, Szczecin, Poland. He has over 10 years of experience working in the area of knowledge discovery including data mining, decision making, data representation, biometric data retrieval, and biometric data manipulation. Dr. Popel participated as the PI and as a senior person on a number of decision making and pattern recognition projects, and published articles and papers in these areas. He also runs a consulting business that helps small and medium companies with research and development projects.

Nalini K. Ratha is a Senior Member of IEEE and co-chair of IEEE Robotics and Automation Society's AutoID TCIBM's. He is organizing the third IEEE AutoID Workshop, and is co-organizing a Tutorial on Biometrics at ICAPR 2001. Before joining IBM, Dr. Ratha completed his Ph.D. (1996) in Computer Science, Michigan State University, East Lansing. He was a member of the Pattern Recognition and Image Processing Lab of Dept.
of Computer Science. His thesis involved mapping computer vision algorithms onto FPGA-based custom computers. Mr. Ratha's advisor was Prof. Anil K. Jain. He recently co-chaired a special session on robust biometrics at KES 2000. He co-organized tutorials on biometrics at CVPR'99 and ICPR'98.

Bill Rogers is the Publisher of the Biometric Digest and Biometric Media Weekly newsletters (U.S.A.) reporting on biometric identification tech nology covering fingerprint ID, voice, hand geometry, facial recognition, iris scan and other methods. Bill moderates several national conferences on biometric technology and is a frequent speaker on this subject. He is also associate Editor of Credit Union tech-talk and CU InfoSECURITY newsletters. He has more
than 40 years experience working with financial institutions nationwide. Bill most recently was CEO for a national data processing firm serving credit unions. He was previously a Vice President with CUNADATA Corp., a national credit union data processor, and a Regional manager with Electronic Data Systems. Bill was with EDS for six years prior to forming his own company in 1988. Bill moderates and sponsors several technology conferences for credit unions. Prior to joining EDS, he was with the Missouri Credit Union League for over 19 years as Vice President of Credit Union Support Services and Manager of the League's data processing division for more than 14 years.

Faramarz F. Samavati received his Master and PhD degrees from Sharif University of Technology in 1992 and 1999 respectively. He is currently an Assistant Professor in the Department of Computer Science at the University of Calgary. His research interests are Computer Graphics, Geometric Modeling, Visualization, and Computer Vision. He has authored papers in Subdivision Surfaces, Sketch based Modeling, Multiresolution and Wavelets, Surface Modeling, Scientific Visualization and Biometric.

Vlad P. Shmerko received both M.S. (1976) and PhD (1984) degrees in electrical and computer engineering from the State University of Informatics and Radioelectronics, Minsk, Belarus, and Habilitation (1990) in computer science from the Institute of Electronics, Riga, Latvia. He joined the Department of Electrical and Computer Engineering, University of Calgary, Canada, in 2001. He is a Senior Member of the IEEE, a Fellow of the IEE (UK), an IAPR Member, and an IEE (UK) Chartered Engineer.

Vlad was a James Chair visiting Professor at St. Francis Xavier University in 1997 (Canada), Honorary Professor of the Technical University of Szczecin (Poland) and of the State University of Informatics and Radioelectronics (Belarus). He has been a team-leader for about 20 industrial projects related to CAD design of hardware and software systems for recognition, identification, and training of personnel. In particular, his *Controlled Stochastic Process Generator* was presented at industrial exhibitions worldwide.

His interest in biometrics began during the 1980s when he was developing simulators for training pilots and astronauts. Vlad brought his experience to the Biometric Technologies Laboratory at the University of Calgary which specialises in synthetic biometrics. His current research interests include biometric-based security system design using artificial intelligence, decision making support, and logic design of nanostructures. He has served as a Guest Editor and Board Editor of several International journals and as the general chair and co-chair for about 20 International conferences, symposia, and workshops. He is the author and co-author of more than 300 technical papers and research reports, including 45 patents and 11 books.

Sargur (Hari) Srihari's research and teaching interests are in pattern recognition, data mining and machine learning. He is involved in three research projects: (i) interpreting handwriting patterns to determine writership — which is a project of relevance to forensics where there is a need for scientific methods of examining questioned documents, (ii) studying friction ridge patterns, such as those found in latent fingerprints, to determine extent of  individualization — which s a topic of relevance to both forensics and biometrics where the degree to which a fingerprint is unique is being established, and (iii) developing methods for searching manuscripts written in the Latin, Arabic and Devanagari scripts — a topic of relevence to questioned document analysis and historic manuscript analysis.

Srihari joined the faculty of the Computer Science Department at the State University of New York at Buffalo in 1978. He is the founding director of the Center of Excellence for Document Analysis and Recognition (CEDAR). Srihari is a member of the Board of Scientific Counselors of the National Library of Medicine. He is chairman of CedarTech, a corporation for university technology transfer. He has been general chairman of several international conferences and workshops as follows: Third International Workshop on Handwriting Recognition (IWFHR 93) held in Buffalo, New York in 1993, Second International Conference on Document Analysis and Recognition (ICDAR 95), in Montreal, Canada, 1995, Fifth ICDAR 99 held in Bangalore, India and Eighth IWFHR 2002 held in Niagara-on-the-lake, Ontario, Canada.

Srihari has served as chairman of TC-11 (technical committee on Text Processing) of the International Association for Pattern Recognition (IAPR). He is presently chair of the IAPR Publicity and Publications committee.

Srihari received a New York State/United University Professions Excellence Award for 1991. He became a Fellow of the Institute of Electronics and Telecommunications Engineers (IETE, India) in 1992, a Fellow of the Institute of Electrical and Electronics Engineers (IEEE) in 1995, and a Fellow of the International Association for Pattern Recognition in 1996. He was named a distinguished alumnus of the Ohio State University College of Engineering in 1999.

Srihari received a B.Sc. in Physics and Mathematics from the Bangalore University in 1967, a B.E. in Electrical Communication Engineering from the Indian Institute of Science, Bangalore in 1970, and a Ph.D. in Computer and Information Science from the Ohio State University, Columbus in 1976.

Harish Srinivasan is currently a Ph.D. candidate at the department of computer science and engineering at the State University of New York at Buffalo. His research is in: (i) applications of statistical machine learning to biometric verification using fingerprints, handwriting and signatures, as the modalities, (ii) content-based image retrieval from scanned handwritten documents and (iii) use of conditional random fields in automatic segmenting and labelling of scanned documents. His research contributions to the above fields have been published at several conferences.

Harish Srinivasan received his B.Tech in Information Technology from the Madras University in 2002 and an M.S. in Electrical Engeneering from the State University of New York at Buffalo.

Adrian Stoica (Dipl Ing. TUI Iasi, Romania, PhD VUT Melbourne, Australia) is a Principal in the Biologically Inspired Technology and Systems Group in the Autonomous Systems Division at the NASA Jet Propulsion Laboratory (JPL), California Institute of Technology, Pasadena, California. He leads JPL research in the areas of Evolvable Systems, Humanoids Robotics and Integrated Circuits Security. He has over 20 years of management of advanced R&D and leadership in innovation and development of new technology in electronics and information systems, robotics and automation, learning and adaptive systems. His research interests and expertise are in the areas of advanced electronics (extreme environment electronics, self-reconfigurable, adaptive and evolvable hardware, adaptive computing devices, sensor fusion hardware), secure electronics (obfuscation techniques for protection from reverse engineering, trusted integrated circuits and biometrics protected ICs), biometrics (embedded and multi-modal), artificial intelligence algorithms (automated circuit design, search/optimization techniques, learning techniques, genetic algorithms, neural networks, fuzzy systems), and robotics (robot learning, humanoid robots, autonomous systems). He has over 100 papers and 4 patents in these areas, is serving in the editorial board of several journals in the field, gave invited keynote addresses and tutorials at conferences. He taught the first Evolvable Hardware short course (UCLA Extension, 2003) and started two annual conferences: the NASA/DOD Conference in Evolvable Hardware in 1999, and the NASA/ESA Conference on Adaptive Hardware and Systems in 2006. He received the Lew Allen Award for Excellence (highest NASA/JPL award for excellence in research) in 1999, and the Tudor Tanasescu Prize of the Romanian Academy in 2001.

Issa Traoré received an Aircraft Engineer degree from Ecole de l'Air in Salon de Provence (France) in 1990, and successively two Master degrees in Aeronautics and Space Techniques in 1994, and in Automatics and Computer Engineering in 1995 from Ecole Nationale Superieure de l'Aeronautique et de l'Espace (E.N.S.A.E), Toulouse, France. In 1998, Traoré received a Ph.D. in Software Engineering from Institute Nationale Poly-

technique (INPT)-LAAS/ CNRS, Toulouse, France. From June–Oct. 1998, he held a post-doc position at LAAS-CNRS, Toulouse, France, and Research Associate (Nov. 1998–May 1999), and Senior Lecturer (June-Oct. 1999) at the University of Oslo. Since Nov. 1999, he has joined the faculty of the Department of ECE, University of Victoria, Canada.

Dr. Traoré, is currently and Associate Professor and holding since Oct. 2003 the position of Computer Engineering Programme Director. His research interests include Behavioral biometrics systems, intrusion detection systems, software security metrics, and software quality engineering.

Patrick S. Wang is IAPR Fellow, professor of Computer Science at Northeastern University, research consultant at MIT Sloan School, and adjunct faculty of computer science at Harvard University Extension School. He received Ph.D. in C.S. from Oregon State University, M.S. in I.C.S. from Georgia Institute of Technology, M.S.E.E. from National Taiwan University and B.S.E.E. from National Chiao Tung University. He was on the faculty at University of Oregon

and Boston University, and senior researcher at Southern Bell, GTE Labs and Wang Labs prior to his present position. Dr. Wang was elected Otto-Von-Guericke Distinguished Guest Professor of Magdeburg University, Germany, and serves as Honorary Advisor Professor for Xiamen University, China, and Guangxi Normal University, Guilin, Guangxi, China. In addition to his research experience at MIT AI Lab, Dr. Wang has been visiting professor and invited to give lectures, do research and present papers in a number of countries from Europe, Asia and many universities and industries in the U.S.A. and Canada, including the recent iCore (Informatics Circle of Research Excellence) Visiting Professor position at university of Calgary Dr. Wang has published over 120 technical papers and 20 books in Pattern Recognition, A.I. and Imaging Technologies and has 3 OCR patents by US and Europe Patent Bureaus. As IEEE Senior member, he has organized numerous international conferences and workshops including conference co-chair of the 18th IAPR ICPR (International Conference on Pattern Recognition) to be held in 2006, at Hong Kong, China, and served as reviewer for many journals and NSF grant proposals. Prof. Wang is currently founding Editor-in-Charge of Int. J. of Pattern Recognition and A.I., and Editor-in-Chief of Machine Perception and Artificial Intelligence by World Scientific Publishing Co. and elected chair of IAPR-SSPR (Int. Assos. for P.R.). In addition to his technical achievements and contributions, Prof. Wang has been also very active in community service, and has written several articles on Du Fu, Li Bai's poems, Verdi, Puccini, Bizet, and Wagner's operas, and Mozart, Beethoven, Schubert and Tchaikovsky's symphonies.

Svetlana N. Yanushkevich received M.S. (1989) and PhD (1992) degrees in electrical and computer engineering from the State University of Informatics and Radioelectronics, Minsk, Belarus, and she received her Habilitation degree (1999) in engineering science from the University of Technology, Warsaw, Poland. She joined the Department of Electrical and Computer Engineering, University of Calgary, Canada, in 2001. She is the founding director of the "Biometric Technologies: Modeling and Simulation" Laboratory at Calgary. Dr. Yanushkevich is a Senior Member of the IEEE, a Member of IAPR, and a Member of the Institute of Electronics, Information and Communication Engineers (IEICE), Japan.

Dr. Yanushkevich took part in several industrial projects on the development of (i) high-performance parallel computing tools for image processing applications, including document understanding, (ii) systems for forensic expertise support based on behavioral biometrics, and (iii) bank staff training systems for signature and handwriting verification. Her current research interests include problems in direct biometrics (analysis) and inverse biometrics (synthesis), computer aided design of systems, nanocomputing in 3D, and artificial intelligence in decision making. She also acts as a consultant to several government and commercial institutions.

Dr. Yanushkevich is the Chair of the IEEE-Industrial Electronics Society Technical Subcommittee for Biometric Technologies, and a Member of the IEEE Technical Committee on Multiple-Valued Logic. She serves as a Guest Editor of several international journals, and she was a general chair and co-chair of about 20 international conferences, symposia, and workshops.

Dr.Yanushkevich published over 200 technical papers, including 15 patents, and 8 authored, co-authored, and 3 edited books. She co-authored (with D. M. Miller, V. P. Shmerko, and R. S. Stanković) the first handbook on logic design of nanocircuits, "Decision Diagram Techniques for Micro- and Nanoelectronic Design, Taylor&Francis, CRC Press, 2006. The book "Biometric Inverse Problems", Taylor&Francis, CRC Press, 2005, written together with A. Stoica, V. Shmerko, and D. Popel, is the first book that focuses on the generation of synthetic biometric data.

Xinge You received the B. S. and M. S. degrees in mathematics from the University of Hubei, Wuhan, China and the Ph.D. degree in computer science from the Hong Kong Baptist University, Hong Kong, in 1990, 2000, and 2004 respectively. He is presently an Associate Professor in the Faculty of Mathematics Computer Science at Hubei University, China. And he is currently work as postdoctoral fellow in the Department of Computer Science at Hong Kong Baptist University, Hong Kong. His current research interests include wavelets and its application, signal and image processing, pattern recognition, and computer vision.

Dan Zhang received the BSc degree in mathematics from Huanggang Normal College, Hubei, China, in 2004. Currently she is pursuing her master degree in the Faculty of Mathematics and Computer science, Hubei University, China. Her research interests include image processing, pattern recognition and biometrics technologies and applications. pattern recognition and biometrics technologies and applications.

David Zhang graduated in Computer Science from Peking University in 1974. In 1983 he received his MSc in Computer Science and Engineering from the Harbin Institute of Technology (HIT) and then in 1985 his PhD from the same institution. From 1986 to 1988 he was first a Postdoctoral Fellow at Tsinghua University and then an Associate Professor at the Academia Sinica, Beijing. In 1994 he received his second PhD in Electrical and Computer Engineering from the University of Waterloo, Ontario, Canada.

Currently, he is a Chair Professor, the Hong Kong Polytechnic University where he is the Founding Director of the Biometrics Technology Centre. Professor Zhang's research interests include automated biometrics-based authentication, pattern recognition, and biometric technology and systems. He is the Founder and Editor-in-Chief, International Journal of Image and Graphics; Book Editor, Kluwer International Series on Biometrics; Chairman, Hong Kong Biometric Authentication Society and Program Chair, the First International Conference on Biometrics Authentication (ICBA), Associate Editor of more than ten international journals including IEEE Trans on SMC-A/SMC-C, Pattern Recognition, and is the author of more than 140 journal papers, twenty book chapters and eleven books. Professor Zhang holds a number of patents in both the USA and China and is a Croucher Senior Research Fellow and Distinguished Speaker of IEEE Computer Society.

Author Index

Subject Index

physiological, 106, 244, 340
synthetic, 6, 7, 370, 401
system, 21, 107, 245, 339
verification, 139, 140, 272
Blood
flow rate, 369, 378
pressure, 376, 398
Borda count, 160, 174

C

Cancelable biometrics, 22, 323
Caricature, 19
Classification, 106, 165, 171, 208
Classifier, 160, 166, 220, 235, 310
Computational geometry, 108
Computer aided
design, 17
training, 384
Confusion matrix, 165
Convergence map, 198, 199
Cooperation
human-human, 379
human-machine, 380, 381
machine-machine, 380
Crossover, 291
single-point, 299
two-point, 299
uniform, 299, 300

D

Data
acquisition, 7, 56
representation, 114
Database, 104, 271
centralized, 273
distributed, 273
Decision
diagram, 397
fusion, 235
making, 282, 309, 368
support, 22, 370, 382, 394
threshold, 281
Decomposition, 67
coefficients, 234
filter, 68

Delaunay
tessellation, 111
triangulation, 107, 109
Digram, 51
model, 52
Discriminant
function, 236
Discriminating
element, 139–141
Dissenting-weighted majority vote
(DWMV), 160, 172
Distance
distribution, 109, 118
Euclidean, 110, 119
transform, 117, 118, 120, 162
vector, 236
Distribution
Gamma, 146, 148
Gaussian, 202
Divergence, 183, 197
Domain
frequency, 187, 228
knowledge, 374, 376
signal, 400

E

Ear
biometrics, 183
signature, 204
topology, 186
Early warning, 390
paradigm, 367
Eigenfaces, 209
Electromagnetic spectrum, 377
Emotion, 9, 321
Energy
distribution, 228
field, 183, 190
transform, 190, 192
Enrollment, 244, 271
Environment
real-world, 371
virtual, 392, 402
Equal error rate (EER), 140, 151
Ergonomics, 343, 373